Soil Science: Properties, Fertility and Management of Soil

Soil Science: Properties, Fertility and Management of Soil

Edited by Bruce Collins

SYRAWOOD
PUBLISHING HOUSE

New York

Published by Syrawood Publishing House,
750 Third Avenue, 9th Floor,
New York, NY 10017, USA
www.syrawoodpublishinghouse.com

Soil Science: Properties, Fertility and Management of Soil
Edited by Bruce Collins

International Standard Book Number: 978-1-68286-569-9 (Hardback)

Cataloging-in-Publication Data

Soil science : properties, fertility and management of soil / edited by Bruce Collins.
    p. cm.
Includes bibliographical references and index.
ISBN 978-1-68286-569-9
1. Soil science. 2. Soil fertility. 3. Soil management. I. Collins, Bruce.
S591 .S65 2018
631.4--dc23

TABLE OF CONTENTS

PREFACE

It is often said that books are a boon to mankind. They document every progress and pass on the knowledge from one generation to the other. They play a crucial role in our lives. Thus I was both excited and nervous while editing this book. I was pleased by the thought of being able to make a mark but I was also nervous to do it right because the future of students depends upon it. Hence, I took a few months to research further into the discipline, revise my knowledge and also explore some more aspects. Post this process, I begun with the editing of this book.

Soil science is the study of the characteristics and properties of soil. It studies the texture, water content, fertility, biota and acidity of soil. The two main branches of soil science are pedology and edaphology. With an increase in issues such as water crisis and land use, soil science studies ways of preserving soil. This book provides comprehensive insights into the field of soil science. As this field is emerging at a fast pace, this book will help the readers to better understand the concepts of soil science.

I thank my publisher with all my heart for considering me worthy of this unparalleled opportunity and for showing unwavering faith in my skills. I would also like to thank the editorial team who worked closely with me at every step and contributed immensely towards the successful completion of this book. Last but not the least, I wish to thank my friends and colleagues for their support.

Editor

Dust Reduction in Bauxite Waste: Role of Gypsum, Carbonation, and Microbial Decomposition

Mark Anglin Harris[1]

[1] College of Natural & Applied Sciences, Northern Caribbean University, Mandeville, WI, Jamaica

Correspondence: Mark Anglin Harris, College of Natural & Applied Sciences, Northern Caribbean University, Mandeville, WI, Jamaica. E-mail: mark.harris@ncu.edu.jm

Abstract

Producing alumina by the Bayer Process creates fine air borne red dust which devalues property and causes irritation to the human respiratory system. Aggregation of such inorganic particles was proposed as a dust-inhibiting corrective. Resistance to breakdown under simulated rain suggests a lower number of dust-size particles after rain. Samples of red mud waste (1) treated 10 years before the study at the 0-15 cm depth zone with 40 t ha^{-1} of gypsum (2) from the subjacent 15-30 cm zone, were collected, crushed and passed through a 0.5 mm diameter sieve. Leaves from *Acacia senensis* (a legume) were finely chopped to < 1 mm and thoroughly mixed with the sieved bauxite waste at 25- and 50%, and the samples incubated for 6 weeks at ambient room temperatures, at 60% soil water-holding capacity. To determine the fraction of potential dust, the treated samples were submerged in de-ionized water for several days until there was no change in discoloration (due to clay dislocation) of the water. The samples were removed from the water and the water evaporated and the residues dried and weighed. In total, the dust-reducing capabilities of the treatments in descending order of proficiency were: 50% phytogenic > 25% phytogenic > 0-15 cm soil depth non-phytogenic > 15-30 cm-depth non-phytogenic. The 50% phytogenic-treatment reduced potential particles of fugitive dust by 70% over the untreated controls and 95% over the crushed-only (subjacent red mud; no organics added) samples. All in all, phyto-organics increased average particle size to > 100 µm by flocculation, thereby creating stable agglomerates which resisted disintegration and breakdown under simulated rain. Reducing the concentration of < 75 µm particles in the air will decrease morbidity due to respiratory illnesses in surrounding populations, harmful effects on vegetation, and the defacement of buildings. This treatment promises the use of gypsum + phyto-organics for reducing the emanation of surface dust from red mud waste sites onto surrounding areas.

Keywords: aggregation, dust-amendments, organic remediation, phytogenic amendment

1. Introduction

Red mud waste (RMW) surfaces (crusts) of bauxite mine spoils are potential sources of dust emission. During storage, the surface gets dried, even though the consistency may be still liquid at a few millimeters below (Graham & Fawkes, 1992). A strong wind can then blow a red dust cloud over the surroundings at a considerable financial and environmental cost. Buildings have been defaced and vegetation damaged by dust accumulation on leaves (Graham & Fawkes, 1992).

According to the International Standardization Organization (1994), "Dust: small solid particles, conventionally taken as those particles below 75 µm in diameter, which settle out under their own weight but which may remain suspended for some time". A Glossary of Atmospheric Chemistry Terms (IUPAC, 1990), characterizes dust as: "Small, dry, solid particles projected into the air by natural forces, such as wind, volcanic eruption, and by mechanical or man-made processes such as crushing, grinding, milling, drilling, demolition, shovelling, conveying, screening, bagging, and sweeping. Dust particles are usually in the size range from about 1 to 100 µm in diameter, and they settle slowly under the influence of gravity." It is at the micro-aggregate size level that excess sodium from the Bayer process produces separation of clay particles. Since micro-aggregates bind < 5 µm particles into larger entities at the < 250 µm level, and wind-blown dust particles do not exceed 100 µm in diameter, it follows that production of stable micro-aggregates from < 5 µm particles to form aggregates > 100 µm should reduce the rate of wind-blown dust.

Bauxite red mud consists primarily of the insoluble fraction of the bauxite ore that remains after extraction of the aluminium-containing components (Zhang et al., 2001). Iron oxides (10-30%), titanium dioxide (2-15%), silicon oxide (5-20%) and undissolved alumina (0-20%) make up the residue (Bardossy, 1982), together with a wide range of other oxides which will vary according to the initial bauxite source (Jones & Haynes, 2011). The high concentration of iron compounds in the bauxite gives the by-product its characteristic red colour, and hence its common name "Red Mud" (AAC, 2012).

The sodium carbonate is precipitated on the surface of residue as entrained moisture evaporates (Alcoa, 2007). Therefore, dry red mud easily spreads into the air and causes dust pollution (Hai et al., 2014).

Though particle size analysis shows that Jamaican Terra Rossa bauxites contain 20% sand (O'Callaghan, 1998), it is the secondary (clay) fraction which contains the bauxite. Similar processing techniques in Jamaica (Bayer process) to that of the Weipa (West Australia) kaolinites indicate similar post-beneficiation clay particle sizes. Above 11 m s^{-1} (40 km h^{-1}), such dust emissions from the residue area can increase rapidly and wind speeds above 14.5 m s^{-1} (50 km h^{-1}) are predicted to be the largest source of dust (Alcoa, 2007). Attaining speeds often exceeding 13m s^{-1} during winter months (Macpherson, 1991), unobstructed trade winds in Jamaica and the northern Caribbean should, on the above basis, normally carry fugitive dust released from the dried surface of red mud dumps.

Freisen et al. (2009) examined the associations between alumina and bauxite dust exposure and cancer incidence and circulatory and respiratory disease mortality among bauxite miners and alumina refinery workers. Their preliminary findings of the very few cases in the limited population study suggest that cumulative inhalable bauxite exposure may be associated with an excess risk of death from non-malignant respiratory disease and that cumulative inhalable alumina dust exposure may be associated with an excess risk of death from cerebrovascular disease. Nevertheless, neither exposure appeared to increase the risk of incident cancers. Further, there was no apparent danger and no association between every bauxite exposure and any of the outcomes (Freisen et al., 2009), though there was a borderline significant association between every alumina exposure and cerebrovascular disease mortality.

Wagner (1997, 2009) also found that exposures to bauxite dust, alumina dust, and caustic mist in contemporary best-practice bauxite mining and alumina refining operations have not been demonstrated to be associated with clinically significant decrements in lung function. Exposures to bauxite dust and alumina dust at such operations were also not associated with the incidence of cancer (Wagner, 1997).

Despite these favourable findings, adverse health effects of bauxite dust are possible. This is because red mud residue and sand consist primarily of alumina, silica and iron oxides (Alcoa Air Assessments, 2007) and silicosis is an irreversible condition with no cure (Wagner, 1997). Silicosis is the most common form of pneumoconiosis (lung-related diseases), which is caused by occupational exposure of free silica dust (ACGIH, 1999). Treatment options currently focus on alleviating the symptoms and preventing complications (ACGIH, 1999). According to Pattajoshi (2006), dust is inevitable in mineral industries. Also, Nouh (1989) reports that a non-occupational form of silicosis has been described that is caused by long-term exposure to sand dust in desert areas, with cases reported from the Sahara, Libyan desert and the Negev, and that the disease is caused by deposition of this dust in the lung.

Cohesiveness between clay particles can be significantly increased by inorganic cementing agents such as CaCO$_3$, and Fe and Al oxides (Zhang, 2015). In a study of African Ultisols and Oxisols, Ahn (1979) observed highly stable micro-aggregates not dependent upon organic matter. Therefore, binding agents need not be from organic sources. After adding gypsum to two different red-brown earths, Shanmuganathan and Oades (1984) noted a reduction in the amount of dispersible clay, an increase in the proportion of water-stable aggregates sized 50-250 μm diameter, and an increase in soil friability. Nevertheless, mechanical strength was reduced by gypsum addition (Aylmore & Sills, 1982) before strengthening by carbonation occurred.

Without dislocation there is no dust. The treatment of this study therefore aimed to "fix" the clay particles into soil aggregates. These were to bind the clay particles against dislocation by the wind. This procedure was also aimed at resisting natural field impacts such as rain and running water which dislocate fine clay particles prior to them becoming airborne dust.

2. Materials and Methods

Twenty kg of (1) gypsum-treated red mud which subsequently underwent a 10-yr lithification by atmospheric carbonation (2) red mud beneath the gypsum-treated layer (Table 1) were randomly collected from the Kirkvine Pond 6 Bauxite residue storage area in Jamaica. Whereas the gypsum-lithified red mud (G) exists in rock-hard form in the 0-15 cm zone, the non-gypsum fraction (R) exists adjacently below 15 cm as a semi-viscous constituent, having been seemingly unaffected by the gypsum which O'Callaghan et al. (1998) had ploughed into the zone (G) directly above, in 1996. Both materials are of the same age but very different physico-chemically. The top layer is non-dispersive while the 15-30 cm layer is highly dispersive. Prior to air-dry, this 15-30 cm depth layer had the physical consistency of untreated red mud waste that supplies wind-blown dust. Samples were crushed at air-dry in a mortar and pestle and passed through a 1-mm aperture sieve. Decomposable organic material as finely chopped (< 2 mm) leaves from *Acacia senensis* ("Kasha", an invasive legume of southern Jamaica) was thoroughly mixed in with either above-mentioned bauxite waste in dry, grinded homogenous (< 1 mm) form and incubated at room temperature for 42 days at a water content of 60% of the field capacity of each soil. This "mix" is based on improved aggregation of similarly treated crushed particles at the < 2 mm diameter size by Harris (2009). Of the two crushed red muds, field capacity (water-holding capacity) was substantially higher for the gypsum-treated samples.

Table 1. Properties of red mud subjacent (15-30 cm) to 0-15 cm depth gypsum-treated red muds

Properties	Values
Red mud pH (saturated extract)	12
EC (saturated extract: dS m^{-1})	.25
Organic carbon (%)	.3
CEC (c molc kg^{-1})	40
Al$_2$O$_3$ (%)	16
CaO (% w/w)	7
Fe$_2$O$_3$ (%)	47
Na$_2$O (%)	3
P$_2$O$_5$ (%)	2
SiO$_2$ (%)	3
TiO$_2$ (%)	6
Bound H$_2$O (%)	14
Particle size	.001-mm

Note. There were six treatments, each having three replicates: (1) G, (2) G25 (*i.e.*, gypsum-treated red mud + 25% phyto-organics), (3) G50 (*i.e.*, gypsum-treated red mud + 50% phyto-organics) (4) R (red mud), (5) R25 (red mud + 25% phyto-organics), (6) R50 (red mud + 25% phyto-organics). Phyto-organic treatments are referred to hereafter as GP or RP.

2.1 Measuring Potential Wind-Blown Dust

Clay dislocation: 1st submergence

After incubation for six weeks, the samples at air dry were subjected to submergence in de-ionized water which simulated rain water. Deionized water was included because Khattab and Othman (2013) noted a general reduction of strength of rocks with an increase in the number of wet-dry cycles when using distilled water. Replicates of treated aggregates having a diameter of approximately one cm were each placed in a watch glass, after which water was slowly added because the weak structure of bauxite waste is subject to breakdown by electrolytes. The relative rates of clay dislocation from the aggregates signified the amount of dust particles which would have emerged on drying. Without clay dislocating forces on the aggregates there would be no potential dust. The extent of clay particle dislocation from the aggregates was determined by measuring the weight of the dislocated particles. The clay dislocation process was monitored for the ensuing days until no further change was observed. At air dry, aggregates were removed from the watch glasses and the masses of air-dry residues of fine dust weighed.

Clay dislocation: 2nd submergence

After clay removal from watch glasses at air-dry, the clay dislocation procedure was repeated. Each aggregate was then mechanically agitated in an end-over-end shaker for a pre-determined time period. At air-dry, all particles were again weighed. The separated fractions were air-dried. They were weighed after sieving on a nest of three sieves: 2 mm-, 500 μm-, and 100 μm-diameter.

3. Results

Table 2 shows that after incubation plus air-dry and 5 days of submergence in deionized water, relatively little observable clay particle dislocation occurred from any of the non-phyto-organic treatments as compared to that of the phyto-organic treatments (Table 2, Figure 1). The R50 (R samples are from 15-30 cm depth) dislocated 29% of the mass of the sample as clay, and the R25 dislocated 33% of the sample mass for an average of 31% dislocation by the subjacent phyto-treated samples. The R50 dislocated less clay than did the R25. This was expected because these soil particles were considered bound by organic matter due to the treatment.

Further, it can be seen that the quantity of clay dislocated was much greater from the 15-30 cm depth compared with the 0-15 cm depth, *i.e.*, in a ratio approximating 15:1 (Table 3). For the 0-15 cm depth the samples containing 25% phyto-organics dislocated no clay, while the samples containing 50% phyto-organics produced a 4% w/w rate of clay dislocation. Thus for the 0-15 cm depth, rate of phyto-organics determined effectiveness to suppress clay dislocation in deionized water. Similarly, at the 15-30 cm depth (Table 3), where the 25% phyto-organics treatment dislocated less clay than the 50% phyto-organic treatment. This was surprising. Clearly, the factor which produced the initial dislocation in deionized water existed in the phyto- samples, and the greater concentration of phyto- material the lesser was the clay dislocation. This pattern is examined later.

After 8 days of submergence in deionized water, non-phyto-organic treatments from all samples dislocated very little clay above the level of detection (Table 3, Figure 1).

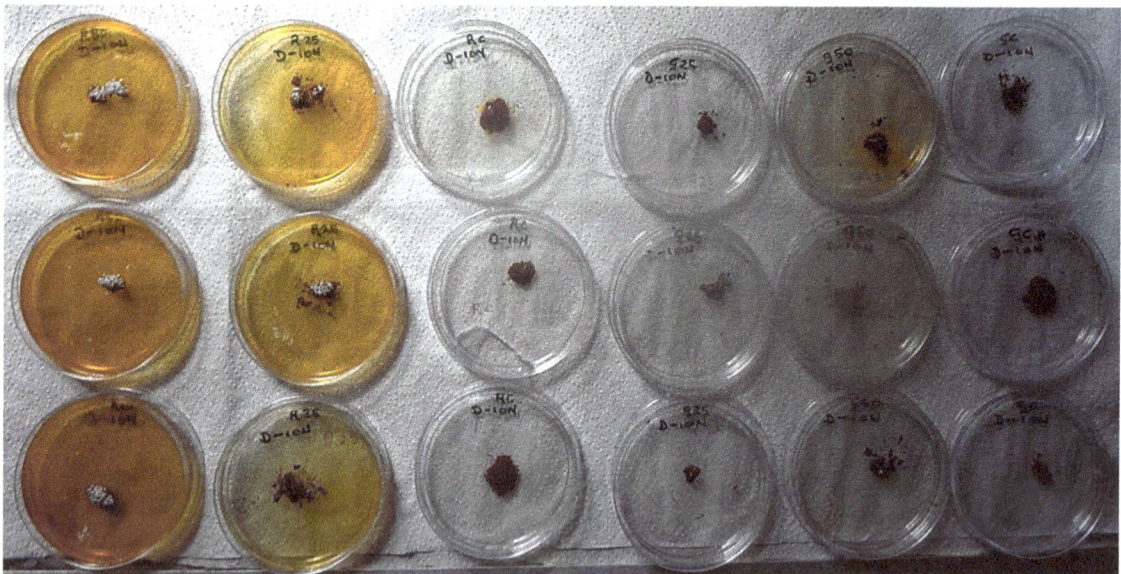

Figure 1A. Clay dislocation in deionized water for remolded aggregate samples from the 15-30 cm depth in carbonated red mud

Note. At extreme left are the R50 (phytogenic-treated) samples; second from left are the R25 (phytogenic) samples with less clay dispersion depicted by the lighter colour.

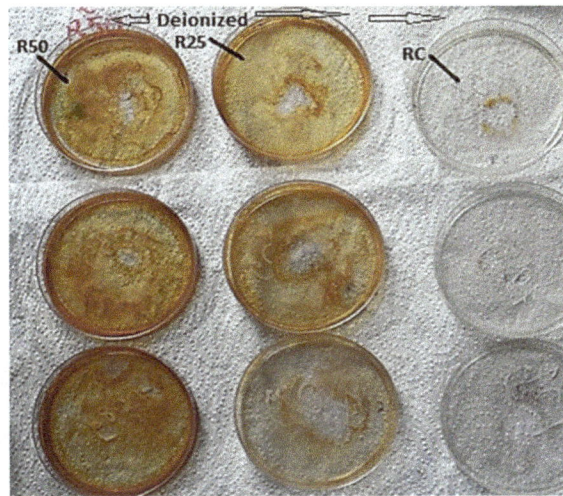

Figure 1B. Dried samples of dislocated clay from the 15-30 cm depth after aggregates were removed from deionized water

Note. R50 = 50% decomposable phyto-organics R25 = 25% phyto-organics, RC = no phyto-organics. Each small circle in background paper is 1-mm in diameter.

Table 2. Clay (% of mass) dislocated after incubation treatments and submergence in de-ionized water for 8 days

Red mud (depth in cm)	0 % phyto-organics	25% phyto-	50% phyto-	Average
0-15	0	0	4	2
15-30	0.01	29	33	31

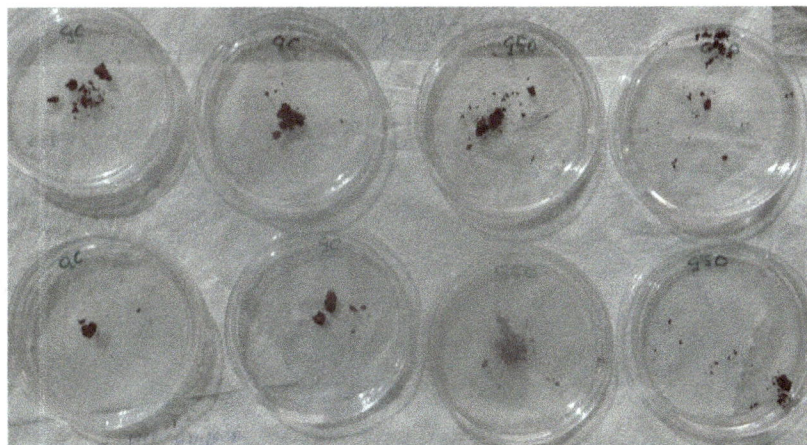

Figure 2A. Post-incubation gypsum-treated samples in distilled water after 1st submergence

Note. GC = 0-15 cm and no phto-organics; G50 = 0.15 cm + 50% organics. No discoloration suggests effective flocculation of red mud individual particles.

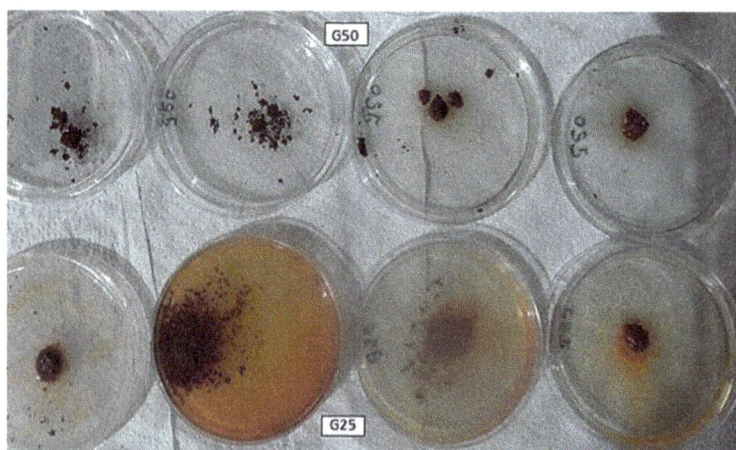

Figure 2B. Post-incubation gypsum-treated samples in non-deionized water after 1st submergence, in contrast to the reaction with deionized water (Figure 2A) where the colorless water indicated no clay dislocation

Non-de-ionized water also dislocated clay from the samples but there were minor but important differences. Firstly, the total mass of clay dislocated was higher for the deionized water (Table 3), where the 15-30 cm level dislocated almost twice as much clay under the deionized water compared to the non-deionized water. On the other hand, for the 0-15 cm level, the non-deionized water caused a 4-fold increase in clay dislocation compared to the deionized water. Again, only the phyto- treatments caused substantial clay dislocations.

Table 3. Clay (% of mass) dislocated by non-deionized water after incubation treatments

Red mud (depth in cm)	0 % phyto-organics (controls)	25% phyto-organics	50% phyto-organics	Average
0-15 (G)	0	8	8	8
15-30 (R)	0	16	16	16

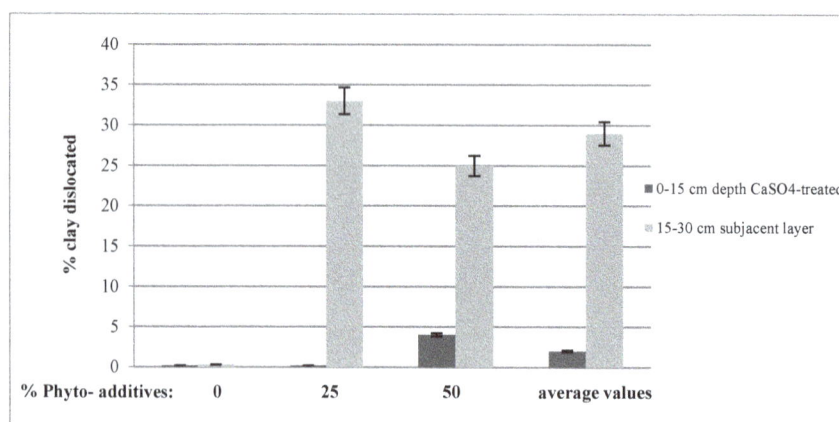

Figure 3. Mass of clay dislocated after incubation and soaking in de-ionized water for 8 days

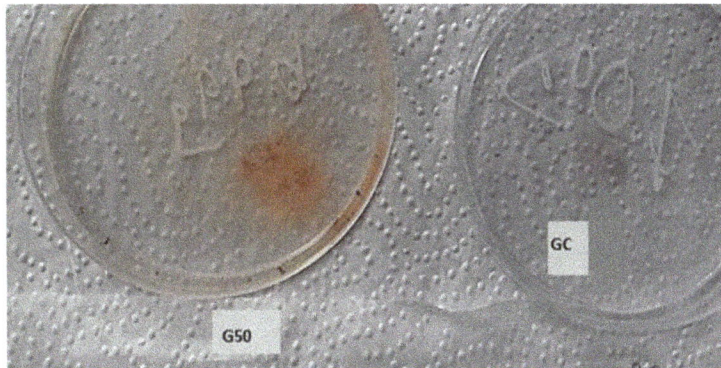

Figure 4. Post-incubation gypsum-treated samples in non-distilled water after the 1st submergence

Note. Dislocated clay particles can be observed from the 50% phyto- treatment at left (G50). Stippled background contains 1-mm diameter circles.

For the G samples (no phyto-), submergence in non-deionized water produced small, barely visible levels of clay dislocation (Table 4, Figures 3 and 4). This nevertheless was greater than that for de-ionized water, where no dispersion appeared from any of the non-phyto samples. Similarly, from the 0-15 cm depth, GP25 samples increased clay dislocation up to half the level of the RP samples (*i.e.*, RP25 or RP50). Of all the phyto-organic treatments, GP samples treated with 50% phyto-organics exhibited the lowest levels of clay dislocation from the aggregates (Table 4, Figures 1 and 5). The possible reasons are discussed later.

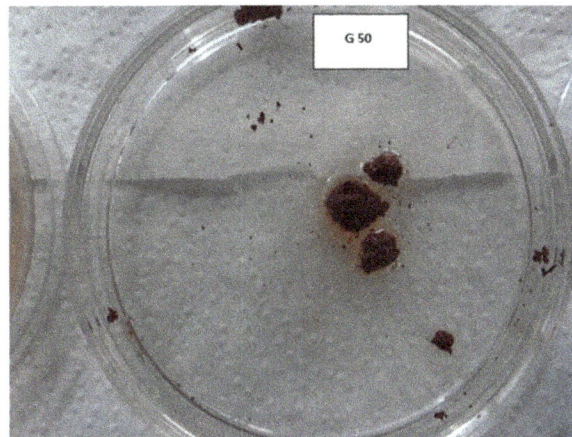

Figure 5. G50 (50% phytogenic additive) dislocated very little clay even in non-de-ionized water

The clay dislocations in deionized water were less severe for the G samples. Only 2% of the GP mine-waste dispersed, whereas 29% of the R samples dislocated in the form of clay particles. This is almost a 15-fold difference. As electrical stresses on soil aggregates vary directly with purity of water, a high concentration of electrolytes (*i.e.*, > EC) in a soil should increase clay dislocation in deionized water. Thus the R samples, with a higher electrical conductivity than the G samples, released more clay particles in deionized water than in non-deionized water.

The relative rates of clay dislocation from the aggregates signified the amount of dust particles which would have emerged on drying. After incubation for 6 weeks, the result for either the G or R samples was that little or no observable clay dislocation occurred in deionized or non-deionized water.

In deionized water, gypsum-treated bauxite waste without phyto-organics (GC) showed no dispersion of clay-particles at the microstructural level (< 100 μm diameter) for the 0-15 cm-depth. For samples taken from the subjacent level, a very slight clay dislocation effect was visible. This was below the detection level of the weighing equipment (Figure 3). However for the RP samples a large amount of dust averaging 25% w/w would have been generated by phyto-organic treatments. This was unexpected. This is discussed later. Based on these results, the GP samples were overwhelmingly less dispersive than the RP samples.

After removal from Petri dishes at air-dry, and after the second episode of clay dislocation in water the remolded aggregate masses were weighed and measured. Even after the 2nd submergence, the phyto-organically treated samples in general retained stability in water to a far greater degree than that of the 1st submergence. Clay dislocation procedure was repeated on the aggregates. The results show that no dislocation occurred at all from any sample. It can be seen that the 2nd submergence produced no dislocation of clay particles (Table 4), because the total mass for each sample submerged was zero.

Table 4. Mass of clay (grams) dislocated from aggregates after 2nd submergence for 8 days

Red mud (depth in cm)	0 % phyto-organics (controls)	All phyto-organics	Total mass
0-15	0	0	0
15-30	0	0	0

In comparison to the low levels of dislocation observed from the G, R, and G50 treatments, the level of clay dislocation exhibited by all RP treatments was more than 10 times as high. However, for all phyto-organic treatments, especially in the case of the RP, the second submergence produced far less clay dislocation (> 10-fold decrease) than the first submergence (Table 4). On the established fact that totally untreated Bayer red muds are (1) totally dispersive (AAC, 2012) and (2) hence totally disaggregated (Bardossy, 1982), clay dispersion would therefore have been far greater in such totally untreated red muds than had occurred with the treatments of this study. Overall, greater clay dislocation was more evident from the de-ionized samples, especially from the R and RP samples.

The lack of dislodged clay particles from the 2nd submergence could have meant that all the dispersible clay had been released at the first submergence (Table 4). To test this assumption, the aggregates were again air-dried and subjected to dry sieving in a shaker where the aggregates were rolled and bounce on a sieve having a 100 micron aperture (the diameter of micro-aggregates). Each aggregate was then mechanically agitated in an end-over-end shaker for a pre-determined time period. At air-dry, all particles were again weighed.

In the rating scheme for this study, the clay dislocation ratio for the RP as compared with the GP samples was > 10:2.6. Nevertheless, the aggregates of this study, which were: (1) totally suffused with the water and (2) 1-cm diameter sub-rounded spheres had a much larger surface area than the same volume in field conditions (where there is just a sub-aerial surface of exposure from which dust could have potentially emanated).

Therefore, as the 2nd submergence of this study caused no observable clay dislocation, the maximum total potential dust for any treatment remained at 29% w/w. After the 2nd submergence, samples were air-dried and shaken in a sieve having apertures 100 μm in diameter. Results are shown in Table 5.

Table 5. Proportion of < 125 μm particles after dry-sieving of aggregates following two submergences in water for a total of 15 days after incubation

Red mud (depth in cm)	0 % phyto-organics (controls)	25% phyto-organics	50% phyto-organics	Total
0-15 (G)	4	2	1	7
15-30 (R)	60	2	0	62

It can be seen that, whereas the phyto-treatments continued to dislocate very little clay even during dry sieving, the non-phyto- treatments dislocated at least twice as much clay as the least efficient phyto-treatment, during this procedure. In fact, R samples (non-phytogenic-) dropped 60% of their mass from the dry-sieve as dislocated < 100 μm clay particles (Table 5). As noted in section 1.0 above, among the world's bauxite ores Jamaican bauxite texturally contains by far the largest clay fraction at 75% (Wehr et al., 2006). Therefore, had this been a field study, it is reasonable to assume that the R (non-phyto-organic) samples would have potentially released 42% of its mass as wind-blown clay. Again, as stated above, several variables determine the actual concentration of wind-blown dust at any given location. Therefore no attempt is made here to quantify the potential rate of wind-blown dust. Nevertheless, it can be seen that despite being by far the least effective treatment of this study, the R (non-phyto-) samples reduced potential wind-blown dust by 33% (*i.e.*, 75 – 42 = 33). Using Tables 2-5, similar calculations estimate the potential dust reduction by other treatments as follows:

G50 = 75 – (4 + 8 + 0 + 1) = 67%

G25 = 75 – (8 + 0 + 2 + 3) = 67%

R50 = 75 – (16 + 0 + 0 + 13) = 56%

R25 = 75 – (16 + 0 + 2 + 3) = 54%

G = 75 – (0 + 0 + 0 + 4) = 71%

R = 75 – (.01 + 0 + 0 + 60) = 14.99%

Prior to the test of aggregate stability, the leading treatment for dust reduction here is still gypsum (71%), with gypsum + phytogenic being second at 67%. Predictably, without gypsum, incubation yielded only 15% dust reduction for non-gypsum treatments (R samples).

In a physical sense, resistance to shattering impacts reduces dust production. After 15 days, the samples were subjected to falling drops of water which simulated raindrops from a pre-determined height. The results showed that the phyto-organically treated samples at 50% from either the 0-15 cm depth or the subjacent red mud withstood more than twice as many impacts before shattering, as each of the non-phyto-organically treated samples (Table 6). Additionally, the G50- proved substantially more resilient in this respect than the R50 samples (Table 6). Without organics, the G aggregates withstood three times as many raindrop impacts as the R aggregates. The overall G treatment produced 50% greater resiliency under simulated raindrop impacts than the R treatments (Table 7). However, the CND resistance of the R25 exceeded that of the G25 samples by 50%. This apparent anomaly is discussed later.

Table 6. Impact of large falling water drops on remolded aggregates (days) after incubation and submergence in water for 15 days as depicted by the counted number of drops (CND) before shattering

Red mud (depth in cm)	0 % phyto-organics	25% phyto-organics	50% phyto-organics	Total
0-15	150	30	430	670
15-30	50	70	330	450

Aggregates treated with 50% phyto-organics were the most resilient in still water for 15 days (Table 7, Figure 6).

Table 7. Longevity of remolded aggregates (days) after incubation and submergence in water for 15 days

Red mud (depth in cm)	0 % phyto-organics (controls)	25% phyto-organics	50% phyto-organics	Total
0-15	2	3	13	21
15-30	12	3	13	28

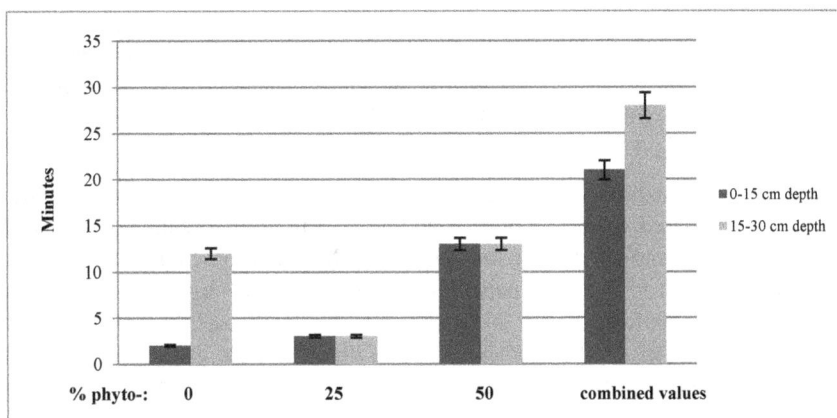

Figure 6. Longevity of remolded aggregates from two depths after incubation and soaking for 15 days

A vast difference of water-stable longevity amongst treated samples can be seen, because samples from the 15-30 cm depth with either 50% phyto-organics (R50) or without phyto-organics (R only) are by at least a 5-fold margin the most resilient under wet sieving (Figure 7, Table 8). Aggregates subjected to wet-sieving maintained their integrity in the following order of decreasing longevity: R50 > R > G > R25 > G25 = G50 (Figure 7, Table 8).

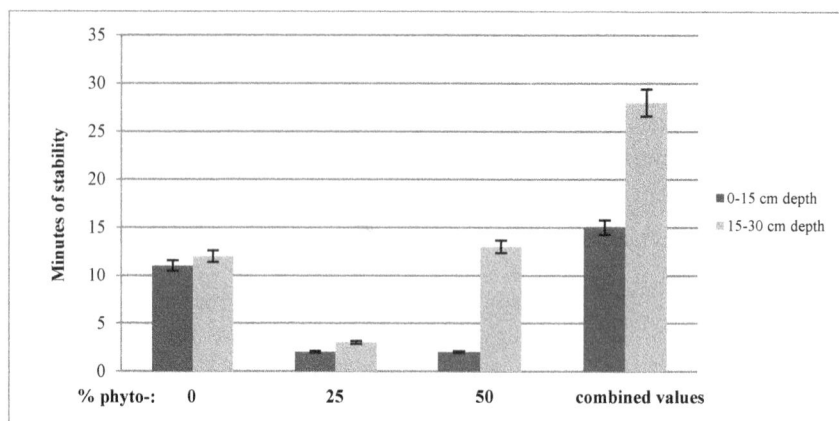

Figure 7. Stability of remolded aggregates from two depths after incubation and soaking

Table 8. Aggregate longevity in minutes of stability during wet-sieving after incubation treatments

Red mud (depth in cm)	0 % phyto-organics	25% phyto-organics	50% phyto-organics	Total
0-15	11	2	2	15
15-30	12	3	13	28

4. Discussion

As dust results from dislocated clay, and contact with water weakens soil aggregates and dislocates clay (Rengasamy et al., 1984), the most resilient aggregates are the lowest potential producers of dust. It can be seen that despite being submerged in still water for 15 days, the phyto-organic (P) samples outlasted all others. In fact, before the end of day 1, several non-P samples had already disintegrated (Table 7), some having done so within a few minutes of submergence.

For aggregates without phyto-organics from the 0-15 cm previously gypsum-treated layer, relatively few measurable clay particles (potential dust) were dislocated during submergence in water for several days. At the end of day 2 of submergence, those samples incubated from the 15-30 cm layer (i.e., under the gypsum-treated layer, without phyto-organics) released no visible clay particles. Wind power increases exponentially with wind speed such that the power of the wind varies as the cube of its speed (Jackson & Hunt, 1975). Thus an increase of even two or three km hr^{-1} is a significant increase in terms of energy expended and hence the amount of dust removable from a surface. As adsorbed Na^+ causes mutual repulsion among clay particles, and the 15-30 cm depth contained Na^+ far in excess of the surface layer, the low level of clay dislocation from that depth was unexpected. This is further discussed later.

In the first submergence, the gypsum plus phyto-organics treatment dislocated > 5-times the amount of fine clay particles (potential dust) compared to the amount from the gypsum-only treatments. This also was against expectations because applying similar treatments to sodic (Na^+-rich or high ESP) subsoils, Harris and Rengasamy (2004) used sub-micron particle analysis to show that phyto-organics with gypsum substantially flocculated (agglutinated) clay particles. Thus by increasing the mass sizes, they decreased the release of individual clay particles. Further, for the first submergence of this study, the greatest reduction of potential dust occurred as follows in descending order of efficiency: G > R > G50 > G25 > R50 = R25.

Decomposable phyto-organics have been shown to bind sodic clay particles (Harris & Rengasamy, 2004) and inorganic particles in bauxite waste (Harris, 2009), and as discussed later, wet-sieving results showed that strong binding of inorganic particles also occurred in this study. Yet, for the ultra-fine (highest dust-potential) inorganic colloids, such a binding force from phyto-organics proved, particularly at the lower phyto-organic rates, ineffective initially in this study. Isomorphic atomic substitution causes the negative charge exhibited by all clay particles. Muneer and Oades (1989) noted an increased number of negatively charged particles formed as decomposition and breakdown of phyto-organics advances in remolded aggregates. They concluded that small negative charges from organic colloids which were the same approximate length as clay particles were dispersing the clay particles. In this study, it can be seen that, compared with non-phyto-organic-treatments, it was the phyto-organic samples that produced more clay dislocation in the 1st submergence, especially the phyto-organics at the lower rate (R25, and to a lesser extent, the G25). On the other hand, though they readily broke down even in still water, non-phyto- treated macro-aggregate samples from the 0-15 cm depth (G) did not dislocate ultra-fine clay particles when those macro-aggregates were submerged in water (Table 2). Furthermore, the following observations were made:

➢   No clay dislocation from the G50 or G25 samples occurred in deionized water;

➢   Between G50 & G25 samples, clay dislocation in non-deionized water was greater from G25.

Clay dislocation by the G25 & G50 in the non-deionized water could have been due to a greater number of negatively charged colloids (organic and/or inorganic) intrinsic to that water as compared to the deionized (less contaminated) type. Indeed, the pH of the non-deionized water had a value of 7.8-8.0. The total mass of negative charges/colloids would be greater in the non-deionized system. The G25, with less microbial substrates than the G50 and hence with more competition among microbial decomposers, would have been expected to have undergone more advanced breakdown of organic matter, thereby presenting smaller colloids to the system. As noted earlier, highly comminuted negative colloids of similar sizes mutually repel. It is thus more likely that of the two treatments, the G25 would have been more likely to contain organic colloids fine enough to repel clay particles of similar sizes. Such electronic forces could have dislocated the clay particles at a higher rate from the G25 than from the G50 aggregates.

For similar reasons, in a study incubating gypsum with sodic sub-soils, Harris and Rengasamy (2004) observed no stable macro-aggregates (*i.e.*, > 250 μm) at any stage of incubation. They concluded that gypsum alone cannot produce stable macro-aggregates, though stabilization occurred at the < 250 μm size level. They also reported that phyto-organic additions caused a slight increasing trend.

The ability of negatively charged clay colloids to disperse clay particles is illustrated by the immediate settling out by flocculation of clay colloids or organic colloids whenever small amounts of multivalent cations from $Al_2(SO4)_3$ (alum) are added to such suspensions. It is very likely that after the first submergence, most or all of the organic colloids became trapped in the dried residue. This may explain why, after dislocating such high levels of clay in the first submergence, at submergence #2, the phyto treatments failed to dislocate any clay at all.

The natural flushing by rainfall for more than a decade (Harris, 2009) would have removed much sodium sulfate in the 0-15 cm layer. This relative increase in the ratio of Ca^{2+} to Na^+ ions promoted the binding action among clay particles. However, annual rainfall also subsequently depleted the unprotected Ca^{2+} ions in the upper (0-15 cm) layer. It was expected that phyto-organic gums from the decomposing phyto-organics, would have induced stable aggregation among the G25 and G50 samples. This did not occur. On the other hand, with the same levels of gypsum in the same bauxite waste, Harris (2009) observed stabilization of macro-aggregates. The explanation may be that Harris (2009) incubated the samples for a much longer period of time (more than 24 weeks) compared to the 6-week incubation of the present study. With more advanced levels of decomposition, stronger binding between inorganic particles may have occurred in that earlier study.

Yet, paradoxically, in this same study of just six weeks, underneath the surface, the 15-30 cm layer, despite having massively accumulated Na^+ ions from the prolonged eluviation from the 0-15 cm layer plus residual Na^+ from the Bayer Process, produced stable macro-aggregates of 6-fold the longevity under wet-sieving than those produced by the Na^+-deficient 0-15 (surface) layer. Moreover, as stated above, Harris and Rengasamy (2004) concluded that with or without gypsum applications, phyto-organics cannot macro-aggregate sodic clays to a level of water-stability, though stable micro-aggregation (< 250 μm) can be achieved. The objective of flushing is to remove the Na^+ in the form of Na_2SO_4 from the exchange reaction with $CaSO_4$ to provide greater opportunities for the adsorption of Ca^{2+} onto clay particles. Yet despite the obvious absence of flushing at the 15-30 cm level (indicated by high Na^+ levels), strong Ca^{2+} adsorption as proven by high aggregate stability, had clearly occurred subsequently (*i.e.*, after gypsum addition to the superjacent layer). The explanation may be that with such unusually high levels of Ca^{2+} in the 15-30 cm level, exchange sites became dominated by the more strongly bound Ca^{2+}. Purely from the standpoint of stability, this nullified the need for Na_2SO_4 flushing. Interestingly, as stated above, the R25 samples dislocated more clay than all other treatments. It is therefore unsurprising that the R25, perhaps for reasons stated above, exhibited also the least resistance to wet sieving (Table 8).

On the other hand, the lower resistance to wet-sieving of the 0-15-cm gypsum-treated phytogenic macro-aggregates suggests an inactivation by carbonation to $CaCO_3$ of a substantial quantity of the Ca^{2+} released in the early stages of the gypsum treatment applied several years before this study. Being exposed sub-aerially, greater atmospheric CO^2 would have been available in the 0-15 cm depth compared to the subjacent layer.

Another possible explanation for the longevity of the R samples to wet sieving entails thixotropically affected entities, referred to by Coughlan et al. (1979) as "fortuitous agglomerates." These fortuitous agglomerates survive wet-sieving but are not true aggregates, being case-hardened during storage possibly in this case by oxidation of iron oxides (Ly, 2001; Zhang, 2015).

As stated above (section 1.2), based on Wehr (2006), the total clay fraction for the Jamaica bauxite waste is 75%. On the assumption that all clay particles are < 100 μm in diameter, potential dust in Jamaica bauxite waste, according to his table, is at least 84% w/w. As the best treatments of this study increased more than 95% of the particle size of the bauxite waste to well over the < 100 μm- diameter threshold after submergence in de-ionized water, it is concluded that the phyto-organic treatments substantially decreased potential dust not just from Jamaica bauxite wastes, but potentially for all other wastes listed above. This includes red muds from St. Croix (Virgin Islands), Guyana, and Haiti.

The relatively high level of aggregate stability under wet-sieving observed for the crushed G (no phyto-organics) samples in this study was not expected because an opposing result occurred in a study of the same lithified red mud waste by Harris (2009). In yet another study, Harris and Rengasamy (2004) also found that no macro-aggregate stability in a sodic sub-soil when treated with gypsum. However, in both of those studies, they used < 2 mm particle sizes, whereas in the current study the particle size used was < 1 mm. Particles must be fine enough to provide a sufficient reactive surface area for the solid-state chemical reactions (Mehta & Monteiro, 1993; Malhorta & Mehta, 1996). It is thus postulated that the larger surface area (more than twice as large) of the smaller particles of this study afforded far greater opportunities for cohesive forces of the Ca^{2+} ions among inorganic particles.

It is the binding action of organic cements and electronic attraction of cations that produce aggregation and, by extension, can reduce dust formation in sodic soils and spoil heaps. It is interesting that both the mine tailings of the present study and the soil studied by of Harris and Rengasamy (2004) reacted similarly to phyto-organic additives. Neither sodic entity was influenced by phyto-organics acting alone. In other words, without gypsum, sodic clays do not react significantly with phyto-organics to form stable macro-aggregates. This is primarily because organic colloids and sodic clays are mutually repulsive, being both negatively charged. Without multi-valent cations there are few positive charges to link the clay particles. Harris and Rengasamy (2004) found that the role of gypsum is to flocculate the clay particles, as has been found by Baldock and others (1994); and Muneer and Oades (1989), where Ca^{2+} in soil solution did not improve macro-aggregation. In this study, however, gypsum acting alone stabilized macro-aggregates from finer inorganic particles.

Applying similar treatments to sodic subsoils, Harris and Rengasamy (2004) found that clay particles had been aggregated by decomposing phyto-organics into a size range not exceeding 30 μm. In this study, agglomerates from the RM + phyto-organics treatment remained stable under physical pressure, and did not release dust particles. Yet, these agglomerates of varying sizes up to 4 mm from the RM + phyto-organics treatments were readily unstable under wet-sieving (Table 3), as was the case for those produced in an experiment conducted by Harris and Rengasamy (2004). The implications here are that under field conditions, such stability would be destroyed during the first rain shower, and that dust would be produced on drying of the crumbled masses. However, destruction of unstable macro-aggregates produce progressively smaller aggregates with an inverse change in soil strength (Dexter, 1988) caused by stronger inter-particle binding mechanisms (Seguel et al., 2006). Thus below a particular wind speed, micro-aggregates above dust-size produce less dust per volume than macro-aggregates. As reported above, dust particles of up to 100 μm can be airborne. As soil micro-aggregates range up to 250 μm (diameter), an appreciable proportion can be airborne. Though micro-aggregate size was not measured in this study, Harris and Rengasamy (2004) found that micro-aggregates which formed under similar conditions had an average size of 30 μm (diameter). It is therefore reasonable to conclude that although dust was reduced in this study by creation of aggregates exceeding 100 μm, a proportion of micro-aggregates created would be below 100 μm, and hence susceptible to creating dust hazards.

Based on the above, it is clear that (1) the water-stability of red mud waste with or without added phyto-organics varies directly with addition of gypsum (2) water-stability of red mud waste determines dust production levels.

Under specific conditions of this study, water-stability was achieved not only during wet-sieving trials, but prior to that, during two episodes of several days of submergence.

5. Conclusions

Gypsum treatment decreased dust-sized clay particles by > 80%. At a rate of 7% CaO and 3% Na_2O (Table 1), average dust particle sizes of < 100 μm in original samples aggregated to larger particles and showed a > 4-fold increase in resistance to clay dislocation over that of the controls. Tuffour et al. (2015) found that finer particles were highly effective in altering soil properties even at low concentrations. Therefore, decreasing the original particle size of the remolded aggregates from < 2 mm in previous studies to < 1mm of this study is also likely to have increased stability of aggregates against clay dislocation. The size of dislocated particles was < 100 μm. Under the influence of simulated de-ionized water, decomposable phyto-organic additives initially increased small quantities of the finest clay particles. Nevertheless, increasing the size of particle clusters in this study potentially produced greater resistance against wind saltation and airborne movements than that which occurs for individual clay particles, or sub-100 μm domains.

References

AAC (Australian Aluminium Council). (2012). *Bauxite Residue in Australia.* Retrieved May 12, 2015, from http://www.alluminium.org.aubauxite.world-aluminium.org/refining/bauxite-residue-management.html

ACGIH. (1999). In J. H. Vincent (Ed.), *Particle Size-Selective Sampling for Health-Related Aerosols.* American Conference of Governmental Industrial Hygienists, Air Sampling Procedures Committee, Cincinnati, OH, USA.

Alcoa. (2007). *Pinjarra Alumina Refinery Efficiency Upgrade.* Alcoa World Alumina Australia. Dust Management Plan for the Alcoa Pinjarra Bauxite Residue Disposal Area.

Aylmore, L. A. G., & Sills, I. D. (1982). Characterization of soil structure and stability using modulus of rupture–exchangeable sodium percentage relationships. *Australian Journal of Soil Research, 20,* 213-224. http://dx.doi.org/10.1071/SR9820213

Baldock, J. A., Aoyama, M., Oades, J. M., Susanto, & Grant, C. D. (1994). Structural amelioration of a South Australian Red Brown Earth using calcium and organic amendments. *Australian Journal of Soil Research, 32,* 571-594. http://dx.doi.org/10.1071/SR9940571

Bardossy, G. (1982). *Karst Bauxites.* Elsevier Scientific Publishing Co.

Coughlan, K. J. (1979). *Influence of micro-structure on the physical properties of cracking clays.* Report to the Reserve Bank, Qld Dept. Primary Industries.

Dexter, A. R. (1988). Advances in characterization of soil structure. *Soil Tillage Research, 11,* 199-238. http://dx.doi.org/10.1016/0167-1987(88)90002-5

Friesen, M. C., Fritschi, L., Del Monaco, A., Benke, G., Dennekamp, M., de Klerk, N., ... Sim, M. R. (2009). Relationships between alumina and bauxite dust exposure and cancer, respiratory and circulatory disease. *Occupational & Environmental Medicine, 66,* 615-618. http://dx.doi.org/10.1136/oem.2008.043992

Graham, G. A., & Fawkes, R. (1992). Red muds disposal management at QAL. *Proceedings of an International Bauxite Tailings Workshop, Pereth, Western Australia.* Australian Bauxite and Alumina Producers.

Hai, L. D., Nguyen, M. K., Tran, V. Q., & Nguyen, X. H. (2014). Material composition and properties of red mud coming from alumina processing plant tanrai, lamdong, Vietnam. *International Journal of Research in Earth & Environmental Sciences.* Retrieved from http://www.ijsk.org/ijrees.html

Harris, M. A. (2009). Structural improvement of age-hardened gypsum-treated bauxite red mud waste using readily decomposable phyto-organics. *Environmental Geology, 56,* 1517-1522. http://dx.doi.org/10.1007/s00254-008-1249-5

Harris, M. A., & Rengasamy, P. (2004). Sodium affected subsoils, gypsum, and green-manure: Inter-actions and implications for amelioration of toxic red mud wastes. *Environmental Geology, 45*(8), 1118-1130. http://dx.doi.org/10.1007/s00254-004-0970-y

ISO (International Organization for Standardization). (1994). *ISO 4225: 1994 Air quality – General aspects – Vocabulary.* Retrieved June 5, 2015, from http://webstore.ansi.org/RecordDetail.aspx?sku=ISO4225:1994

IUPAC. (1990) *Glossary of Atmospheric Chemistry Terms.* Retrieved June 8, 2015, from http://old.iupac.org/reports/1990/6211calvert/glossary.html

Jackson, P. S., & Hunt, J. C. R. (1975). Turbulent wind flow over a low hill. *Quarterly Journal of the Royal Meteorological Society, 101,* 929-955. http://dx.doi.org/10.1002/qj.49710143015

Jones, B. E. H., & Haynes, R. J. (2011). Bauxite Processing Residue: A Critical Review of Its Formation, Properties, Storage, and Revegetation. *Environmental Science and Technology, 41,* 271-315. http://dx.doi.org/10.1080/10643380902800000

Khattab, S., & Othman, H. (2013). Durability & Strength of Limestone Used in Building. *Al Rafidain Engineering, 21*(3).

Ly, L. (2001). A study of iron mineral transformation to reduce red mud tailings. *Waste Management, 21*(6), 525-534. http://dx.doi.org/10.1016/S0956-053X(00)00107-0

Malhorta, V. M., & Mehta, P. K. (1996). *Pozzolanic and Cementitious Materials.* Lightning Source Inc.

McPherson, J. (1991). *The Making of the West Indies.* Longmans.

Mehta, P. K., & Monteiro, O. J. M. (1993). *Concrete: Microstructure, Properties and Materials.* McGraw Hill.

Muneer, M., & Oades, J. M. (1989). The role of Ca^{2+} organic interactions in soil aggregate stability. II. Field studies with 14C-labelled straw, $CaCO_3$, and $CaSO_4$-$2H_2O$. *Australian Journal Soil Research, 27,* 401-409. http://dx.doi.org/10.1071/SR9890401

Nouh, M. S. (1989). Is the desert lung syndrome (non-occupational dust pneumoconiosis) a variant of pulmonary alveolar microlithiasis? Report of 4 cases with review of the literature. *Respiration, 55*(2), 122-6. http://dx.doi.org/10.1159/000195715.PMID 2549601

O'Callaghan, W. B., McDonald, S. C., Richards D. M., & Reid, R. E. (1998). *Development of a topsoil-free vegetative cover on a former red mud disposal site.* Alcan Jamaica Rehabilitation Project paper.

Pattajoshi, P. K. (2006). Assessment of airborne dust associated with chemical plant: A case study. *Indian Journal of Occupational & Environmental Medicine, 10,* 32-4. http://dx.doi.org/10.4103/0019-5278.22893

Rengasamy P, Greene, R. S. B., Ford, G. W., & Mehanni, A. H. (1984). Identification of dispersive behaviour and the management of Red-brown earths. *Australian Journal of Soil Research, 22,* 413-431. http://dx.doi.org/10.1071/SR9840413

Seguel, O., Horne, & Rainer, F. (2006). Structure properties and pore dynamics in aggregate beds due to wetting-drying cycles. *Journal of Plant Nutritonal & Soil Science, 169*(2). http://dx.doi.org/10.1002/jpln.200521854

Shanmuganathan, R. T., & Oades, J. M. (1984). Influence of anions on dispersion and physical properties of the A horizon on a red brown earth. *Geoderma, 29,* 257-277. http://dx.doi.org/10.1016/0016-7061(83)90091-5

Tuffour, H. O., Thomas, A.-G., Awudu, A., Caleb, M., David, A., & Abdul, A. K. (2015). Assessment of changes in soil hydro-physical properties resulting from infiltration of muddy water. *Applied Research Journal, 1*(3), 137-140.

Wagner, G. R. (1997). Asbestosis and Silicosis. *Lancet, 349*(9061), 1311-1315. http://dx.doi.org/10.1016/S0140-6736(96)07336-9

Wehr, J. B., Fulton, I., & Menzies, N. W. (2006). Revegetation strategies for bauxite refinery residue: A case study of Alcan Gove in Northern Territory, Australia. *Environmental Management, 37,* 297-306. http://dx.doi.org/10.1007/s00267-004-0385-2

Zhang, X. W., Kong, L. W., Cui, X. L., & Yin, S. (2015). Occurrence characteristics of free iron oxides in soil microstructure: Evidence from XRD, SEM and EDS. *Bulletin of Engineering Geology and the Environment,* 1-11. http://dx.doi.org/10.1007/s10064-015-0781-2

Zhang, Y., Qu, Y., & Wu, S. (2001). Engineering geological properties and comprehensive utilization of the solid waste (red mud) in aluminium industry. *Environmental Geology, 41,* 249-256. http://dx.doi.org/10.1007/s002540100399

Crop Suitability Mapping for Rice, Cassava, and Yam in North Central Nigeria

Roland Clement Abah[1,2] & Brilliant Mareme Petja[3]

[1] College of Agriculture and Environmental Sciences, Department of Environmental Sciences, University of South Africa, Pretoria, South Africa

[2] National Agency for the Control of AIDS, Abuja, Nigeria

[3] College of Agriculture and Environmental Sciences, Department of Environmental Sciences, University of South Africa, Pretoria, South Africa

Correspondence: Roland Clement Abah, National Agency for the Control of AIDS, Plot 823 Ralph Shodeinde Street, Central Area, Abuja, Nigeria. E-mail: rolann04@yahoo.com

Abstract

Agricultural production has contributed over time to food security and rural economic development in developing countries particularly supporting the countryside. Evidence of crop yield decline exist in the Lower River Benue Basin. This was a crop suitability mapping for rice, cassava, and yam to guide policy makers in strategic planning for sustainable agricultural development. Data was collected on various themes including climate, drainage, soil, satellite imagery, and maps. Remote Sensing was used to analyse satellite imagery to produce a digital elevation model, land use and land cover map, and normalised difference vegetation index map. GIS was used to produce thematic maps, weighted percentages of attribute data, and to produce crop suitability maps through weighted overlay. Soils in the study area require fertility enhancement with inorganic fertilisers for better crop yield. Soils in the Lower River Benue Basin are suitable for yam, cassava, and rice cultivation on maps of suitable areas. Some areas were found to be highly suitable for the cultivation of rice (34.22%), cassava (17.08%) and yam (16.08%). Some other areas were found to be moderately suitable for the cultivation of cassava (48.18%), rice (45.46%), and yam (48.85%). Areas with low suitability were 14.99% (rice), 33.68% (cassava), and 29.57% (yam). This study has demonstrated the importance of crop suitability mapping and recommends that farmers' cooperative societies and policy makers utilise the information presented to improve decision making methods and policies for agricultural development.

Keywords: GIS, remote sensing, precision agriculture, crop suitability mapping, sustainable agriculture

1. Introduction

Nigerian cassava production is by far the largest in the world, and Benue and Kogi state in the North Central zone are the largest producers of cassava (IITA, 2004). The country produces about 50 million metric tons a year within a cultivated area of about 3.7 million hectares. Nigeria accounts for 20% of world produce, 34% of Africa's produce, and 46% of West Africa's produce (FAO, 2016). Nigeria accounts for 71% (over 37 million tons) of the 94% of world production of yams which comes from West Africa (IITA, 2009). Nigeria is Africa's largest consumer of rice. Rice production in Nigeria is mainly for market value as rice generates more income than most agricultural produce. Nigeria is one of the leading importers of rice in the world. Most agricultural produce in Nigeria including cassava, yam and rice is by small-scale farmers (FAO, 2016).

Suitable parameters for the cultivation of cassava, yam and rice exist in many areas of Nigeria. Conditions for cassava cultivation in savannah regions are documented in Titus et al. (2011) and Ande (2011). Cassava can grow on a wide variety of soils within a temperature of between 25 °C and 29 °C, and with a rainfall range of 500 to 1500 mm. Cassava can grow on level to moderate slope and does not require much water for growth. The conditions for rice cultivation in southern guinea savannah is presented in Aondoakaa and Agbakwuru (2012), and rice requires a temperature range of 20 °C to 27 °C and a rainfall range of 1150mm to 3000mm. The main ecologies for rice cultivation in West Africa include rain-fed upland, rain-fed lowland, and irrigated lowland with water control. Conditions for yam cultivation as discussed in Kutugi (2002) and Eruola et al. (2012) are similar to that of cassava but yam has less tolerance for water stress.

These conditions are prevalent in Benue state which is predominantly made of small-scale farmers heavily involved in the cultivation of cassava, yam, and rice. Through an integrated scientific planning approach which is aimed at enhancing small-scale farm activities, the aim of development which is centred on enriching quality of life in all segments of the population particularly the rural population can be achieved (M. Ghosh & S. K. Ghosh, 2013).

As part of efforts towards enhancing the production of cassava, yam and rice, it is beneficial to accurately match agricultural practice with appropriate spatial information on adequate conditions. Geographic Information System (GIS) and remote sensing has been extensively used in other sectors of national development but the use of such technology to support decisions for sustainable agricultural development in rural settings is still evolving. The use of these technologies is therefore encouraged towards improvements in the standards and quality of rural life (Petja et al., 2014).

Agricultural land use patterns are highly dynamic features of a cultural landscape and social and economic factors are the most prominent factors that influence land use change in rural areas (Ortserga, 2012). The Food and Agricultural Organisation framework for land evaluation (FAO, 1976) has provided guidance for land suitability assessment in developing countries where data scarcity often constrains modelling. Riveira and Maseda (2006) revealed that there is a shortage of models focused on rural land use and that designing a rural land use planning model should involve the integration of different computer tools. According to Kumara (2008), the principal application of GIS in rural development are land and resource mapping, integration of local and scientific spatial knowledge, community-based natural resource management, area planning, environmental management, and management of pests and natural hazards. The integration of local knowledge into GIS makes analysis more participatory and enhances ownership and utilisation of information.

The utilitarian value of GIS and remote sensing provides robust analytical and manipulative capabilities which can enable modelling for rural agricultural enhancement (Enete & Amusa 2010). Various studies (Nuga, 2001; Rilwani & Ikuoria, 2006; Rilwani & Gbakeji, 2009; Uchua et al., 2012) have revealed the need to adopt geo-informatics methods to improve agricultural productivity to meet the nutritional need of the teeming Nigerian masses as well as for export income.

A study was conducted by Ashraf (2010) which involved land suitability analysis for wheat using multi-criteria evaluation and GIS. The study by Ashraf (2010) used GIS to provide information at local level for farmers to select their cropping patterns. In a large study by Stickler et al. (2007), the biophysical potential for three major crops (soybean, sugar cane, oil palm) in the tropics were mapped globally. Stickler et al. (2007) identified growth requirements for these crops and used the data to develop spatially-explicit variables and identified regions where these crops can be profitably grown. Heumann et al. (2013) embarked on land suitability modelling using a geographic socio-environmental niche-based approach in north-eastern Thailand. The study by Heumann et al. (2013) tried to understand the land suitability for crops and utilised data on the built environment, natural abiotic conditions, and household social factors which were responsible or externally influenced the human modification of the niche.

This study aimed to produce crop suitability maps for cassava, rice, and yam by utilising a broad range of quantified data on physical aspects of the local community for the improvement of agriculture in Benue state. Cassava, rice and yam are widely cultivated crops in Benue state but produce have not appreciably increased over the years with indications of low productivity, low yields, and high post-harvest losses owing a subsistence culture of farming. This study explores the most suitable areas for the cultivation of cassava, rice, and yam which can lead to sustainable increase in yield.

2. Method

2.1 Study Area

The area of study falls within Benue State, north central Nigeria, between Latitudes 7°13′N and 8°00′N and Longitudes 8°00′E and 9°00′E (Figure 1). There are thirteen Local Government Areas (Makurdi, Gboko, Tarka, Gwer west, Gwer east, Guma, Buruku, Otukpo, Agatu, Ushongo, Ohimini, Obi, and Konshisha) covered either in whole or in part by the study area.

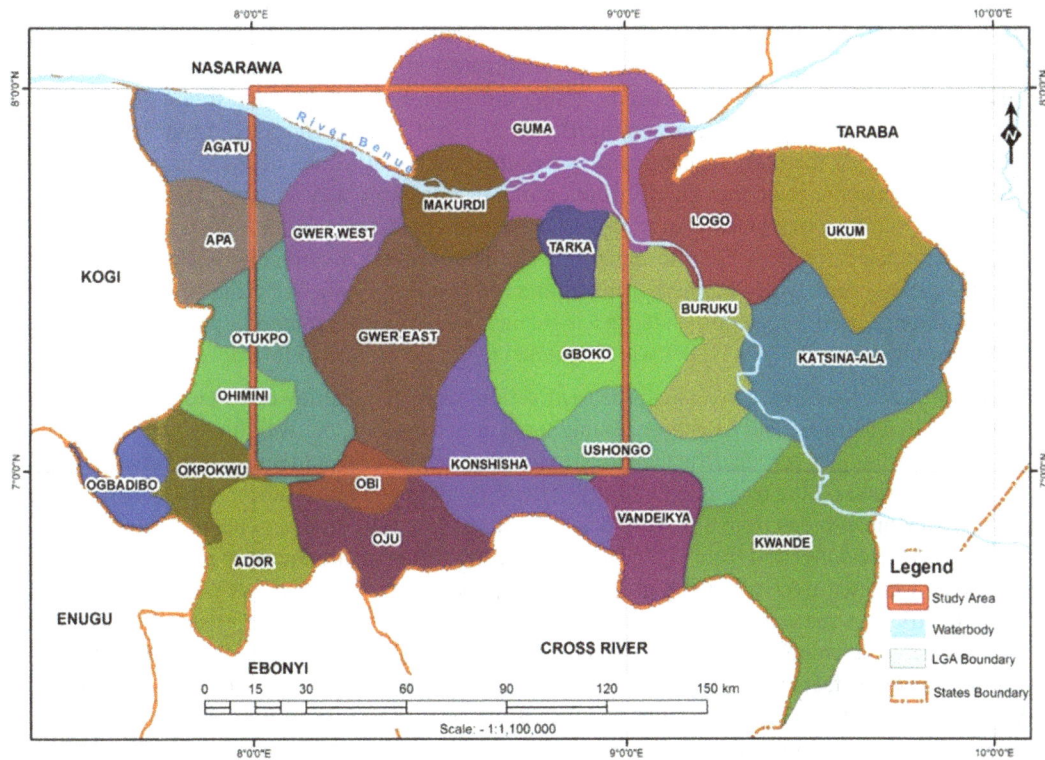

Figure 1. Map of Benue state showing study area

2.2 Data Collection

The methodology utilised for the study involved primary and secondary data for physical, remote sensing and GIS analysis. Data on climatic parameters (1973-2014) were collected from the Nigerian Meteorological Agency. Drainage data (1955-2012) on the River Benue from Makurdi and Umaisha hydrological stations and on River Katsina Ala from the Nigerian Hydrological Service Agency. Physical and chemical analysis of 36 top and subsurface soil samples collected from Makurdi, Tarka, and Gboko Local Government Areas located in Benue State and supported by secondary soil data from other studies to cover the study area.

The Landsat Satellite imagery was obtained from Global Land Cover Facility (GLCF) through the earth explorer platform. Landsat 7 ETM+ data was obtained for the study area for the year 2015, which had ortho-rectified the systematic radiometric, atmospheric and geometrical distortions of the imagery to a quality level of 1G before delivery (USGS, 2015). The Landsat scenes covered a region of approximately 182 km × 185 km and had a spatial resolution of 30 metres. The Landsat scenes covering the study area were Path 187 and 188 of Row 055. The Topographic maps of the study area were obtained from the Office of the Surveyor General of the Federation, Nigeria in Abuja. The topographic map sheets were at a scale of 1:50,000 for more details to be captured. The NASA Shuttle Radar Topographic Mission (SRTM) has provided digital elevation data (DEMs) for over 80% of the globe. This data was downloaded from the National Map Seamless Data Distribution System, or the USGS ftp site. The elevation details was obtained from the SRTM using the Global Mapper 15 software and compared with the contour extracted from the topographic map using the ArcMap 10.3 software, to have a full understanding of the topography. Maps were derived for themes such as climate, drainage, and soil.

2.3 Data Processing

The remote sensing analyses for the research included Land use land cover analysis and NDVI. These two (2) analyses were achieved using a combination of software (Idrisi 17.0 Selva edition and ArcMap 10.3) and geoprocessing operations. The spatial analyses were done in Idrisi while the cartographic finishing was achieved using the ArcMap 10.3 software. Classification involved labelling the pixels belonging to particular spectral classes using the spectral data available. The supervised method of classification was used which gave rise to the training sets. The Landsat imagery was first mosaicked using the geo-reference properties of both imagery and a feathering of two (2) was applied to reduce the edging. The various bands from 1-4 was independently mosaicked.

After which, a subset of the study area was made from the two (2) scenes of Landsat imagery downloaded. This subset was done using the Idrisi 17 Selva edition software. From empirical analysis and Principal Component Analysis, it has been proven that the bands that carry the greatest information about natural environment are the visible (Red, Blue and Green) wavelength bands. Using the Idrisi Selva software a true colour composite was made in Red, Green and Blue (RGB) representing Bands 3, 2 and 1 respectively. The tool considered both the variance and covariance of the class signatures as it assigned each cell to one of the classes represented in the signature file. With the assumption that the distribution of a class sample was normal, classes were characterised by the mean vector and the covariance matrix. Given these two characteristics for each cell value, the statistical probability was computed for each class to determine the membership of the cells to the class. The NDVI is expressed as the difference between the near infrared and red bands normalised by the sum of those bands. This is the most commonly used vegetation index as it retains the ability to minimise topographic effects while producing a linear measurement. The NDVI was calculated using the empirical format by Rouse et al. (1973).

Operations such as vector to raster conversion, reclassification, weighted overlay etc. were performed at this stage using the ArcMap 10.3 software and its geoprocessing tools in ArcToolbox. A "Weighted Overlay Operation" was adopted using GIS techniques for identification of areas of the various crop suitability depending on a number of thematic layers and based on the principle of Multi-Criteria Evaluation. The ArcMap 10.3 software was used to create the various thematic maps from available data. The maps (rainfall, drainage, temperature, DEM, Land use land cover and soil) were converted from vector format to raster format using the conversion tools in ArcToolbox for use in the GIS weighted overlay operation. Using the spatial analyst tools in ArcToolbox, the various raster maps were reclassified. A scale of 1 to 5 was adopted to indicate the level of importance. Value 5 represented extreme importance while value 1 represented not important. The scaling of the criteria was done in line with the level of contribution of the factors to the growth of rice, yam, and cassava from literature and conditions obtainable in the study area. Given the requirements for the growth of rice, yam and cassava from literature, the range requirements of extreme importance for each crop was ranked within the biophysical results obtained in this study. All the parameters were compared against each other in a pair-wise comparison matrix which was a measure of the relationship between the parameters in order to rule out bias. Subsequently, a numerical value expressing the level of importance of one parameter against another was assigned. After the preparation of all the thematic layers, reclassification as well as preparation of the table of weights, the weighted overlay operation was performed on the ArcMap 10.3 software. The crop requirements used and assigned weights are presented in Table 1 and 2. The crop suitability maps were created through the weighted overlay geoprocessing tool in ArcMap 10.3 ArcToolbox by using the weights assigned to each of the parameters (climate, soil, land cover, and DEM). Using five classes, the various layers were classified from very high suitability to very low suitability. Suitability maps were created for rice, yam, and cassava. Each raster was assigned a percentage of influence according to its importance derived for each crop. Similar GIS and remote sensing models have been used elsewhere (Stickler et al., 2007; Ashraf, 2010; Petja et al., 2014).

3. Results

3.1 Physical Conditions

The annual average rainfall amount recorded for the period 1973-2013 was 1194.1 mm, and the median was 1207.9 mm. The year with the highest amount of rainfall was 1999 (1617.1 mm). Other years with high amounts of rainfall were 1984 (1572 mm), 1998 (1537.6 mm), 1975 (1508.6 mm), 2012 (1466.7 mm), 1980 (1425.5 mm), and 2009 (1407.5 mm). The year with the lowest amount of rainfall was 2003 (761.5 mm). The average annual temperature calculated for the period January 1973 to December 2014 was 27.84 °C. The highest annual temperature averages were recorded in 2005 (28.6 °C), 1998 (28.55 °C), 2010 (28.5 °C), and 2003 (28.43 °C). The lowest temperature values were recorded in 2012 (26.8 °C) and 1974 (27.2 °C). Relative humidity in Makurdi was quite high annually with an annual average of 67.8% calculated for the period 1974-2008. The most extreme value (99.5%) was recorded on August 13 in 1997.

Table 1. Requirements of extreme importance for cultivation of rice, yam, and cassava

Parameters	Rice	Yam	Cassava
Rainfall (mm)	> 1500	1000-1250	750-1000
Temperature (°C)	23-26	26-29	26-29
Soil classes	Clay loam	loamy sand	loamy sand
Soil pH	5.0-5.5	6.0-6.5	5.5-6.0
Soil organic carbon	1.5 < 2.0	2.0 >	2.0 >
Soil Phosphorus (mg kg^{-1})	5-10	10-15	10-15
Soil Potassium (cmol kg^{-1})	0.8-1.0	0.5-0.7	0.5-0.7
Land Cover classes	Wetland	Scattered vegetation	Scattered vegetation
DEM (metres)	0-100	100-200	100-200

Note. Kutugi (2002), Titus et al. (2011), Ande (2011), Aondoakaa and Agbakwuru (2012), Eruola et al. (2012).

Table 2. Weighted index of parameters

Parameters	Weights (%)		
	Rice	Yam	Cassava
Rainfall	23.08	33.33	23.08
Temperature	10.25	15.00	10.38
Humidity	5.14	10.00	5.00
Soil class	12.82	4.45	7.69
pH	10.26	1.11	1.54
Organic carbon	5.13	5.56	4.62
Phosphorus	7.69	3.33	6.15
Potassium	2.56	2.22	3.08
Land cover	7.69	8.33	7.69
DEM (slope)	15.38	16.67	30.77
Total	**100%**	**100%**	**100%**

The average discharge of River Benue at Umaisha hydrological station for the period was 4,919.47 cubic metres per second (m^3/s). The maximum discharge for the period was 19,120 m^3/s which was recorded on the 15th October 2012. Average discharge of River Benue at Makurdi hydrological station was 3,468.24 m^3/s. The peak flow discharge of 16,400 m^3/s was recorded in three days 19th, 29th, and 30th in the month of September 2012 while the peak flow of 2011 was 9,436 m^3/s. At River Katsina Ala hydrological station, the average discharge from January 1955 to May 2014 was 933.12 m^3/s. The maximum discharge for the period was 4,401 m^3/s which was recorded on the 20th October 1977.

Soils in Makurdi were mostly loamy sand. Loamy sand soils have low water holding capacity, good drainage and aeration. Soils from Tarka, and Gboko were mostly sandy loam. Loamy sand and sandy loam soils appear moderately suitable for irrigation, but may be drought prone (Utsev et al., 2014). A summary of the chemical composition of analysed soil samples from Makurdi, Tarka, and Gboko is presented in Table 3.

Table 3. Descriptive summary of chemical properties of soils in the study area

Parameters	Makurdi		Tarka		Gboko	
	Mean	Std. Dev.	Mean	Std. Dev.	Mean	Std. Dev.
pH	5.9083	.28110	5.6500	.30896	5.3083	.18809
C (%)	.5200	.06769	1.1275	.57089	1.2725	.49393
N (%)	.0342	.00515	.0817	.04324	.0983	.03810
P (mg kg^{-1})	5.01	11.79	4.56	7.45	13.73	11.05
Ca (cmol kg^{-1})	2.7283	.62098	2.8517	.91714	3.9967	.77787
Mg (cmol kg^{-1})	2.2408	.45821	1.4733	.27988	2.2742	.72055
K (cmol kg^{-1})	.0892	.02353	.0708	.01379	.0967	.02425
Na (cmol kg^{-1})	.0592	.01084	.0550	.01168	.0675	.01055
Al3+ (cmol kg^{-1})	.2033	.08083	.3758	.59220	.1242	.09811
H+ (cmol kg^{-1})	.6092	.11389	.8542	.33288	.7425	.20877
ECEC (cmol kg^{-1})	5.9025	.80093	5.6808	.93675	7.3025	1.44385
Base saturation (%)	86.0833	2.71221	78.5000	14.78021	88.0000	2.79610

3.2 Land Use and Land Cover of Study Area

The study area had a predominance of scattered cultivation which supported the finding that the study area has a preponderance of agrarian peasants. Scattered cultivation covered a total area of 4,691.18 km^2 which made up 38.28% of the total study area which affirmed the field findings. The Built-up area accounted for 2,343.14 km^2 (19.12%) of the total area under study. Wetland and Waterbody (including rivers) covered a total area of 1,645.84 km^2 (13.43%) and 1,523.29 km^2 (12.43%) respectively. Bareland surfaces covered an area of 1,388.48 km^2 (11.33%) while Rock outcrops accounted for the least area occupying 662.99 km^2 representing 5.41% of the total area under investigation. The generated land use and land cover map is presented in Figure 2.

3.3 Normalised Difference Vegetation Index

The NDVI results showed that the study area is appreciably vegetated which buttressed the finding from the Land Use Land Cover. The NDVI analysis showed values ranging from -1 to +1 (Figure 3). After the reclassification operation, areas without vegetation were found to occupy a total area of 601.72 km^2 which represented 4.91% of the study area. Sparsely vegetated areas covered 8,312.51 km^2 (67.83%) which was the highest vegetal cover class. This was followed by 27.26% (3,340.69 km^2) which was covered by high vegetation. The result of the NDVI showed the general vegetation condition of the study area (Figures 4). These results further attest to the general suitability and potential of the study area for crop cultivation.

3.4 Rice Suitability Classes

The total area of 4,193.65 km^2 representing 34.22% of area under investigation was found to be highly suitable for rice cultivation (Table 4). Most of the other parts of the study area are moderately suitable for rice cultivation 45.46% (5,570.51 km^2). Very high suitable areas covered only 500.00 km^2 (4.08%). The suitability map is presented as Figure 5.

3.5 Cassava Suitability Classes

Cassava suitability classes showed that moderate suitability covered the largest part of the study area occupying 5,904.52 km^2 (48.18%). It was closely followed by areas of low suitability covering an area of 4,127.49 km^2 representing 33.68% of the total area (Table 5). Highly suitable areas occupied 2,093.13km^2 (17.08%) of the total area under investigation. The least area was occupied by the very low suitability class covering 96.90km^2 (0.79%) of the study area. Cassava is a crop that can survive on many soil types and usually copes with adverse weather conditions. It is therefore not surprising that given these fringe suitability classes (moderate and low), cassava seems to be a thriving crop in Benue state (Figure 6).

3.6 Yam Suitability Classes

Moderately suitable areas for yam cultivation made up 48.85% (5,986.41 km^2) of the study area and spread across the entire area under investigation (Table 6). The closest to moderate suitability was low suitability covering 29.57% (3,623.86 km^2) of the study area. The areas marked with very low suitability for yam cultivation was 598.54 km^2 representing 4.88% of the total study area (Figure 7).

Table 4. Suitability classes for rice

Suitability classes	Area (km^2)	Percentages %
Very low suitability	153.25	1.25
Low suitability	1,837.52	14.99
Moderate suitability	5,570.51	45.46
High suitability	4,193.65	34.22
Very high suitability	500.00	4.08
Total	12254.92	100%

Table 5. Suitability classes for cassava

Suitability classes	Area (km^2)	Percentages %
Very low suitability	96.90	0.79
Low suitability	4,127.49	33.68
Moderate suitability	5,904.52	48.18
High suitability	2,093.13	17.08
Very high suitability	32.89	0.27
Total	12254.92	100%

Table 6. Suitability classes for yam

Suitability classes	Area (km^2)	Percentages %
Very low suitability	598.54	4.88
Low suitability	3,623.86	29.57
Moderate suitability	5,986.41	48.85
High suitability	1,994.80	16.28
Very high suitability	51.31	0.42
Total	12254.92	100%

4. Discussion

In this paper, the use of remote sensing and GIS techniques allowed for the inclusion of various attributes specific to the study area which enhanced the accuracy and presentation of suitability maps for rice, cassava, and yam. These maps have revealed most suitable areas where cultivation of these crops should be focused. These results are in line with the assertion by Lingjun et al. (2008) that GIS and remote sensing has allowed for a transition from qualitative to quantitative assessment of land suitability based on relevant natural, economic, social and technical data. Similar modelling techniques have been documented in literature (Joss et al., 2008; Twumasi et al., 2012) which attest to the utilitarian value of this approach to suitability mapping for improved crop cultivation. The Lower River Benue Basin is known for high amounts of agricultural produce especially cereals, roots and tubers, and legumes. It is therefore not surprising that most parts of study area was found to be moderately suitable for the cultivation of rice, yam, and cassava. Notwithstanding, the suitability maps indicated that areas highly suitable and very highly suitable for these crops are not as predominant except for rice which had an appreciable percentage marked as highly suitable. The suitability map for rice (Figure 5) showed a high variation.

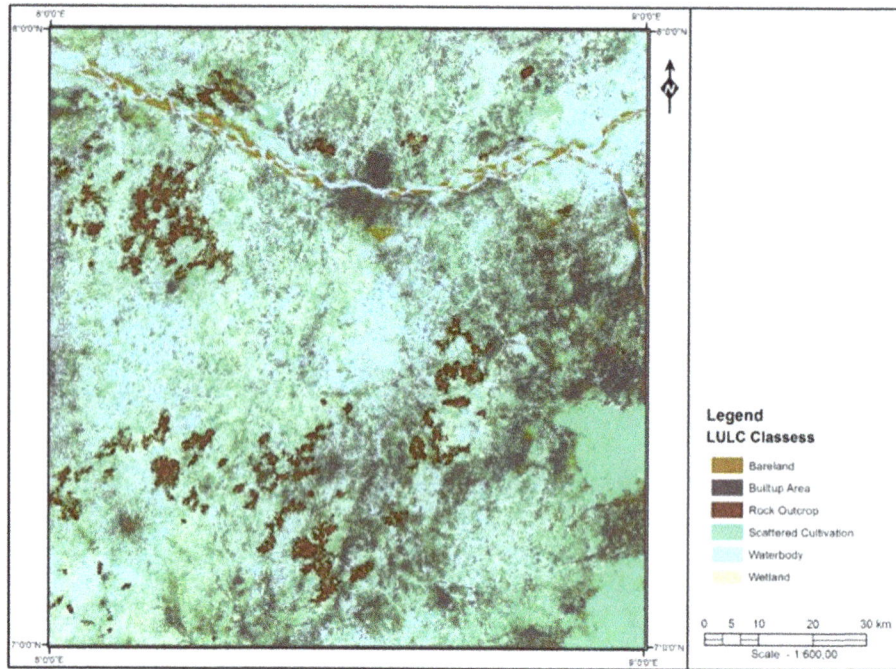

Figure 2. Land use land cover of the study area

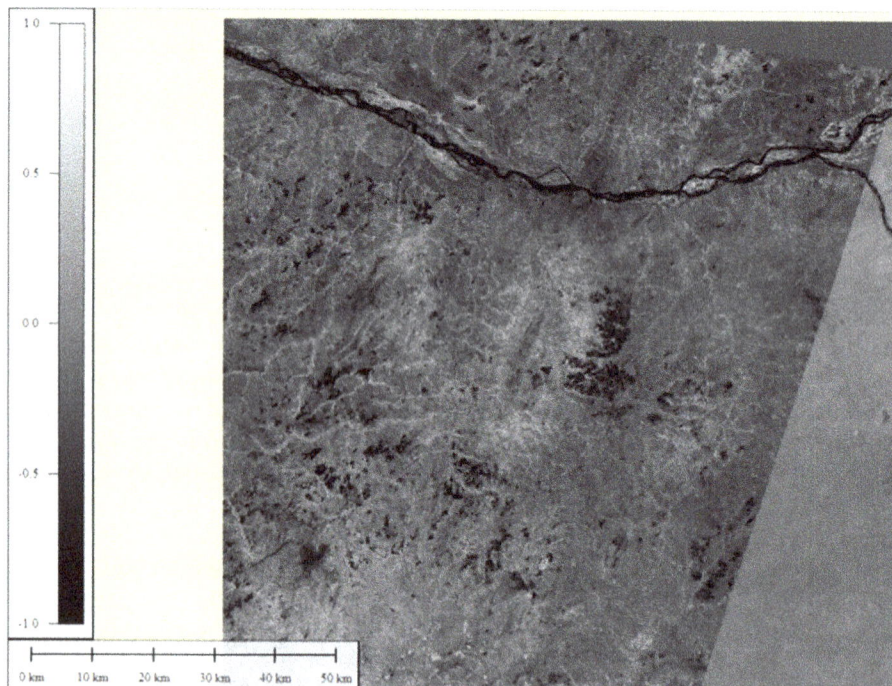

Figure 3. NDVI map of the study area

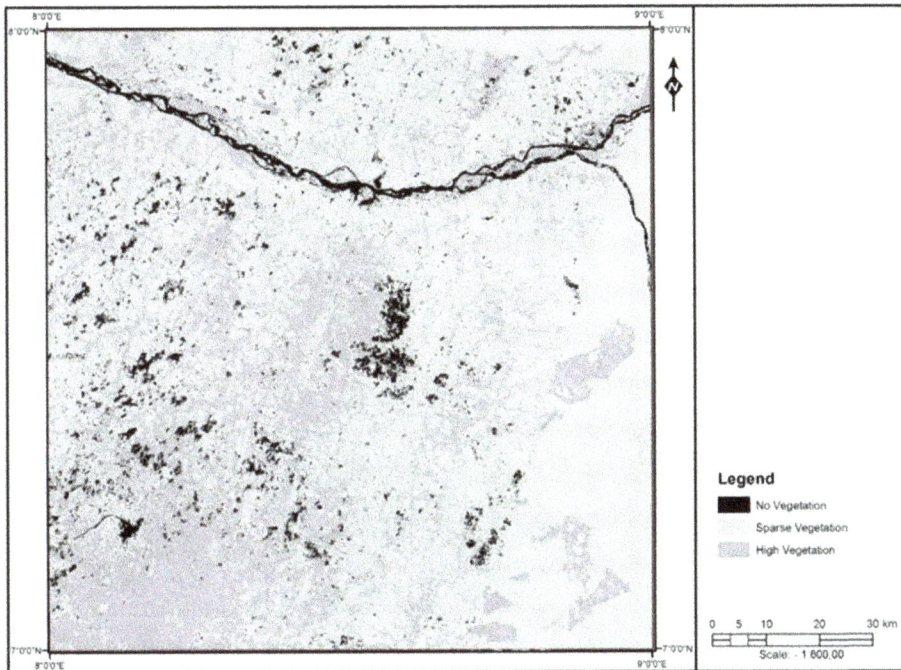

Figure 4. Reclassified NDVI map of the study area

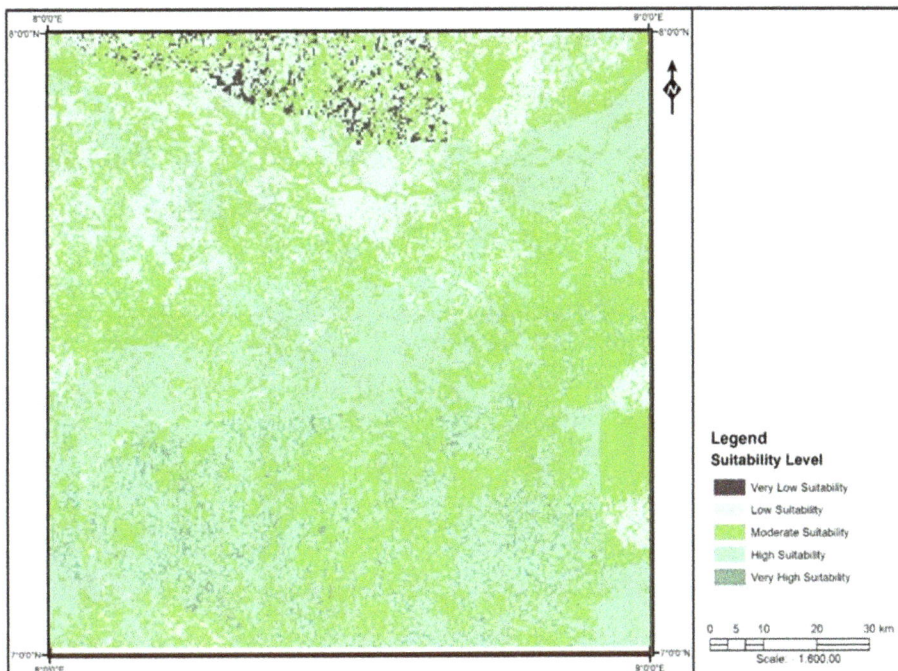

Figure 5. Rice suitability classes map

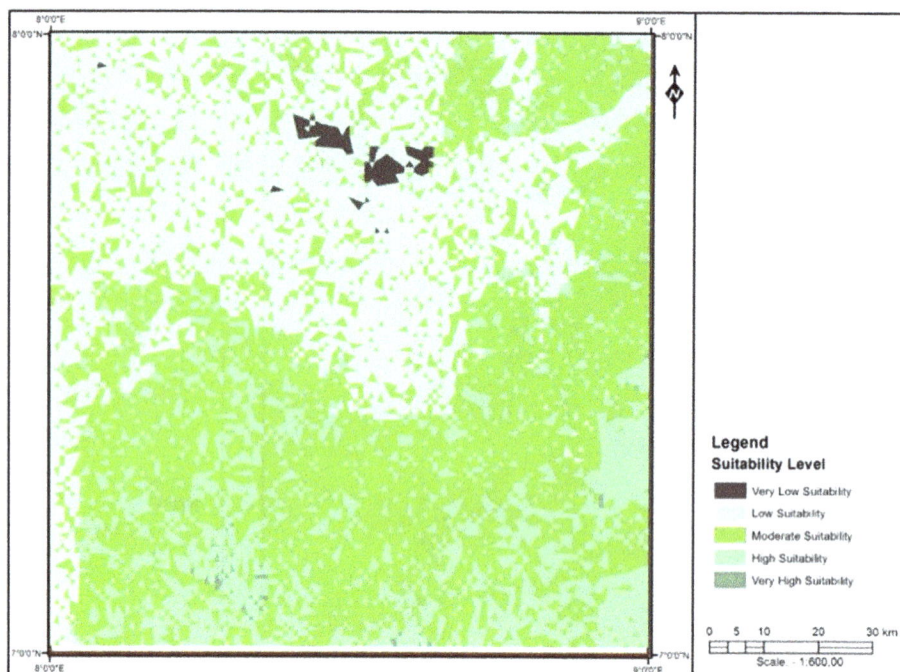

Figure 6. Cassava suitability classes map

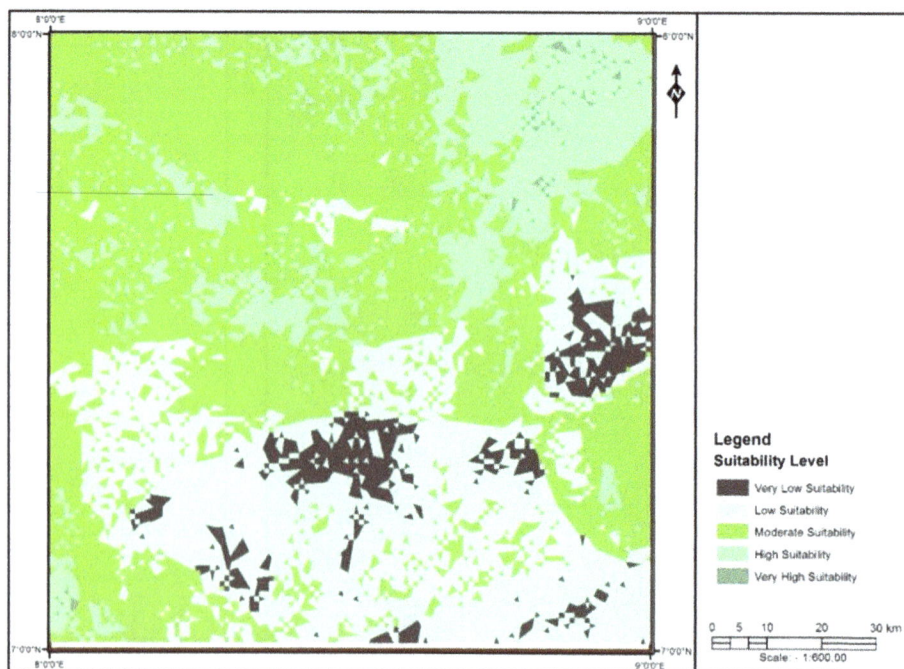

Figure 7. Yam suitability classes map

The areas found to be highly suitable and very highly suitable for rice cultivation fall under four Local Government Areas (L.G.As) including Gboko, Konshisha, Gwer east, and Otukpo (Figure 1). The streams in these areas are not tributaries of the River Benue but actually flow southwards from the River Benue. These areas experience the highest rainfall amounts and the least temperature. The areas marked as very highly suitable, highly suitable and moderately suitable for rice cultivation fall under the soil type Ferric Acrisol and Distric Notosol. One of the LGAs (Otukpo) used to have the largest rice mill in Nigeria but was neglected by

Government and succumbed to issues of illegal levies within the premises and obscure activities of middlemen. The areas marked high and very high suitability for rice cultivation have the highest concentration of built up areas.

The suitability map for cassava (Figure 6) showed that cassava is more successfully cultivated in areas of moderate elevation. The areas of very high suitability fell under Gwer east LGA. Areas marked as moderately suitable which were predominant fell under Tarka, Gboko, Gwer east, Konshisha, and Otukpo (Figure 1). Most parts of Benue however are known to produce large amounts of cassava annually. The rainfall and temperature of these areas are similar to those of rice suitability. The areas of very low suitability were areas close to Makurdi.

The suitability map for yam (Figure 7) was quite different from that of rice and cassava. The areas marked with very low suitability and low suitability were in parts of Gboko, Konshisha, Ushongo and parts of Gwer east which were quite suitable for rice cultivation. This may be a function of the relief system in the area. The suitability map for yam showed the least variation. The preferred areas for yam suitability appeared more in areas of lesser rainfall, higher temperature and moderate relief in contrast to that of rice.

Overall, rice had the highest suitability percentages for both the very highly suitable and the highly suitable categories (Figure 8). This is an indication that more areas are quite suited for rice cultivation than for yam and cassava. However, Figure 8 showed that the study area was moderately suitable for either of the crops examined with an average of more than 40% for each crop. Cassava has the least suitability for the combined low suitability classes (34.47%), and was followed by yam (34.45%). Generally, the areas of suitability for these crops potentially provides a good population of farmers for these crops given that the major occupation in Benue State is crop farming. However, the predominant mode of farming is a mix of traditional and semi-traditional. The suitable areas presented have potential to provide adequate physical conditions necessary to optimise the produce of cassava, rice and yam. Associated benefits with cultivation of crop in most suitable areas include reduced cost of inputs, minimal labour efforts, and availability of close markets.

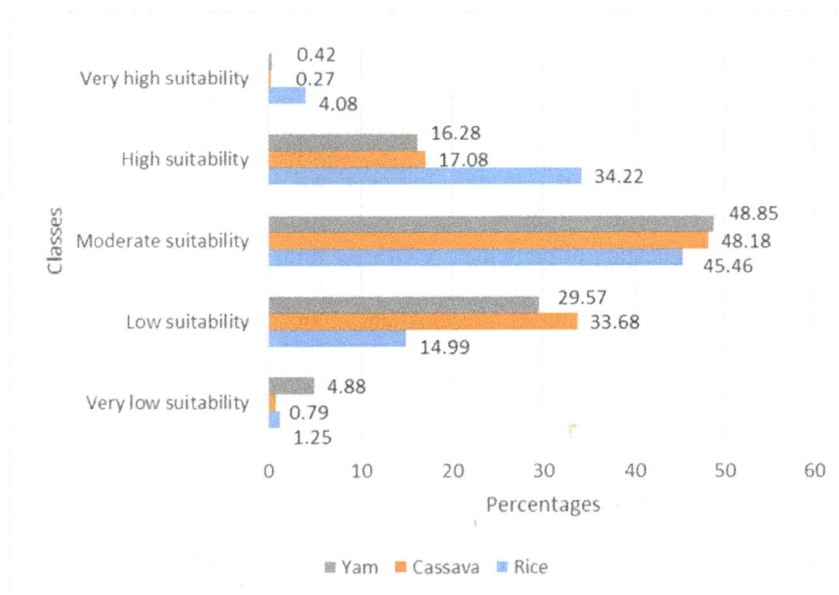

Figure 8. Suitability classes for rice, yam, and cassava in the study area

The study has dealt with crop suitability mapping for the promotion of rice, cassava, and yam cultivation in Benue State. The study utilised GIS and remote sensing methodology including an assessment of climate, soil, and crop cultivation variables. In order to draw focus to the improvement of crop cultivation in the study area, suitability maps have been produced to highlight areas most suitable for crop cultivation and especially the cultivation of rice, cassava, and yam. It is expected that farmers' cooperative societies and policy makers will utilise the information presented to improve decision making methods on choice of crop cultivation.

References

Ande, O. T. (2011). Soil suitability evaluation and management for cassava production in the derived savanna area of southwestern Nigeria. *International Journal of Soil Science, 6*, 142-149. http://dx.doi.org/10.3923/ijss.2011.142.149

Aondoakaa, S. C., & Agbakwuru, P. C. (2012). An assessment of land suitability for rice cultivation in Dobi, Gwagwalada area council, FCT Nigeria. *Ethiopian Journal of Environmental Studies and Management EJESM, 5*(4), 444-451. http://dx.doi.org/10.4314/ejesm.v5i4.S2

Ashraf, S. (2010). Land suitability analysis for wheat using multi-criteria evaluation and GIS method. *Research Journal of Biological Sciences, 5*, 601-605. http://dx.doi.org/10.3923/rjbsci.2010.601.605

Enete, A. A., & Amusa, T. A. (2010). Challenges of agricultural adaptation to climate change in Nigeria: A synthesis from the literature. *Field Actions Science Reports 4*. Retrieved from http://factsreports.revues.org/678

Eruola, A. O., Bello, N. J., Ufoegbune, G. C., & Makinde, A. A. (2012). Effect of variety selection on growth, development and yield of white yam in southwestern Nigeria. *International Journal of Agriculture and Forestry, 2*(3), 101-104. http://dx.doi.org/10.5923/j.ijaf.20120203.04

FAO. (1976). A framework for land evaluation. *Soils Bulletin 32*. Rome: FAO.

FAO. (2016). *Nigeria at a glance.* Retrieved from http://www.fao.org/nigeria/fao-in-nigeria/nigeria-at-a-glance/en/

Ghosh, M., & Ghosh, S. K. (2013). *Rural planning models using GIS and remote sensing: A review.* Retrieved from http://www.gisdevelopment.net/application/lis/rural/mi0599.htm

Heumann, B. W., Walsh, S. J., Verdery, A. M., McDaniel, P. M., & Rindfuss, R. R. (2013). Land suitability modeling using a geographic socio-environmental niche-based approach: A case study from northeastern Thailand. *Annals of the Association of American Geographers, 103*. http://dx.doi.org/10.1080/00045608.2012.702479

IITA. (2004). *Nigerian's Cassava Industry.* Statistical Handbook.

IITA. (2009). *Crops: Yam.* Retrieved from http://www.iita.org/yam

Joss, B. N., Hall, R. J., Sidders, D. M., & Keddy, T. J. (2008). Fuzzy-logic modelling of land suitability for hybrid poplar across the Prairie Provinces of Canada. *Environmental Monitoring and Assessment, 141*, 79-96. http://dx.doi.org/10.1007/s10661-007-9880-2

Kumara, B. A. U. I. (2008). Application of participatory GIS for rural community development and local level spatial planning system in Sri Lanka. *Geospatial Application Papers.* Retrieved from http://www.gisdevelopment.net/application/lis/rural/srilanka.htm

Kutigi, A. D. (2002). Comparative analysis between indigenous and modern tillage practices in the production of Dioscorea rotundata and Dioscorea cayenensis in the middle belt of Nigeria. *Journal of Agriculture Environment, 2*(1), 53-66.

Lingjun, L., Zong, H., & Yan, H. (2008). *Study on land use suitability assessment of urban-rural planning based on remote sensing—A case study of Liangping in Chongqing.* The International Archives of the Photogrammetry, Remote Sensing and Spatial Information Sciences, XXXVII, B8. Beijing. Retrieved from http://www.isprs.org/proceedings/XXXVII/congress/8_pdf/1_WG-VIII-1/22.pdf

Nuga, B. O. (2001). Application of geographic information systems (GIS) for sustainable land resource management in Nigeria: A review. *Journal of Applied Chemistry and Agricultural Research, 7*, 104-111. http://dx.doi.org/10.4314/jacar.v7i1.41132

Ortserga, D. S. (2012). Towards a model explaining change in agricultural land use patterns in Ute districts of Vandeikya local government area, Benue State, Nigeria. *International Journal of Humanities and Social Science, 2*, 251-264.

Petja, B. M., Nesamvuni, E., & Nkoana, A. (2014). Using geospatial information technology for rural agricultural development planning in the Nebo Plateau, South Africa. *Journal of Agricultural Science, 6*, 10-18. http://dx.doi.org/10.5539/jas.v6n4p10

Rilwani, M. L., & Gbakeji, J. O. (2009). Geoinformatics in agricultural development: challenges and prospects in Nigeria. *Journal Social Sciences, 21*, 49-57.

Rilwani, M. L., & Ikuoria, I. A. (2006). Precision farming with geoinformatics: A new paradigm for agricultural production in a developing country. *Transactions in GIS, 10*(2), 177-197. http://dx.doi.org/10.1111/j.1467-9671.2006.00252.x

Riveira, I. S., & Maseda, R. C. (2006). A review of rural land use planning models. *Environment and Planning B: Planning and Design, 33*, 165-183. http://dx.doi.org/10.1068/b31073

Rouse, J. W., Haas, R. H., Schell, J. A., & Deering, D. W. (1973). *Monitoring vegetation systems in the Great Plains with ERTS* (SP-351(I), pp. 309-317). Third ERTS Symposium, NASA.

Stickler, C., Coe, M., Nepstad, D., Fiske, G., & Lefebvre, P. (2007). *Readiness for REDD: A preliminary global assessment of tropical forested land suitability for agriculture*. United Nations Framework Convention on Climate Change (UNFCCC) Conference of the Parties (COP), Thirteenth session, December 3-14, 2007. Bali, Indonesia.

Titus, P., Lawrence, J., & Seesahai, A. (2011). *Commercial cassava production, technical bulletin*. Retrieved from http://www.cardi.org/.../2011/.../commercial-cassava-production-technical

Twumasi, Y. A., Manu, A., Coleman, T. L., Merem, E. C., & Osei, A. (2012). Visualization of rangeland resources from space: A case of Sahel region, West Africa. *International Journal of Geomatics and Geosciences, 3*, 259-268.

Uchua, K. A., Olowolafe, E. A., & Nduke, G. E. (2012). *Mapping and analysis of agricultural systems using GIS in the lower river Benue basin Nigeria, Conference paper*. Tropical and Subtropical Agricultural and Natural Resource Management (TROPENTAG), September 19-21, Germany.

United States Geologic Survey (USGS). (2015). *Landsat Missions: Landsat processing details*. US Department of Interior. Retrieved from http://landsat.usgs.gov/Landsat_Processing_Details.php

Utsev, J. T., Ajon, A. T., & Ugama, T. I. (2014). Irrigation Soil Analysis in River Katsina-Ala Catchment Areas of North-Central Nigeria. *International Journal of Engineering and Technology, 4*(6), 387-393.

3

Knowledge and Adoption of Conservation Agriculture Technologies by the Farming Community in Different Agro-Climatic Zones of Tamilnadu State

M. R. Ramasubramaniyan[1], J. Vasanthakumar[2] & B. S. Hansra[3]

[1] National Agro Foundation, Chennai, Tamilnadu, India

[2] Faculty of Agriculture, Annamalai University, Tamilnadu, India

[3] School of Agriculture, Indira Gandhi National Open University, New Delhi, India

Correspondence: M. R. Ramasubramaniyan, National Agro Foundation, CSIR Road, Taramani, Chennai 600 113, Tamilnadu State, India. E-mail: nafsree@gmail.com

The research is self sponsored as part of doctoral thesis of the first author.

Abstract

Tamilnadu State in India is one of the earlier beneficiaries of Green Revolution which contributed to multifold increase in agricultural productivity. Though the State has been progressive, it has also experienced the ill effects of over exploitation of natural resources through intensive agriculture. There is an urgent need to shift its focus from over exploitative intensive farming to more sustainable farming with optimal use of resources without causing imbalances. Conservation Agriculture (CA) offers potential solution which not only enhances the productivity but also maintains the environmental safety and ecological sustainability. With this at the backdrop, the present study was conducted during 2013-2014 in all the seven agro-climatic zones of Tamilnadu State in India covering 350 respondents to understand the knowledge and adoption levels of Conservation Agriculture among the farming community in the State. Three Conservation agricultural technologies namely, Minimum Tillage, Crop Rotation and Permanent Soil Cover were identified. Knowledge of the farmers about these technologies and their adoption by the farmers were studied. As regards the awareness and knowledge level of respondents majority of them do not have knowledge on minimum tillage (72.6%) and permanent soil cover (75.1%) but a vast majority is knowledgeable on crop rotation (71.1%). Farmer characteristics such as age, educational status and innovativeness of farmers played a significant impact on the knowledge levels of CA whereas number of years of experience in farming and land holding pattern did not have significant influence on the knowledge levels of farmers on CA. Among the knowledgeable farmers only 11.5% of farmers adopted minimum tillage, 27.6% of farmers adopted permanent soil cover and 78% adopted crop rotation. None of the farmers adopted CA as a whole comprising all the three components.

Keywords: minimum tillage, permanent soil cover, crop rotation, knowledge, adoption, conservation agriculture, agro-climatic zones, Tamilnadu, India

1. Introduction

The urge to increase the food grain production to feed the growing population has led to Intensive agriculture in India through Green Revolution. While the agriculture productivity increased multifold, the associated problems like over exploitation of natural resources, environmental degradation and imbalance in biodiversity etc have caused serious concerns about sustainability of such production systems. Moreover, there has been a paradigm shift from mere sustenance and production oriented agriculture in early 60's and 70's to systems oriented agriculture after 2000. A search for alternative solution which is environmentally safe, ecologically sustainable and yet economically profitable was made by all the stakeholders like farmers, scientists, government and civil society organizations. A set of soil-crop-nutrient-water-landscape system management practices known as Conservation Agriculture (CA) offers potential solution which not only enhances the productivity but also maintains the environmental safety and ecological sustainability. Conservation Agriculture (CA) is an approach to manage agro eco systems for improved and sustained productivity, increased profits and food security while preserving and enhancing the resource base and the environment (FAO, 2011).

The key elements or practices which characterize CA include: (i) minimum soil disturbance by adopting minimum tillage and traffic for agriculture operations (ii) leave and manage the crop residues on the soil surface and (iii) adopt spatial and temporal crop sequencing/crop rotations to derive maximum benefits from inputs and minimize adverse environmental impacts (Abrol & Sangar, 2006). The adoption of CA practices however faces many barriers viz., mindset, know how, machineries and policies (Derpsch et al., 2012).

Tamilnadu state, one of the earlier adopters of intensive agriculture through Green Revolution has also become a victim of it in terms of environmental degradation, loss of natural resources and the other associated ill effects. There is an urgent need to correct this in order to ensure the long term sustainability of farming for which CA offers potential solution. Though CA is relatively new concept, it has found an important place in global agriculture including India. It has also been practiced extensively in Indo Gangetic Plains which was the seat of success of Green Revolution. Similarly CA is relevant to Tamilnadu, which was one of the early adopters of Green Revolution. However, not much of work has been reported on CA in Tamilnadu. In view of the above, the present study was intended to explore the penetration of CA technologies among the farming community across Tamilnadu State and also to study the Knowledge of Conservation Agriculture technologies as a package, its suitability and adoption at all the seven agro-climatic zones of Tamilnadu.

2. Methodology

Tamilnadu State has been classified into seven agro-climatic zones. The study was conducted in all the seven agro-climatic zones covering entire Tamilnadu State which included North Eastern Zone, North Western Zone, Western Zone, Cauvery Delta Zone, Southern Zone, High Rainfall Zone and Hilly Zone. It is assumed that CA is common irrespective of different agro-climatic zones and is hypothesized that there has not been much awareness and knowledge on CA in Tamilnadu and the present study is intended to investigate the same. Since, each of the agro-climatic zones is characterized by unique climate, rainfall, cropping systems and cropping patterns the present research was intended to study the knowledge and adoption of conservation agriculture. The study was conducted in 2013-2014. The study area was selected in such a way that in each of the agro-climatic zones, the blocks where annual crops are predominantly cultivated were selected as conservation agriculture is more applicable and relevant to those cropping pattern where intensive agriculture throughout the year is practiced.

From the selected blocks, study villages were selected by simple random sampling. Since the cropping pattern was almost uniform across each of the agro-climatic zones as per the secondary data, one block per agro-climatic zone was randomly selected irrespective of the number of blocks present in the zone. The respondents were selected using simple random sampling method. Totally three hundred and fifty respondents were randomly selected in seven agro-climatic zones at fifty respondents in each of the agro-climatic zone.

Data was collected with the use of a well structured and pre- tested interview schedule. The data thus collected were statistically analysed using SPSS package.

2.1 Brief about the Study Area

The districts where the study was conducted in each of the agro-climatic zones is summarized below.

2.1.1 Kancheepuram District–North Eastern Zone

Kancheepuram district is situated on the North East coast of Tamil Nadu. It lies between 11°00′ to 12°00′ latitudes and 77°28′ to 78°50′ longitudes. The district has a total geographical area of 4,43,210 hectares and a coastline of 57 km. the district has the maximum temperature of 45 °C and a minimum of 21.1 °C during summer and a maximum of 28.7 °C and a minimum of 14 °C in winter. The district is mainly dependent on the monsoon rains. During normal monsoon, the district receives a rainfall of 1200 mm.

2.1.2 Dharmapuri District–North Western Zone

Dharmapuri district is situated in the North western Corner of Tamil Nadu. It is located between latitudes N 11°47′ and 12°33′ and longitudes E 77°02′ and 78°40′. The total geographical area of Dharmapuri District is 4497.77 Sq Kms. The climate of the Dharmapuri District is generally warm with the maximum temperature touching 38 °C. and minimum of 17 °C. On an average the District receives an annual rainfall of 895.56 mm. The District economy is mainly agrarian in nature. It has a total gross cropped area of 1.69 lakh hectares. Millets, pulses and Paddy are the main crops

2.1.3 Theni District–Western Zone

The district lies at the foot of the Western Ghats between 9°39' and 10°30' North latitude and between 77°00' and 78°30' of East Longitude. In the plains, the temperatures range from a minimum of 13 °C to a maximum of 39.5 °C. In the hills the temperatures can range from as low as 4-5 °C to 25 °C. The district is known for its salubrious climate, hills and lakes. Its economy is mostly agricultural. Utilization of land area for cultivation in Theni district is 40.33%. The principal crops include sugarcane, cotton, paddy, millets and pulses.

2.1.4 Sivaganga District–Southern Zone

The district of Sivagangai, extending over an area of 4468.11 Sq. Km, is situated in the southeastern portion of the state. The district lies between 9°43' and 10°2' north latitude and 77°47' and 78°49' east Longitude. Temperature is low during the month of January and the lowest mean daily temperature is 19.8 °C. The hottest month in the district is July during which period the maximum temperature is 33.83 °C. Mean humidity varies from 65% in July to 77% in November. The average rainfall of the District is below 800 mm. The principal crop of Sivaganga district is paddy. The other crops that are grown are millets, cereals, pulses, sugarcane, and groundnut.

2.1.5 Thanjavur District–Cauvery Delta Zone

Thanjavur District lies in the East Coast of Tamil Nadu. It is located between 90°50' and 110°25' of the northern latitude and 70°25' and 78°45' of the Eastern longitude. In Thanjavur district brown coloured soil was the maximum constituting nearly 65%. Red soil and black soils were found in 19.30 and 15.97 percent of the area respectively. The climate of Thanjavur can be termed as a fairly healthy one like other coastal areas. The South-West monsoon sets in June and continues till September followed by North-East monsoon in October upto January. The total gross cropped area is 2.41 lakh hectares. The major crops of the district are Paddy, Pulses, Sugarcane, Groundnut, Gingelly, Cotton and Coconut.

2.1.6 Dindigul District–Hilly Zone

Dindigul District is located between 10°05' and 1°09' North Latitude and 77°30' and 78°20' East Longitude. The district has a net sown area of 2.53 lakh hectares. The major crops of the district are Paddy, Maize, Sugarcane, Pulses and Cotton.

2.1.7 Kanyakumari District–High Rainfall Zone

The district lies between 77°15' and 77°36' of the eastern longitudes and 8°03' and 8°35' of the northern latitudes. The district has a gross cropped area of 0.92 lakh hectares. The major soil type in the district is Red soil (65,608 ha), which constituted about 67% of the total cultivated area. Lateritic soil (20,003 ha) is the next major soil type, which formed 22% of the total cultivated area in the district.

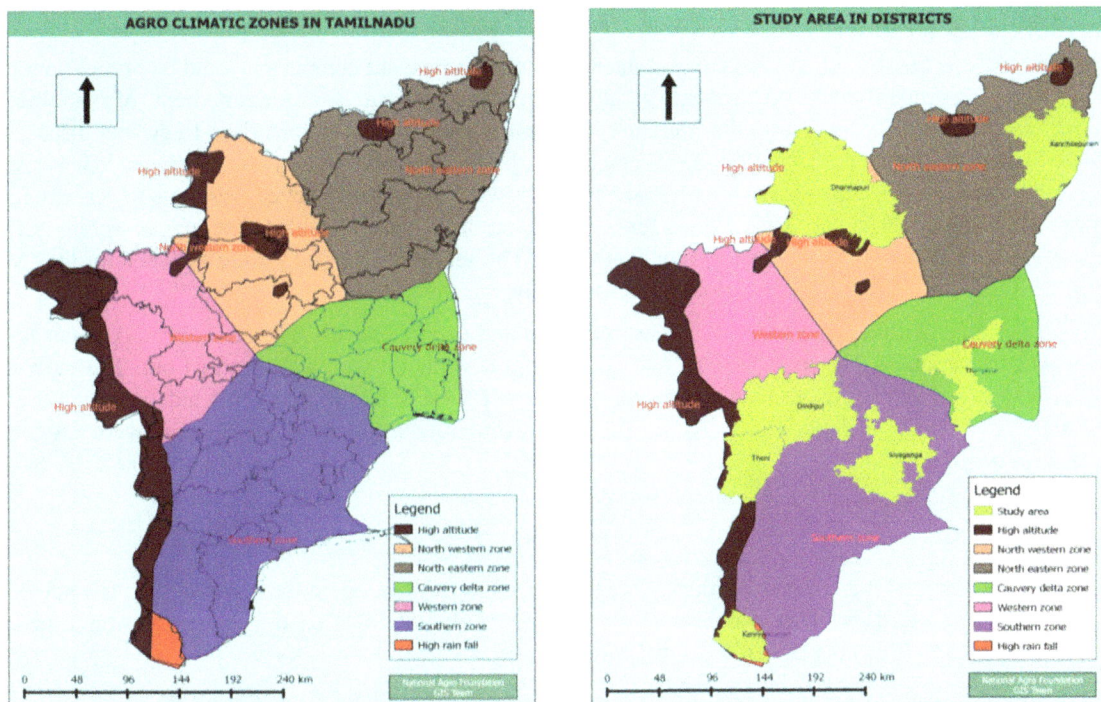

Figure 1. Map of agro-climatic zones of tamilnadu and study area districts

Table 1. Agro-climatic zone wise distribution of districts and the blocks

S No	Name of the agro-climatic zone	Districts covered	No. of blocks
1	North Eastern Zone	Kancheepuram, Tiruvallur, Cuddalore, Vellore, Villupuram and Tirunvannamalai	100
2	North western Zone	Dharmapuri, Krishnagiri, salem and Namakkal (Part)	53
3	Western Zone	Erode, Coimbatore, Tiruppur, Theni, Karur (part), Namakkal (part), Dindigul, Perambalur and Ariyalur (part)	60
4	Southern zone	Madurai, Sivagangai, Ramanathapuram, Virudhunagar, Tirunelveli and Thoothukudi.	78
5	Cauvery Delta Zone	Thanjavur, Nagapattinam, Tiruvarur, Trichy and parts of - Karur, Ariyalur, Pudukkottai and Cuddalore	74
6	Hilly zone	The Nilgiris and Kodaikanal (Dindigul)	5
7	High rainfall zone	Kanyakumari	9

Table 2. Agro-climatic zone wise distribution of districts and the blocks under study

S No	Name of the agro- climatic zone	Name of the district under study	Name of the block under study
1	North Eastern Zone	Kancheepuram	Chitamur block
2	North Western Zone	Dharmapuri	Pappireddypatti
3	Western Zone	Theni	Vadugapatti and Cumbum
4	Southern Zone	Sivaganga	Thirupathur
5	Cauvery Delta Zone	Thanjavur	Ammapettai
6	Hilly Zone	Dindigul	Poomparai
7	High Rainfall Zone	Kanyakumari	Karungal

Note. The map of the study area is depicted in Figure 1.

2.2 Selection of Dependent and Independent Variables

The critical analysis of the study, its meaningful interpretation and relevant conclusion could be brought out only when relevant dependent and independent variables are selected and the measurement of variables is appropriately followed. By reviewing the literature, discussion with extension scientists, freewheel discussions conducted and by observations made by the researcher in the initial stages twenty three independent variables which may influence the dependent variables of the study were selected. A letter was sent to ten extension scientists of the State Agriculture Universities for the selection of relevant variables. The judges were requested to indicate the relevancy rating on a three-point continuum. The scores of 3, 2 and 1 were assigned for the "More relevant", "Relevant" and "Irrelevant" responses respectively.

Based on the rating and judgments by the judges, the mean and coefficient of variation were worked out for all the independent variables. The overall mean and coefficient of variation were also worked out. The individual variables with mean scores greater than the overall mean score were selected. Thus five independent variables were finally selected and included for the study. The dependent variables selected for the study were knowledge and extent of adoption.

2.3 Measurement of Variables

2.3.1 Age

The chronological age of the respondents during the time of interview was operationalized as age. The age of the respondents in completed years was considered for analysis. The classification developed by Arunmozhidevi (2004) and adopted by Vijayalakshmi (2012) was followed.

Table 3. Age

S No	Category	Score
1	Young (18-35 years)	1
2	Middle (36-50 years)	2
3	Old (above 50 years)	3

2.3.2 Educational Status

It referred to the level of literacy of the respondent during the interview. The categories were illiterate, functionally literate, primary education, middle education, secondary education and college education. The respondents who could not read and write were categorized as illiterate. The respondents who could read and write were considered as functionally literate. Primary education refereed to the formal school education upto fifth standard. Middle education referred to formal school education upto eighth standard. Secondary education referred to the formal education upto plus two or junior college level. Collegiate education meant the education as diploma/degree after schooling. The following scoring procedure developed by Mansingh (1993) was followed to arrive at a score on educational status of the respondents.

Table 4. Educational status

S No	Category	Score
1	Illiterate	1
2	Functionally literate	2
3	Primary education	3
4	Middle education	4
5	Secondary education	5
6	College education	6

2.3.3 Experience in Farming

It was operationalized as the number of years of experience in farming possessed by the respondents during interview. The scoring procedure adopted by Puthiraprathap (2003) was used.

Table 5. Experience in farming

S No	Category	Score
1	Upto 10 years	1
2	11 – 20 years	2
3	Above 20 years	3

2.3.4 Farm Size

Farm size referred to the extent of land cultivated by an individual at the time of enquiry. The area was directly taken as a measure and categorized into three by using the following scoring procedure as adopted by Puthirapratap (2003).

Table 6. Farm size

S No	Category	Area	Score
1	Marginal farm	Less than 2.5 acres	1
2	Small farm	Between 2.51-5.0 acres	2
3	Big farm	More than 5.0 acres	3

2.3.5 Innovativeness

Innovativeness was operationalized as the degree to which an individual is relatively earlier in adopting a new idea or technology. The scoring procedure developed by Singh (1972) and used by Santhi (2006) was followed to measure innovativeness.

Table 7. Innovativeness

Question: When would you prefer to adopt any improved technology?		
S No	Answer	Score
1	As soon as it was brought to my knowledge	3
2	After seeing other farmers have done it successfully	2
3	I prefer to wait and take my own time	1

3. Results and Discussion

3.1 Profile of the Respondents

3.1.1 Age

Table 8. Distribution of respondents according to their age (n = 350)

S No	Category	Frequency	Percentage
1	Young (< 35 years)	74	21.1
2	Middle (35-45 years)	101	28.9
3	Old (> 45 years)	175	50.0
4	Total	350	100.0

The categorization of respondents as per their age is presented in Table 8. The respondents were categorized as young (18 to 35 years of age), middle aged (36-45 years) and old aged (more than 45 years of age). Old aged farmers form the majority (50%) closely followed by middle aged (28.9%). Young farmers were only 21.1% which shows that majority of the youth are not attracted to farming.

3.1.2 Educational Status

Table 9. Distribution of respondents according to their educational status (n = 350)

S No	Category	Frequency	Percentage
1	Functionally Literate	63	18.0
2	Primary	113	32.3
3	Middle	88	25.1
4	Secondary	52	14.9
5	Graduate	34	9.7
	Total	350	100.0

The categorization of respondents according to their educational status is presented in Table 9 as educational status tend to influence the knowledge and adoption of new technologies like CA. Accordingly, majority of the respondents had primary education (32.3%) followed by middle level education (25.1%). The proportion of respondents under secondary and college education category was 14.9% and 9.7% respectively.

3.1.3 Experience in Farming

Table 10. Distribution of respondents according to their experience in farming (n = 350)

S No	Category	Frequency	Percent
1	Low (up to 10 years)	86	24.6
2	Medium (10-20 years)	90	25.7
3	High (> 20 years)	174	49.7
	Total	350	100.0

Experience in farming is an important farmer characteristic which helps farmers in evaluating the usefulness of any new technology based on their past experiences and thus influences the decision making of farmers about a particular technology. The categorization of respondents according to their experience in farming is presented in Table 10. Accordingly, about half of the farming community (49.7%) had rich experience of farming over 20 years. Respondents with low and medium level of farming experience were almost equal which is 24.6% and 25.7% respectively.

3.1.4 Land Holding Pattern

Table 11. Distribution of respondents according to their farm size (n = 350)

S No	Category	Frequency	Percent
1	Marginal & Small (< 2.5 ac)	164	46.9
2	Medium (2.5-5 ac)	109	31.1
3	Large (> 5 ac)	77	22.0
	Total	350	100.0

The categorization of respondents as per the land holdings are presented in Table 11. Accordingly, majority of the respondents (46.9%) fall under the category of marginal farmers with land holding upto 2.5 acres whereas 31.1% of the respondents were under medium land holding category of 2.51-5.0 acres. Big farmers formed only 22.0%.

3.1.5 Innovativeness

Table 12. Distribution of respondents according to their innovativeness (n = 350)

S No	Category	Frequency	Percentage
1	Low	177	50.6
2	Medium	124	35.4
3	High	49	14.0
	Total	350	100.0

Innovativeness is a key farmer characteristic which influences the degree of speed with which a new technology is acknowledged and adopted by a farmer. Mostly, innovativeness is an inherent character of the farmer which further decides the risk taking ability as well as decision making ability. Table 12 presents the distribution of respondents based on their innovativeness. It shows that more than half of the respondents (50.6%) had low level of innovativeness as they preferred to take their own time to adopt a new technology. Only 14.0% of the respondents had high level of innovativeness as they wished to adopt any new technology as soon as they knew about it.

The low level of innovativeness may be due to the fear of failure of adoption of innovative technologies. Moreover, since majority of them are small and marginal farmers with agriculture being the prime occupation, the respondents tend to take minimum risk, which is also one of the reasons for low level of innovativeness.

3.2 Knowledge and Adoption of CA

3.2.1 Knowledge Level of the Respondents on Conservation Agriculture Practices

CA Being a new technology, the present study attempted to capture information on level of knowledge of respondents on individual components of CA as well as CA as a whole. The findings of the study are presented below.

1) Knowledge level on minimum tillage

The following table presents the findings on the distribution of respondents based on the knowledge level of minimum tillage as a conservation agriculture practice.

Table 13. Distribution of respondents according to their knowledge level on minimum tillage (n = 350)

S No	Level of knowledge	Frequency	Percentage
1	No Knowledge	254	72.6
2	Knowledgeable	96	27.4
	Total	350	100.0

The data presented above indicate that about three fourth of the respondents (72.6%) had low or no knowledge level on minimum tillage practices. About 27.4% of the respondents had medium level of knowledge on minimum tillage. The respondents with medium level of knowledge were predominantly followers of sustainable agriculture and organic farming practices.

Majority of the respondents followed conventional farming with intensive tillage and they continue to practice the same as they do not have any knowledge on minimum or zero tillage. Usharani and Neelu (2011) had earlier reported similar views with low level of knowledge on minimum tillage.

2) Knowledge of permanent soil cover

The level of knowledge of the respondents on permanent soil cover is presented in below.

Table 14. Distribution of respondents according to their level of knowledge on permanent soil cover (n = 350)

S No	Level of knowledge	Frequency	Percentage
1	No knowledge	263	75.1
2	Knowledgeable	87	24.9
	Total	350	100.0

The results of the data from the above Table reveal that about 24.9% of the respondents had medium knowledge on permanent soil cover with stubbles and crop mulches. A very high proportion of respondents (75.1%) did not have proper on permanent soil cover. Only a small proportion of respondents followed mulching as they were predominantly sugarcane growers and covering the soil with sugarcane trashes as advised by the sugar mills. A portion of respondents were also growers of perennial crops like mango who normally had some cover crops in the initial stages to control the weeds as well as to earn income during the initial gestation periods of the perennial crops.

The reason for very low levels of awareness and knowledge on minimum tillage and permanent soil cover may be due to the fact that these were relatively new technologies and no agency had ever introduced these technologies to them. However, these findings were in contrast with the findings of Mazvimavi et al. (2010).

3) Crop rotation

The level of knowledge knowledge of the respondents on crop rotation is presented in the below.

Table 15. Distribution of respondents according to their knowledge on crop rotation (n = 350)

S No	Level of knowledge	Frequency	Percentage
1	No knowledge	101	28.9
2	Knowledgeable	249	71.1
	Total	350	100.0

As far as the third conservation agriculture practice viz., crop rotation is concerned, the data from Table 15 shows that a vast majority of respondents (71.1%) expressed that they are knowledgeable about crop rotation. However, it was also observed that about one fourth (28.9%) of the respondents did not have knowledge on crop rotation.

The reason for high level of knowledge on crop rotation may be due to the fact that crop rotation has been an age old practice in the study area and is being followed for generations together. Moreover, farmers expressed that following crop rotations is beneficial for maintaining the soil fertility and management of pests. These findings are in line with the findings of Nyanga et al. (2011).

4) Overall knowledge on conservation agriculture

The distribution of respondents based on their knowledge levels on conservation agriculture as a whole is presented below.

Table 16. Distribution of respondents according to their overall knowledge on CA (n = 350)

S No	Category	Frequency	Percentage
1	Low	164	46.9
2	Medium	96	27.4
3	High	90	25.7
	Total	350	100.0

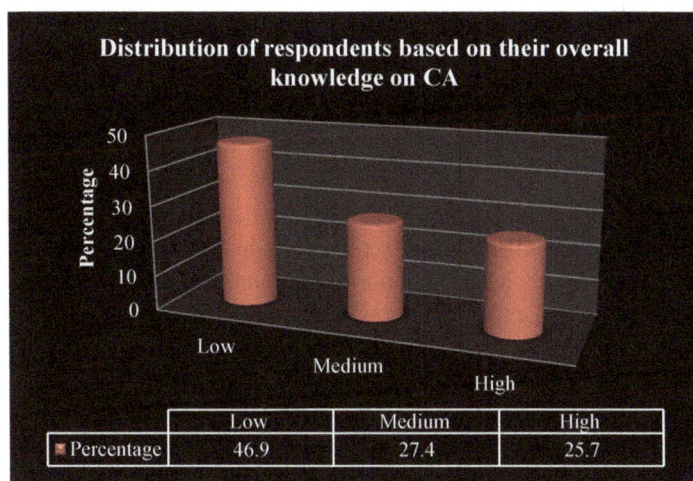

Figure 2. Distribution of respondents based on their overall knowledge on CA

The findings from the Table above give a picture of overall knowledge of respondents on CA comprising minimum tillage, permanent soil cover and crop rotation as a whole. It indicates that about 46.9% of the respondents have low level of knowledge on CA. While 27.4% of the respondents have medium level of knowledge on CA, about 25.4% of the respondents have high knowledge level on CA. The findings presented above clearly indicate that about half of the respondents had only lower levels of knowledge on CA. Only one fourth of the respondents had higher levels of knowledge on CA. Though majority of the respondents do not have knowledge on minimum tillage and permanent soil cover, the high level of knowledge on crop rotation reflects on the overall knowledge on CA as a whole with about 53.1% of respondents knowledgeable about it. These findings find support from the findings of Usharani and Neelu (2011).

3.3 Relationship between the Socio-Economic and Psychological Characteristics of the Respondents and Their Knowledge Level on CA

The following section deals with the relationship of socio-economic and psychological characteristics of respondents and their knowledge level of conservation agriculture.

3.3.1 Distribution of Respondents Based on Their Socio-Economic and Psychological Characteristics and Knowledge Level

1) Age

The age wise distribution of respondents based on their knowledge on CA practices is presented in the Table below.

Table 17. Age wise distribution of respondents based on knowledge level of CA

S No	Category of age	Level of Knowledge on CA			Total	Chi square value	P value
		Low	Medium	High			
1	Young	33	24	17	74		
		(44.6%)	(32.4%)	(23.0%)			
		[20.1%]	[25.0%]	[18.9%]			
2	Medium	50	35	16	101	12.607	0.013*
		(49.5%)	(34.7%)	(15.8%)			
		[30.5%]	[36.5%]	[17.8%]			
3	Old	81	37	57	175		
		(46.3%)	(21.1%)	(32.6%)			
		[49.4%]	[38.5%]	[63.3%]			

Note. The value within ( ) refers to Row Percentage; The value within [ ] refers to Column Percentage; ** denotes significance at 1% level; *denotes significance at 5% level.

The results from the above table shows that P value is significant at 5% indicating that there exists a relationship between age and level of knowledge of CA. About 44.6% of young farmers had low level of knowledge on CA whereas 32.4% and 23.0% of young farmers had medium and high level of knowledge on CA. Among middle aged farmers, 49.5% had low, 34.7% had medium and 15.8% had high level of knowledge of CA. About 46.3% of old aged farmers had low level of knowledge on CA whereas 21.1% and 32.6% had medium and high level of knowledge on CA.

Among various categories of farmers, old aged farmers had significantly higher level of knowledge on CA than other categories at 5% level. This may be due to the fact that the some of the CA practices like crop rotation were part of the traditional Indian agriculture and since majority of the farmers were old aged, their knowledge levels on traditional agriculture would be high and so the knowledge on CA.

Hussain et al. (2010) and Mugo (2012) had reported similar views in the past as that of current findings.

2) Educational status

The distribution of respondents on knowledge of CA based on their educational levels is summarized in the Table below.

Table 18. Distribution of respondents based on their educational status and level of knowledge on CA

S No	Educational Category	Level of Knowledge on CA			Total	Chi square value	P value
		Low	Medium	High			
1	Functionally literate	29	18	16	63		
		(46.0%)	(28.6%)	(25.4%)			
		[17.7%]	[18.8%]	[17.8%]			
2	Primary	63	31	19	113		
		(55.8%)	(27.4%)	(16.8%)			
		[38.4%]	[32.3%]	[21.1%]			
3	Middle	42	25	21	88		
		(47.7%)	(28.4%)	(23.9%)	(100.0%)	17.985	0.021*
		[25.6%]	[26.0%]	[23.3%]			
4	Secondary	20	15	17	52		
		(38.5%)	(28.8%)	(32.7%)	(100.0%)		
		[12.2%]	[15.6%]	[18.9%]			
5	Graduate	10	7	17	34		
		(29.4%)	(20.6%)	(50.0%)	(100.0%)		
		[6.1%]	[7.3%]	[18.9%]			

Note. The value within ( ) refers to Row Percentage; The value within [ ] refers to Column Percentage; ** denotes significance at 1% level; *denotes significance at 5% level.

The data from the Table 18 above indicate that there existed an association between educational status of the respondents and their level of knowledge on CA, since the P value is significant at 5%. Among the functionally literate group of farmers, 46.0% had low level of knowledge on CA whereas 28.6% had medium and 25.4% had high level of knowledge on CA.

Among the respondents with primary education, 55.8% had low level of knowledge on CA and 27.4% and 16.8% had medium and high level of knowledge on CA respectively. Among the respondents with middle level of education 47.7% had low, 28.4% had medium and 23.9% had high level of knowledge in CA. Among the group of respondents with secondary education, 38.5% had low level, 28.8% had medium and 32.7% had high level of knowledge on CA. Among the graduate farmers 29.4% had low, 20.6% had medium and 50.0% had high level of knowledge on CA. The results also indicated that the respondents with higher educational status had significantly higher level of knowledge on CA than the other categories.

Education tended to improve the knowledge level of any new technologies as they get access to information through various sources like print media. Moreover, better educational status of the respondents was likely to increase their contacts with various sources of information like extension agencies. In addition to this, with increased educational levels, farmers tended to explore various sources of information with their inherent ability to assimilate new technologies. This might have resulted in higher knowledge of CA with respect to higher educational status of the respondents though the proportion of highly educated respondents is low. These findings were in line with the findings of Hussain et al. (2010), Chiputwa et al. (2011), Uaiene (2011), and Mugo (2012).

3) Experience in farming

The study on relationship of respondents with respect to their experience and level of knowledge on CA was analysed and respondents were categorized accordingly. The categorization of farmers with respect to their level of knowledge and their experience in farming is presented in the following Table.

Table 19. Distribution of respondents based on their experience in farming and level of knowledge on CA

S No	Category of experience	Level of Knowledge on CA			Total	Chi square value	P value
		Low	Medium	High			
1	Low	42	29	15	86	7.513	0.111
		(48.8%)	(33.7%)	(17.4%)			
		[25.6%]	[30.2%]	[16.7%]			
2	Medium	45	25	20	90		
		(50.0%)	(27.8%)	(22.2%)			
		[27.4%]	[26.0%]	[22.2%]			
3	High	77	42	55	174		
		(44.3%)	(24.1%)	(31.6%)			
		[47.0%]	[43.8%]	[61.1%]			

Note. The value within ( ) refers to Row Percentage; The value within [ ] refers to Column Percentage; ** denotes significance at 1% level; *denotes significance at 5% level.

It could be understood from the data presented in Table 19 that among those respondents with lesser experience in farming, 48.8%, 33.7% and 17.4% of the respondents have low, medium and high level of knowledge on CA respectively. 50.0%, 27.8% and 22.2% of the respondents with medium experience in farming have low, medium and high level of knowledge on CA respectively. Among farmers with rich experience in farming, 44.3% have low level of knowledge on CA whereas 24.1% and 31.6% have medium and high level of knowledge on CA. Since the P value is insignificant (0.111) at all levels, it could be inferred that there is no significant association between experience in farming and level of knowledge on CA among various categories of the respondents. The findings of FAO (2001) which was confirmed by Mugo (2012) are in line with the observations of the current study.

4) Land holding pattern

The distribution of respondents with respect to their knowledge on CA among various categories of respondents based on their land holding pattern is presented in the following Table 20.

Table 20. Distribution of respondents based on their landholding and level of knowledge on CA

S No	Category of farmers	Level of Knowledge on CA			Total	Chi square value	P value
		Low	Medium	High			
1	Marginal	87 (53.0%) [53.0%]	49 (29.9%) [51.0%]	28 (17.1%) [31.1%]	164	18.600	0.163
2	Small	51 (46.8%) [31.1%]	29 (26.6%) [30.2%]	29 (26.6%) [32.2%]	109		
3	Big	26 (44.3%) [15.9%]	18 (24.1%) [18.8%]	33 (31.6%) [36.7%]	77		

Note. The value within ( ) refers to Row Percentage; The value within [ ] refers to Column Percentage; ** denotes significance at 1% level; *denotes significance at 5% level.

From the data from the Table 20, it could be noted that 53.0% of the respondents with marginal land holding have low level of knowledge on CA whereas 29.9% and 17.1% of the marginal farmers category have medium and high level of knowledge on CA. among the small farmers category, 46.8% have low level of knowledge, 26.6% have medium and high level of knowledge on CA respectively. Among big farmers, 44.3% have low knowledge on CA whereas 24.1% of respondents have medium knowledge and 31.6% have high knowledge on CA. Since P value is insignificant, it could be inferred that there is no significant relationship between knowledge on CA and various categories of farmers based on their land holding pattern. The reason for the above trend might be due to the fact that acquisition of knowledge and information is primarily a farmer characteristic and not a farm characteristic.

Though the size of the land holding may influence the acquisition of in depth knowledge and the adoption of acquired knowledge, the overall awareness and knowledge on new technologies like CA may not dependent on the size of the land holding as irrespective of land holding farmers tend to increase their profitability   and long term sustainability of farming. This is in line with the findings of McRobert and Rickards (2010).

5) Innovativeness

The following table presents the distribution of respondents among various categories of their innovativeness with respect to level of knowledge on CA.

Table 21. Distribution of respondents based on their innovativeness and level of knowledge on CA

S No	Category of farmers	Level of Knowledge on CA			Total	Chi square value	P value
		Low	Medium	High			
1	Low	37 (75.5%) [22.6%]	8 (16.3%) [8.3%]	4 (8.2%) [4.4%]	49		
2	Medium	64 (36.2%) [39.0%]	49 (27.6%) [51.0%]	64 (36.2%) [71.1%]	177	32.703	0.001**
3	High	63 (50.8%) [38.4%]	39 (31.5) [40.6%]	22 (17.7%) [24.4%]	124		

Note. The value within ( ) refers to Row Percentage; The value within [ ] refers to Column Percentage; ** denotes significance at 1% level; *denotes significance at 5% level.

It could be noted from the data from the Table 21 that 75.5% of the respondents with low innovativeness have low level of knowledge on CA. 16.3% of respondents had medium level of knowledge on CA among the respondents with low innovativeness whereas a meager proportion of 8.2% of the respondents had higher level of knowledge on CA. Among the respondents with medium level of innovativeness, the proportion of low and high level of knowledge on CA was equal at 36.2% whereas only 27.6% of respondents had medium level of knowledge on CA.

It was surprising to note that among the farmers with high innovativeness, a majority of 50.8% of the respondents with low level of knowledge on CA whereas only a meager proportion of 17.7% of the respondents had high knowledge level on CA. Similar views were expressed by Hussain et al. (2010).

3.4 Extent of Adoption of Conservation Agriculture Practices

3.4.1 Minimum Tillage

The table below shows the distribution of respondents based on the adoption of minimum tillage practices.

Table 22. Distribution of respondents according to their level of adoption of minimum tillage (n = 350)

S No	Category	Frequency	Percentage
1	Adopters	11	3.1
2	Non adopters	339	96.9
	Total	350	100.0

Table 23. Distribution of respondents according to their level of adoption of minimum tillage among the knowledgeable about minimum tillage (n = 96)

S No	Category	Frequency	Percentage
1	Adopters	11	11.5
2	Non adopters	85	88.5
	Total	96	100.0

The results in the table above show very low adoption of minimum tillage by the respondents (3.1%). A whopping majority of the respondents were non adopters of minimum tillage to the tune of 96.9%. The main reason for very low adoption of minimum tillage was the ignorance of respondents about minimum tillage as they had never heard about existence of such technology. Even among the total respondents who were knowledgeable about minimum tillage, only 11.5% adopt it. This may be due to the fact that most of the respondents had strong belief in conventional farming practices including intensive plouging of land using heavy implements like rotary tiller.

3.4.2 Permanent Soil Cover

The following table presents the distribution of respondents based on the adoption of permanent soil cover practices.

Table 24. Distribution of respondents according to their level of adoption of permanent soil cover (n = 350)

S No	Category	Frequency	Percentage
1	Adopters	24	6.9
2	Non adopters	326	93.1
	Total	350	100.0

Table 25. Distribution of respondents according to their level of adoption of permanent soil cover who are knowledgeable about permanent soil cover

S No	Category	Frequency	Percentage
1	Adopters	24	27.6
2	Non adopters	63	72.4
	Total	87	100.0

The data from Table 25 clearly show that only 6.9% of the respondents adopted permanent soil cover whereas the vast majority of the respondents did not adopt permanent soil cover by stubble mulching. The meager percentage of adopters was predominantly sugarcane growers who covered the interspaces of the crop using sugarcane trashes. It could also be noted that among the respondents who were knowledgeable about permanent soil cover, $1/4^{th}$ of them adopted the same whereas about $3/4^{th}$ of them did not adopt it.

It was given to understand from the respondents that they did practice trash mulching as per the advice of extension officials of sugar mills of their respective area. It was also expressed that not many followed this as it required lot of labour and with heavy demand for labour and their increasing wages; it was not profitable for them. However, those respondents who practiced trash mulching expressed that they could control the incidence of weeds and could observe an improvement in soil fertility though the process of improvement was slow.

3.4.3 Crop Rotation

The classification of respondents based on their adoption levels of crop rotation practices are presented below

Table 26. Distribution of respondents according to their level of adoption of crop rotation (n = 350)

S No	Category	Frequency	Percentage
1	Adopters	273	78.0
2	Non adopters	77	22.0
	Total	350	100.0

The results from the above table indicate that more than three fourths (78%) of the respondents practiced crop rotation with at least two crops in rotation. The crops in rotation varied from place to place. However, paddy remained one of the crops in all the locations.

However, majority of the respondents expressed that they adopted crop rotation primarily due to non availability of sufficient irrigation water for cultivation of second crop of paddy. Some of the non adopters also revealed that though they wished to cultivate different crops in rotation, they resorted to mono-cropping of paddy predominantly due to the reason that their lands are low lying and were bound to be under water stagnation.

Adopting CA principles involves shift from (1) excessive tillage with associated runoff to drastically reduced tillage combined with in situ soil and moisture conservation (2) residue burning or incorporation to surface retention of residues and (3) crop based management to cropping system based management (Bhaduri, 2012). Conservation Agriculture is seen as effective technology to increase farmers' resilience to climate variability and address soil degradation. However adoption of conservation agriculture is limited since promotion of CA lacks involvement of farmers (Giller et al., 2009). Most of the farmers are not aware of minimum/zero tillage practices. Farmers are not retaining crop residues in their fields; they are using it for other uses undermining the benefits of retaining crop residues in the field. Different patterns of crop rotations are followed by farmers making full use of land throughout the year (Usharani & Neelu, 2011).

4. Conclusion

Strategies to promote CA will call for moving away from conventional approach and to involve all stakeholders for end-end solutions. The study revealed that among the CA technologies, only Crop Rotation is popular among farmers and is being followed though to a limited extent. Other CA technologies like Minimum tillage and Permanent soil cover is not known to majority of the farmers and is also not being adopted by them. It implies the need for popularizing CA technologies and educating the farmers on the usefulness of CA technologies. Establishing frontline demonstrations, organizing training programs, promotion of community owned CA programs through institutional and governmental support would enhance the adoption of CA which has potential for long term profitability and sustainability of agriculture production systems.

References

Abrol, I. P., & Sunita, S. (2006). General article on Sustaining Indian agriculture – Conservation agriculture the way forward. *Current Science, 91*(8).

Anonymous. (2010). *State Agriculture Plan – Tamil Nadu 2010*. Chennai, Government of Tamil Nadu.

Arunmozhidevi, M. C. (2004). *Research-Extension-Clientale Linkages in Dairy Sector in Karnataka* (Unpublished Ph.D. Thesis). Annamalai University, Annamalainagar

Bhaduri, A. (2012). *Soil health, conservation agriculture and climate change*. Symposium by CASA and SPWD at NASC, New Delhi.

Chiputwa, B., Augustine, S. L., & Patrick, W. (2011). *Adoption of Conservation Agriculture Technologies by Smallholder Farmers in the Shamva District of Zimbabwe: A Tobit application*. Paper presented at the meeting of the Southern Agricultural Economics Association (SAEA) in Texas, USA, Feb. 5-8.

FAO. (2011). *What Is Conservation Agricutlure?* Rome. Retrieved from http://www.fao.org/ag/ca/1a.html

Giller, K. E., Witter, E., Corbeels, M., & Tittonell, P. (2009). Conservation agriculture and smallholder farming in Africa: the heretics' view. *Field Crops Research, 114*, 23-34. http://dx.doi.org/10.1016/j.fcr.2009.06.017

Hussain, M., Saboor, A., Ghafoor, A., Javed, R., & Zia, S. (2010). Factors affecting the adoption of no tillage crop production system. *Sarhad J. Agric., 2*(3), 409-412.

Mansingh, P. J. (1993). *Construction and Standardization of Socio economic Status Scale* (Unpublished Ph.D. Thesis). Tamilnadu Agriculture University, Coimbatore.

Masvimavi, K., Ndlovu, P. V., Nyathi, P., & Minde, I. J. (2010). *Conservation Agriculture practices and Adoption by Smallholder Farmers in Zimbabwe*. Poster presented at the Joint 3rd African Association of Agriculture Economists (AAAE) and 48th Agriculture Economists Association of South Africa (AEASA), Cape Town, South Africa, September 19-23, 2010.

McRobert, J., & Rickards, L. (2010). Social research: Insights into farmers' conversion to no-till farming systems. *Extension Farming Systems Journal, 6*(1), 43-52.

Meena, M. S., & Singh, K. M. (September, 2013). *Conservation Agriculture: Innovations, Constraints and Strategies for Adoption*. http://dx.doi.org/10.2139/ssrn.2318985

Mugo, B., & Jepkorir. (2012). *Factors affecting adoption of Conservation Agriculture among small holder farmers in Ngata Division, Nakuru County*. Retrieved from http://erepository.uonbi.ac.ke:8080/xmlui/handle/123456789/9535

Nyanga, H. P. (2012). Factors Influencing Adoption and Area under Conservation Agriculture: A Mixed Methods Approach. *Sustainable Agriculture Research, 1*(2).

Puthiraprathap, D. (2003). *Relative effectiveness of farm communication through mass media including new media: An experimental approach with rural women* (Unpublished Ph.D. Thesis). Tamilnadu Agriculture University, Coimbatore.

Theodor, F., Rolf, D., & Amir, K. (2012). *Overview of the Global Spread of Conservation Agriculture*. Field Actions Science Reports, Institute Veolia Environnement, France, November, 2012. Retrieved from http://factsreports.revues.org/1941

TSBF-CIAT. (April, 2011). Final Technical Report on Promoting Conservation Agriculture to Improve Land Productivity and Profitability among Smallholder Farmers in Western Kenya. *The African Network for Soil Biology and Fertility (AfNet)*. Tropical Soil Biology and Fertility Institute of CIAT (TSBF-CIAT).

Usha, R. A., & Neelu, N. (2011). Knowledge and Awareness of Conservation Agriculture among the Farmers of Mewat District of Haryana. *International Journal of Current Research, 3*(11), 001-005.

4

Silicon Release from Local Materials in Indonesia under Submerged Condition

Linca Anggria[1,2], Husnain[2], Kuniaki Sato[1] & Tsugiyuki Masunaga[1]

[1] Faculty of Life and Environmental Science, Shimane University, Matsue, Japan

[2] Indonesian Soil Research Institute, Bogor, Indonesia

Correspondence: Tsugiyuki Masunaga, Faculty of Life and Environmental Science, Shimane University, 1060 Nishikawatsu, Matsue, Shimane 690-8504, Japan. E-mail: masunaga@life.shimane-u.ac.jp

Abstract

Five inorganic materials (steel slag, silica gel, electric furnace slag, fly ash and Japanese silica fertilizer) and six organic materials (rice husk-biochar, rice straw compost, media of mushroom, cacao shell-biochar, rice husk-ash and elephant grass), were evaluated as Si fertilizer sources for rice plants (*Oryza sativa* L.) in two soil types (red clayey and sandy soil). Evaluation was carried out by incubating them at 30°C under submerged condition for 70 days. The soil solution was replaced at day 7, 14, 21, 42, 49, 56, 63 and 70 and the amount of silicon (Si) release, pH, Eh, calcium (Ca), magnesium (Mg), iron (Fe) and manganese (Mn) concentrations in soil solutions were determined. The amount of Si release ranged from n.d. (not detected)-32444.7 mg Si kg^{-1} and 105.84-48524.0 mg Si kg^{-1} in red clayey and sandy soil solutions, respectively during 70 days of incubation. Reduction in soil Eh was accompanied with an increase in the solubility of the soil Si especially for silica gel, electric furnace slag, elephant grass and media of mushroom. Higher exchangeable Ca content in soil tended to suppress Si release from rice straw compost, rice husk-ash and cacao shell-biochar. Considering the results of present study and availability of the materials, we concluded that steel slag of the inorganic materials and rice straw/husk and cacao shell-biochar of organic materials had the highest potential as Si fertilizer source in Indonesia.

Keywords: inorganic material, organic material, coexisting element, solubility

1. Introduction

Silicon is not recognized as an essential element, but as a beneficial element, Si enhances diseases resistance, alleviates metal toxicity, improves nutrient balance, prevents lodging and enhanced drought tolerance of rice (Ma & Yamaji, 2006). Silicate minerals liberate dissolved Si (DSi) as monosilicic acid (H_4SiO_4) by chemical weathering (Cornelis et al., 2011). Furthermore, Si is taken up by the roots in the form of H_4SiO_4 (Ma & Yamaji, 2006). However, the soluble Si content of tropical soils, such as highly weathered mineral (Ultisols and Oxisols) is generally less than that in most temperate soils as a result of Si leaching (Foy, 1992). While desilification and fertilization processes are extremely active in red soil (Liang et al., 2015). However, in sandy soil that consists mostly of quartz (SiO_2), the chemical decomposition of this mineral is complex (Marafon & Endres, 2013). One of the most important factors that influence the solubility of Si in soil is redox potential. Low soil Eh as flooding condition, normally leads to an increase in available Si concentration (Liang et al., 2015). The solubility of Si containing mineral is affected by pH, where the soil pH regulates the solubility and the mobility of Si (Tubana & Heckman, 2015). Silica concentration is lowest at pH 8-9 and Si concentration in soil solution may rise sharply when pH value decreases from 7 to 2 (Beckwith & Reeve, 1963).

Most of the land in Indonesia is acidic due to high level of leaching of basic cations. There is around 102,000,000 ha of acidic soil with Ultisols and Oxisols are the dominant soil beside Entisols, Inceptisols and Spodosols (Subagyo et al., 2000; Mulyani et al., 2009). These soils have been used for rice production in Indonesia. Although Si is very abundant element in soil with the range from 25 to 35 %, repeated cropping can reduce the levels of plant-available Si to the point that supplemental Si fertilization is required for maximum production (Datnoff & Rodrigues, 2005). In Indonesia, lower soil Si content was found to be severe in intensive rice field where enormous Si uptake is not followed by sufficient Si replenishment (Husnain et al., 2008). In present, the most common forms of silicate materials used as fertilizer are various industrial by products (Haynes, 2014). Slag from iron and alloy manufacturing that consist of calcium silicate which could be used to meet the demand of Si. Fly ash from coal combustion where the dust-collection system removes the fly ash from the combustion gases before they are discharged in the atmosphere is high in Si content (Ramezanianpour, 2014). These Si rich materials from industrial wastes are also applied to increase soil pH (Haynes et al., 2013).

Besides industrial wastes, potential organic sources of silicate material have been assessed for use as an agricultural amendment. As plants accumulate Si, there is possibility of using crop residues as Si source. For example, rice (*Oryza sativa* L.) husk and sugarcane (*Saccarum* spp.) bagasse have considerable Si concentration (Gascho, 2001). Biogenic amorphous silica is a natural constituent from unicellular organism, compost and crop residue (Rabovsky, 1995; Tubana & Heckman, 2015). However, the demand of Si from crop residues for agriculture is insufficient from plant residues. To address this issue, it is desirable to explore cheap and abundant local materials as Si source.

In Indonesia, there are some potential sources as silicate fertilizer from industrial by product and plant material-based silica. Factories that produce slag as by product of steel with crude steel production was 4.7 million ton on 2011 (Ministry of Industry Republic of Indonesia, 2014). Production of coal in Indonesia is around 437 million ton (Outlook Energy Indonesia, 2014) with fly and bottom ash as waste. Indonesia is the world's third largest cocoa bean producer (FAO, 2010) so it is also possible to use cacao shell and leaf as source of Si fertilizer. Considering the large amount of Si accumulated in rice straw and husk, straw compost and husk burning are an interesting Si source for plants.

Emphasis should be made not only on Si content but also on its solubility. The release of Si from the local materials into soil solution varies in different combination of materials and soils. Factors controlling dissolution of Si include iron (Fe), calcium (Ca), manganese (Mn), pH, particle size of the materials and presence of organic matter (Makabe et al., 2013; Kendrick, 2006). The factors that cause variation Si release from material and soil should be evaluated to improve use of material as Si source. Therefore, the objective of this study is to evaluate Si release from different local materials used as soil fertilizers under submerged conditions in relation with soil chemical properties and other controlling factors.

2. Materials and Methods

The Si release from the local materials was characterized through laboratory incubation experiments.

2.1 Si Source Materials

Eleven materials were collected from Indonesia and Japan. There are two groups of materials, namely (1) five inorganic materials (fly ash from coal company in South Sumatera, steel and electric furnace slag/EFS were obtained from Banten, silica gel, Japanese silica fertilizer (JSF) is a slag-based silicate fertilizer). (2) Six organic materials (elephant grass (*Pennisetum purpureum*), rice (*Oryza sativa* L.) straw compost/RSC, rice husk-biochar/RHB, cacao (*Theobroma cacao* L.) shell-biochar (cacao SB), media of mushroom/MM and rice husk-ash/RHA from West Java. The geographic and climatic condition of South Sumatera is $1°0'-4°0'S$ and $102°0'-106°0'E$ (Badan Pusat Statistik (BPS), 2015), Banten is $5°7'50''-7°1'1''S$ and $105°1'11''-106°7'12''E$ (Banten, 2016). West Java is $5°50'-7°50'S$ and $104°48'-108°48'E$ (Jawa Barat, 2016), with rainfall of about 3409,3573 and 2682 mm on 2013 in South Sumatera, Banten and West Java, respectively (Badan Meteorologi Klimatologi dan Geofisika (BMKG), 2000-2013). The elephant grass and media of mushroom were air-dried for 2-3 days. Materials were ground into fine powder in agate grinding jars, using a mixer mill (MM 200, Retsch GmbH, Haan, Germany).

Samples were oven-dried 12 hours at 80°C. Total element (Ca, Mg, K, Na, Fe Mn, Cu, Zn, Cd and Ni) composition of materials was measured by ICP after digestion in Teflon vessel with HNO_3 at 160°C for 4 hours and diluted with distilled water up to 25 ml after kept resting overnight (Koyama & Sutoh, 1987). Total carbon (C) was assessed using dry combustion methods (Sumigraph NC-22 Analyzer).

Available Si (Table 1) was extracted from materials with 0.5 M HCl (1: 150 ratio for 1h shake at 110 rpm) (Savant el al., 1999) and Na_2CO_3/NH_4NO_3 (10 g L^{-1}/16 g L^{-1}) (1:100 ratio for 1h shake at 110 rpm) (Pereira et al., 2003). The concentration of Si in all extracts was determined using the molybdenum blue method and measured by Spectrophotometry UV 1800 Shimadzu. The wavelength use for the Si detection was 810 nm. pH (H_2O) was determined on 1: 30 (w/v) soil: water suspensions with pH meter (D-51, Horiba).

2.2 Soil Samples

Two types of soil were used, a red clayey soil (Ultisol) and a sandy soil (Entisol) with textural classes of clay loam and fine sand, respectively. Red clayey soil is slightly acidic (pH 5.7) and relatively rich in available Fe and Mn (72.5 and 52.2 mg kg^{-1}). Exchangeable Ca and Mg were 4.3 and 2.4 $cmol_c kg^{-1}$, respectively. Sandy soil is neutral in pH (7.3), has high content of exchangeable Ca (26.4 $cmol_c kg^{-1}$, 1M ammonium acetate extractable Ca) and available Fe (136.4 mg kg^{-1}). The available Si concentration of red clayey and sandy soil was 267.1 and 129.3 mg SiO$_2$ kg^{-1}, respectively. According to Sumida (1992), those were classified to be below critical level of available silica for rice (300 mg SiO$_2$ kg^{-1}). We expected the difference of these properties to influence the dynamics of dissolved Si from the local materials.

The soil sample was air dried and passed through a 2 mm sieve. Exchangeable Ca and Mg were extracted with 1 M ammonium acetate pH 7.0 and measured by Inductive Coupled Plasma Spectroscopy (ICPE-9000 Shimadzu, Kyoto Japan). Available Si was extracted by acetate buffer pH 4 with ratio of 1:10, for intermittent shaking for 5h at 40 °C, determined using the silicate molybdenum blue method (Imaizumi & Yoshida, 1958). Soil pH (H$_2$O) was determined on 1:2.5 (w/v) soil: water suspensions with pH meter (D-51, Horiba). The contents of available Fe and Mn were obtained by extraction with 0.1 N HCl and quantified using the ICP.

2.3 Incubation Experiment

Under submerged condition, the soil was incubated with Si materials as treatment. The experiment was replicated three times. Air dried soil of 10 g was placed in a 50 mL centrifuge plastic tube, 1 g of organic material and 40 mL of distilled water were added into the tube. For inorganic material (steel slag, fly ash, EFS and JSF) 0.02 g of 30% 0.5 M HCl and silica gel 0.02 g of 30% Na$_2$CO$_3$/NH$_4$NO$_3$ as silicon dioxide (SiO$_2$) were added to the centrifuge tube containing 10 g of soil, and then 40 mL of distilled water was added. The tube was covered and mixed thoroughly, incubated at 30°C for 70 days. After incubation, the redox potential (Eh) and pH of soil solution were measured with Eh meter and pH meter (TOA HM-14P and D-51 Horiba, respectively) without disturb the soil. The supernatant was obtained after filtration (paper filter Advantec No. 6). Silica, Ca, Mg, Fe and Mn concentrations in supernatant were measured using ICPE-9000 Shimadzu. To resume the incubation, residue on the filter paper was washed back into tube with distilled water and distilled water was added up to a total volume of 40 mL base on the weight (Makabe et al., 2013). The soil solution was replaced with distilled water at day (d) 7, 14, 21, 42, 49, 56, 63 and 70 assuming field water replacement by drainage / leaching and irrigation in the field.

2.4 Data Analysis

The release of Si in the soil solution after the incubation experiment is symbolized the concentration of Si (mmol L^{-1}). The net release of Si and the other elements (ΔSi, ΔCa, ΔFe, ΔMg, ΔMn) from the materials was estimated based on the difference between the concentration of elements in samples with materials and without material (control). A correlation analysis was conducted to identify relationship between the Si and other elements. Statistical analyses were done using the statistical package SPSS 22.

3. Results and Discussion

3.1 Chemical Composition of Si Material Sources

The chemical composition of 11 materials used in this study is shown in Table 1.

Table 1. Chemical composition of materials

Material	pH	Si		TC	Ca	Mg	Fe	Mn	Cu	Zn	Cd	Ni
		Na$_2$CO$_3$/NH$_4$NO$_3$	HCl 0.5 M									
		------------- mmol kg^{-1} -------------		-- mg kg^{-1} --								
Steel slag	9.8	5.2	801.0	1400	198105.2	34818.5	281649.6	9605.1	147.1	46.1	67	113.6
Silica gel	6.8	234.3	96.6	-	11013.9	3283.6	6124.5	102.9	-	-	-	-
Fly ash	7.7	28.3	300.1	107802.6	18368.1	9006.0	29612.9	341.9	30.7	102.3	18.4	47.2
EFS	9.5	4.7	412.5	7700.0	157214.1	59459.2	151671.3	11589.5	120.9	942.3	56.4	109.9
JSF	12.4	137.1	975.0	20259.0	240126.9	18261.8	8606.2	23003.8	-	-	-	-
RSC	9.5	74.3	44.0	222578.3	14364.9	4340.8	16297.7	1833.5	26.6	87.8	9.7	18.7
RHB	7.6	56.5	2.8	427394.4	1952.0	991.5	2489.0	552.5	3.8	38.7	1.6	4.3
RHA	10.7	98.8	10.3	3650.4	3429.4	793.3	372.6	592.5	2.2	38.1	1.1	3.5
Cacao SB	10.5	5.7	133.5	423490.9	22278.9	15601.3	4189.9	1198.3	54.4	317.9	8.1	27.6
Elephant grass	5.2	192.1	5.4	427055.8	6043.3	2528.2	216.8	59.9	6.2	35.5	1.4	7.2
MM	8.6	52.5	2.6	360115.7	76687.3	2692.8	1594.9	246.3	6.8	27	6.8	15.2

Note. -: not determined, TC: total carbon, EFS: electric furnace slag, JSF: Japanese Si fertilizer, RSC: rice straw compost, RHB: rice husk-biochar, RHA: rice husk-ash, cacao SB: cacao shell-biochar, MM: media of mushroom.

Inorganic materials were alkaline in nature with JSF having the highest pH, except silica gel. According to Savant et al. (1997), the amount of H_4SiO_4 in soil solution is affected by pH and the soil pH regulates the solubility and the mobility of Si (Tubana & Heckman, 2015). Among the elements, Ca, Mg, Mn and Fe were reported to relate to soil Si availability (Makabe et al., 2013; Hansen et al., 1994). Among all the materials used, JSF had highest Ca and Mn content. The highest Fe and Mg content were found in steel slag and EFS, as chemical composition of EFS was $CaSiO_3/MgSiO_3$, while steel slug was $CaSiO_3$ (Tubana & Heckman, 2015). The lowest Ca, Mn, Mg and Fe content were found in silica gel.

The chemical composition of organic materials indicated that RHA had the highest pH (10.7) and elephant grass had the lowest (5.2). Carbon content in organic materials ranged from 222578.3 to 427394.4 mg kg^{-1}, except RHA material. MM had higher Ca content than the other organic materials because in mushroom cultivation, the media was added with $CaCO_3$ to neutralize acid that was released by mushroom. Cacao SB had high content of Mg (15601.3 mg kg^{-1}). RSC compost had the highest Fe and Mn and the lowest was elephant grass. Among materials, except JSF as reference, the available Si content was highest in steel slag.

Concentration of Si was higher with 0.5 M HCl than Na_2CO_3/NH_4NO_3 (alkaline) solution for inorganic materials (300.1-975.0 mmol Si kg^{-1}), except silica gel. For organic material, Si concentration was high with Na_2CO_3/NH_4NO_3 solution (52.5-192.1 mmol Si kg^{-1}), except cacao SB.

Concerning about heavy metal pollution by material application, Cu, Zn, Cd and Ni contents of all the materials were below the regulatory limit of Environmental Protection Agency (EPA) (1993); Cu = 4300, Zn = 7500, Cd = 85, and Ni = 420 mg kg^{-1}.

3.2 Release Pattern of Si from Materials (Δ Si) for 70 Days Incubation

The temporal changes in Si concentrations in the soil solution show that Si release rate and pattern for soil and materials differed. The Si release pattern was different between two type of soil and eleven materials. Release of Si from the materials in red clayey soil during 70 days of incubation is shown in figure 1a. Concentration of Si in soil solution from silica gel was stable and reached different peaks 42-56 days after flooding. Silica release from steel slag in red clayey soil started after day 49 and then increased rapidly until the end of incubation. It may indicate continuous dissolution of Si from steel slag. The dissolution pattern of Si from steel slag was different to dissolution pattern of silicate slag fertilizer reported by Makabe et al. (2013), where Si dissolution increased rapidly during the first 22 days in weakly acidic solution.

In contrast, fly ash released Si on the first 49 days then remained not detected (n.d.). A slightly higher peak of Si release from silica gel and fly ash at day 42 was probably because the soil solution was not changed for three weeks after day 21, this condition resulted in high Si in soil solution. Meanwhile, EFS and JSF did not apparently release Si throughout the incubation period in red soil. It was probably that the bond between Si and the other elements such as Fe-O-Si, $CaO-SiO_2-H_2O$ (Hansen et al., 1994; Flint & Wells, 1934) in these materials was too strong to release Si in the red soil incubation condition.

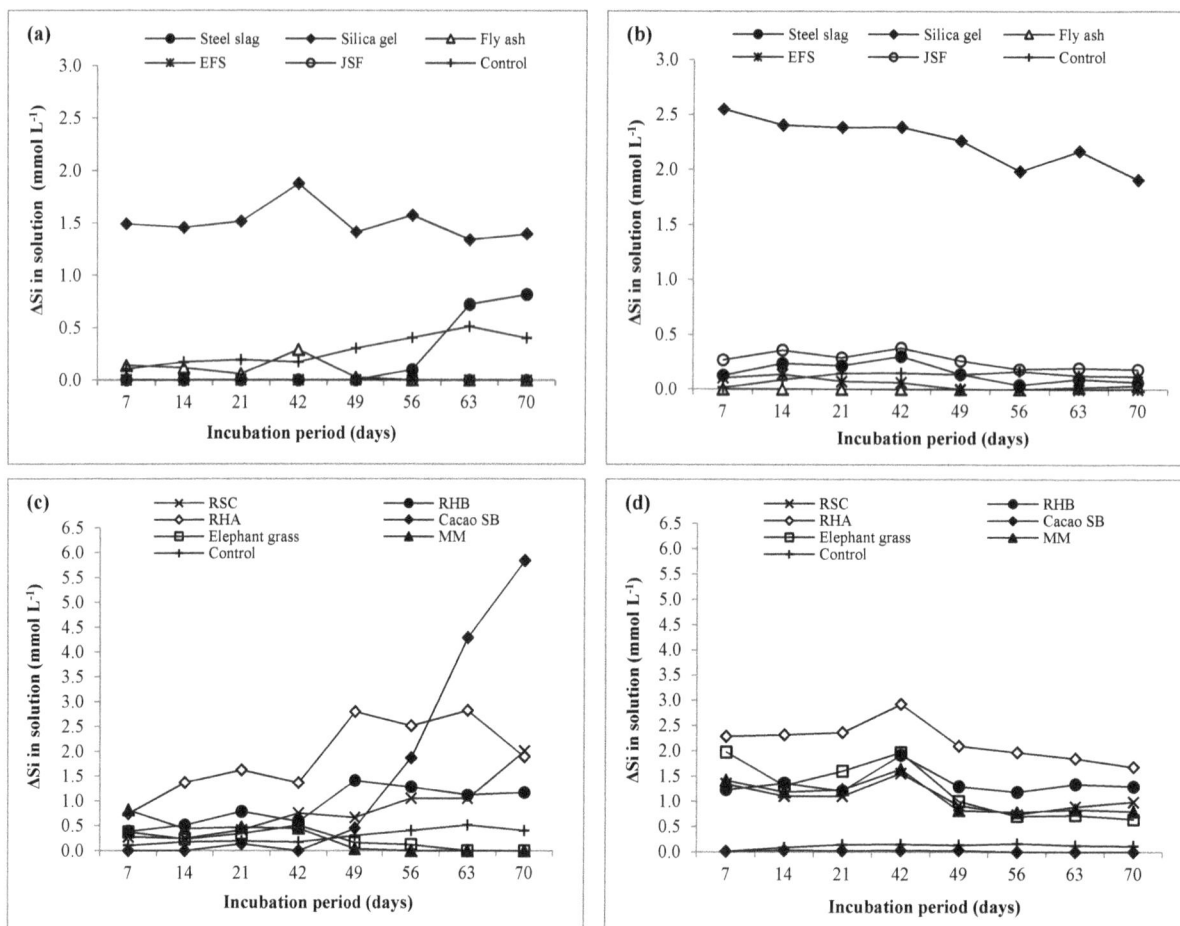

Figure 1. Release pattern of Si from the materials (Δ Si): (a) red clayey and (b) sandy soils with inorganic materials; (c) red clayey and (d) sandy soils with organic materials

Figure 1c shows release of Si from organic materials in red clayey soil. Silica concentration in soil solution with added RSC tend to increase as incubation period increased. Release of Si from RHB and RHA had similar

pattern, where the release of Si was started from day 7. Silica release from cacao SB started on day 21 up to the end of incubation time. Release of Si from MM is described by an initially fast release on day 7 followed by a progressively slight release until 42 days of incubation. Silica release rates from elephant grass tended to be slightly higher on 56 days of incubation.

The effect of EFS application gradually becomes less pronounced and not detected toward the end of experiment in sandy soil (Figure 1b). Concentration of Si in sandy soil with steel slag application increased during first weeks of incubation time. Silica concentration reached the maximum value after 42 days of incubation and then gradually decreased. Silica release from JSF was high during first 49 days, then decreased. The rate of Si concentration from silica gel decrease with the time of incubation. We recorded that Si release from fly ash in sandy soil solution was less (n.d.-0.03 mmol L^{-1}) than red clayey soil, which Si release was only during the last 14 days of incubation.

Si release from silica gel was higher in sandy soil than red clayey soil. As silica gel is made by neutralizing water glass. Thus, according to Bunker et al. (1988) that the tetrahedral SiO_4 sites common to all silicates glasses were susceptible to nucleophilic attack primarily by OH^- to form a reactive five-coordinated intermediate which can be decomposed to rupture the Si-O-Si bond. Therefore, a significant quantity of OH^- could improve the formation of a five-coordinated intermediate which could lead to a great dissolution of the silica gel. Moreover, as Si concentration was higher with alkaline solution (Table 1).

Figure 1d show release of Si from organic material in sandy soil solution. Silica release from RSC and MM had similar rate, where Si concentration in sandy soil solution was high in the first 42 days of incubation. Slightly different with RSC and MM, Si release was high for 49 days after submerged with added cacao SB and elephant grass. Silica release rates from RHB were fluctuating, with the highest Si concentration on day 42 (1.92 mmol Si L^{-1}). The fact that Si release from RHA was highest on day 42, then decreased until the end of incubation.

3.3 The Amounts of Si Release and Coexisting Element from the Materials for 70 Days Incubation

The cumulative amounts of Si and the element expected interact on with Si release from materials during the 70 days of incubation are listed in Table 2.

Table 2. Silicon and other elements release from materials and soil during 70 days.

Materials or soil	Red clayey soil					Sandy soil				
	Si	Ca	Mg	Fe	Mn	Si	Ca	Mg	Fe	Mn
	------------------- mmol kg^{-1} -------------------					------------------- mmol kg^{-1} -------------------				
Steel slag	508.32	377.43	126.72	80.15	0.20	375.97	196.65	n.d	0.15	0.77
Silica gel	1158.74	3.34	10.67	20.63	2.79	1733.00	82.31	4.16	0.39	0.01
Fly ash	79.33	185.16	77.80	0.06	n.d.	6.40	201.61	50.17	0.44	n.d
EFS	n.d.	861.84	471.59	36.35	0.41	64.22	108.13	n.d	0.08	0.43
JSF	n.d.	2050.13	340.69	0.82	2.09	852.50	2217.51	n.d	n.d	0.31
RSC	280.88	192.04	124.31	37.44	25.72	377.73	208.35	169.49	4.01	9.83
RHB	304.86	10.26	6.16	9.01	0.42	456.12	39.84	34.83	0.02	2.82
RHA	611.29	21.47	28.97	8.68	0.55	706.72	n.d	n.d	0.22	0.49
Cacao SB	659.98	52.48	129.25	11.26	1.84	6.61	n.d	63.24	0.30	0.20
Elephant grass	77.59	321.31	208.95	269.13	77.24	442.99	1441.29	309.08	12.13	0.68
MM	98.20	1527.00	226.20	28.28	42.77	388.21	1599.20	227.12	1.02	1.18
Soil	9.66	0.80	0.63	0.91	0.03	3.78	9.50	5.59	0.18	n.d

Note. n.d.: not detected, EFS: electric furnace slag, JSF: Japanese Si fertilizer, RSC: rice straw compost, RHB: rice husk-biochar, RHA: rice husk-ash, cacao SB: cacao shell-biochar, MM: media of mushroom.

The amount of Si release in red clayey and sandy soils ranged from n.d.-1158.74 mmol Si kg^{-1} and 3.78-1733.00 mmol Si kg^{-1}, respectively. The highest Si release in both red clayey and sandy soil was silica gel. Release of Si from red clayey soil (control) was higher than sandy soil, while Si concentration from soil was lower compared to that in the materials.

According to Marxen et al. (2016), Si concentrations in the soil solution from rice straw increased only when the organic matrix surrounding the phytoliths was decomposed and the surface of the phytoliths became exposed to soil solution. The release of Si was higher from RHA than RHB due to higher available Si content in RHA (Table 1), beside that C content in RHB was higher than RHA. According to Xiao et al. (2014), C and Si form in biochar result in the mutual protection between C and Si. Silica in biochar becomes difficult to dissolve, reflecting the protection of Si by C.

Organic materials in this research were high in Si concentration with alkaline solution, except cacao SB, which was high in acid solution. The results were similar to the initial Si concentrations in organic materials (Table 1). It is possible that alkaline solution dissolves organic matter that covers Si and thus Si may release from organic matter. According to Molina (2014), alkaline solution dissolved protoplasmic and structural components from fresh organic tissues.

Release of Ca, Mg, Mn and Fe were different among the materials and two soil types. In red clayey soil, released amounts of Ca and Mg were the highest from JSF (2050.13 mmol kg^{-1}) and EFS (471.59 mmol kg^{-1}), it might be due to high Ca content in both of materials (Table 1). While the lowest of Ca and Mg were silica gel and RHB (3.34 and 6.16 mmol kg^{-1}, respectively) due to Ca and Mg content was low in both materials. The highest Fe and Mn were released from elephant grass (269.13 and 77.24 mmol kg^{-1}, respectively).

Calcium and Mg release from soil (red clayey soil) was lower than in the materials. Red clayey soil has lower Fe solubility compared to in the materials, except fly ash and JSF.

For sandy soil, Ca release was the highest with JSF application due to its high Ca content. Manganese release was the highest with RSC (9.83 mmol kg^{-1}), while Fe and Mg release was the highest from elephant grass.

Calcium release was lower in soil (sandy soil) than in the materials, except RHA and cacao SB. The release of Mg from soil (sandy soil) was lower than in the materials, except steel slag, silica gel, EFS, JSF and RHA. Almost the same with Mg, Fe concentration from sandy soil was also lower than materials, except steel slag, EFS and JSF. Furthermore, Mn concentration from materials was higher than sandy soil, except fly ash. Kato and Owa (1997) reported that the application of the slags increase the Ca concentration in soil solution.

Eight of eleven materials had higher Si release in sandy soil than red clayey soil. According to Dematte et al. (2011), the chemical decomposition of clay mineral is complex, which made sandy soils more responsive on Si release than red clayey soils to the material application.

3.4 Effects of pH and Eh

It is generally stated that Si availability depends on soil types (Wei et al., 1997). In detail, pH, Eh and the type of coexisting metals influence the adsorption of monosilicic acid by oxides (Tubana & Heckman, 2015; Liang et al., 2015).

The increase of pH and decrease of Eh (Table 3) in red clayey soil solution due to added of materials and also effect of submergence. In sandy soil solution, the trend of pH and Eh was different with red clayey soil solution. Steel slag, EFS, JSF, RHA and cacao SB tend to increase soil solution pH. Where steel slag, EFS, JSF, cacao SB, elephant grass and MM decrease soil solution Eh.

Table 3. Mean pH and Eh values of red clayey and sandy soil solution during 70 days of incubation

Material	Red clayey soil				Sandy soil			
	Mean of pH	Δ pH	Mean of Eh	Δ Eh	Mean of pH	Δ pH	Mean of Eh	Δ Eh
			---------- mV ---------				---------- mV ----------	
Steel slag	6.3	0.9	155.9	-28.3	9.8	1.2	-33.8	-14.9
Silica gel	4.8	-0.6	256.5	72.3	8.3	-0.4	21.8	40.7
Fly ash	6.1	0.7	140.9	-43.3	8.6	-0.1	-7.6	11.3
EFS	7.2	1.7	68.0	-116.2	10.2	1.5	-67.6	-48.7
JSF	6.9	1.5	100.9	-83.3	11.0	2.3	-90.8	-71.9
RSC	6.5	1.1	-19.1	-203.4	6.9	-1.8	-13.2	5.7
RHB	5.6	0.2	171.3	-12.9	7.8	-0.9	43.3	62.3
RHA	6.4	1.0	161.3	-22.9	9.4	0.7	-9.8	9.1
Cacao SB	8.0	2.6	-7.7	-191.9	9.8	1.1	-77.3	-58.3
Elephant grass	5.9	0.5	-43.2	-227.4	6.4	-2.3	-92.8	-73.9
MM	6.7	1.3	-91.9	-276.1	6.9	-1.8	-71.9	-53

Note. Δ pH means the change in pH from control (5.4 and 8.7 for red clayey and sandy soil solution, respectively); Δ Eh means the change in Eh from control (184.2 and 18.9 for red clayey and sandy soil solution, respectively). EFS: electric furnace slag, JSF: Japanese Si fertilizer, RSC: rice straw compost, RHB: rice husk-biochar, RHA: rice husk-ash, cacao SB: cacao shell-biochar, MM: media of mushroom.

3.4.1 Red Clayey Soil

The mean value of differences (Δ) during the entire incubation period for soil solution pH ranged from -0.6 to 2.6 units. Where, the increase in soil solution pH was 0.2 to 2.6 units. The soil solution pH was elevated by addition of all the materials, other than silica gel. The lowest soil solution pH was found in silica gel (4.8) and a maximum increase was obtained in cacao SB (8.0). Among inorganic material, the highest pH was gained with EFS application (7.2). Soil solution pH with added steel slag, fly ash, RSC and RHA were almost the same (6.1-6.5). Meanwhile, JSF and MM were close in soil solution pH (6.9 and 6.7, respectively). RHB and elephant grass increased soil solution pH with almost the same value (5.6 and 5.9, respectively).

According to Ponnamperuma (1972) that submerging soil cut off oxygen supply, where aerobic organisms use up the oxygen present in the soil. Kashem and Singh (2001) reported oxygen is reduced at Eh > 300 mV and Mn^{4+} at Eh of 200 mV. We observed that Eh decrease with decreasing pH, especially with organic material treatment. Kashem and Singh (2001) reported that organic material had contributed to low and negative values of Eh resulting into higher increase in pH values.

Decreases in Eh were observed for the soil samples with addition of Si materials after submergence, except for soil with silica gel, where the values rose to 256.5 mV. The submerged condition with the addition of steel slag and fly ash result almost the same value of Eh (155.9 and 140.9 mV, respectively). The negative Eh was found in soil solution with RSC, cacao SB, elephant grass and MM (-7.7 to -91.9 mV). We observed RHB and RHA had almost the same in soil solution Eh (171.3 and 161.3 mV, respectively), it might be because both material were rice husk.

Even though, steel slag and EFS from steel company but the Eh of soil solution Eh with EFS (68.0 mV) was lower than steel slag. It was probably due to Mn content in EFS was higher that influenced Eh. Whereas, soil solution Eh with JSF application was higher than EFS (100.9 mV), but it was also probably due to high Mn content in original material.

3.4.2 Sandy Soil

It is obvious from Table 3 that the increased in sandy soil solution pH occurred following additions of steel slag, EFS, JSF, RHA and cacao SB with the highest increase of 2.3 units was JFS. Soil pH decreased with addition of silica gel, fly ash, RSC, RHB, elephant grass and MM with incubation time. The largest decreasing -2.3 units of pH was elephant grass.

Steel slag, EFS, RHA and cacao SB increased pH around 9.4-10.2. Meanwhile, RSC, elephant grass and MM decreased pH close to neutral (6.4-6.9). Silica gel and fly ash were almost the same in lowering soil solution pH (8.3 and 8.6, respectively) whereas pH 7.8 in soil solution was obtained with RHB application.

The Eh in soil solution with additional of steel slag, EFS, JSF, cacao SB, elephant grass and MM markedly decreased after submergence and the maximum negative value of 92.8 mV was observed for soil solution with elephant grass application. Elephant grass, cacao SB and MM as organic matter increased microbial activity in sandy soil thus decreasing soil solution Eh.

The Eh change was not as large as observed in red clayey soil. Meanwhile, Eh increased with addition of silica gel, fly ash, RSC, RHB and RHA. The highest soil solution Eh was obtained after added RHB (43.3 mV). Interactive the effect of submergence with RSC, RHB and RHA had the same result, which soil solution Eh was not decrease with those material. The same trend was probably due to the same source from rice plant.

3.5 Characteristics of Si Release from the Materials

Generally, soil pH regulates the solubility and the mobility of Si (Tubana & Heckman, 2015). The Eh of soils controls the stability of various oxidized components such as Mn IV and Fe III in submerged soils (Sahrawat, 2005). Reduction in Eh was accompanied with an increase in the solubility of the soil Si, where Si increase in soil solution was attributed to the release from ferric silica complexes under anaerobic conditions (Ponnamperuma, 1965). This is in line with Snyder et al. (2006) who reported that Si release such as monosilicic acid and polysilicic acid that have high chemical activity can react with Fe in the formation of slightly soluble silicate. Makabe et al. (2013) reported that Si concentration had significantly negative correlation with Ca in soil solution.

Figure 2 revealed some significant correlations between Si release with Ca or Eh changes for each material. We characterized the materials as below.

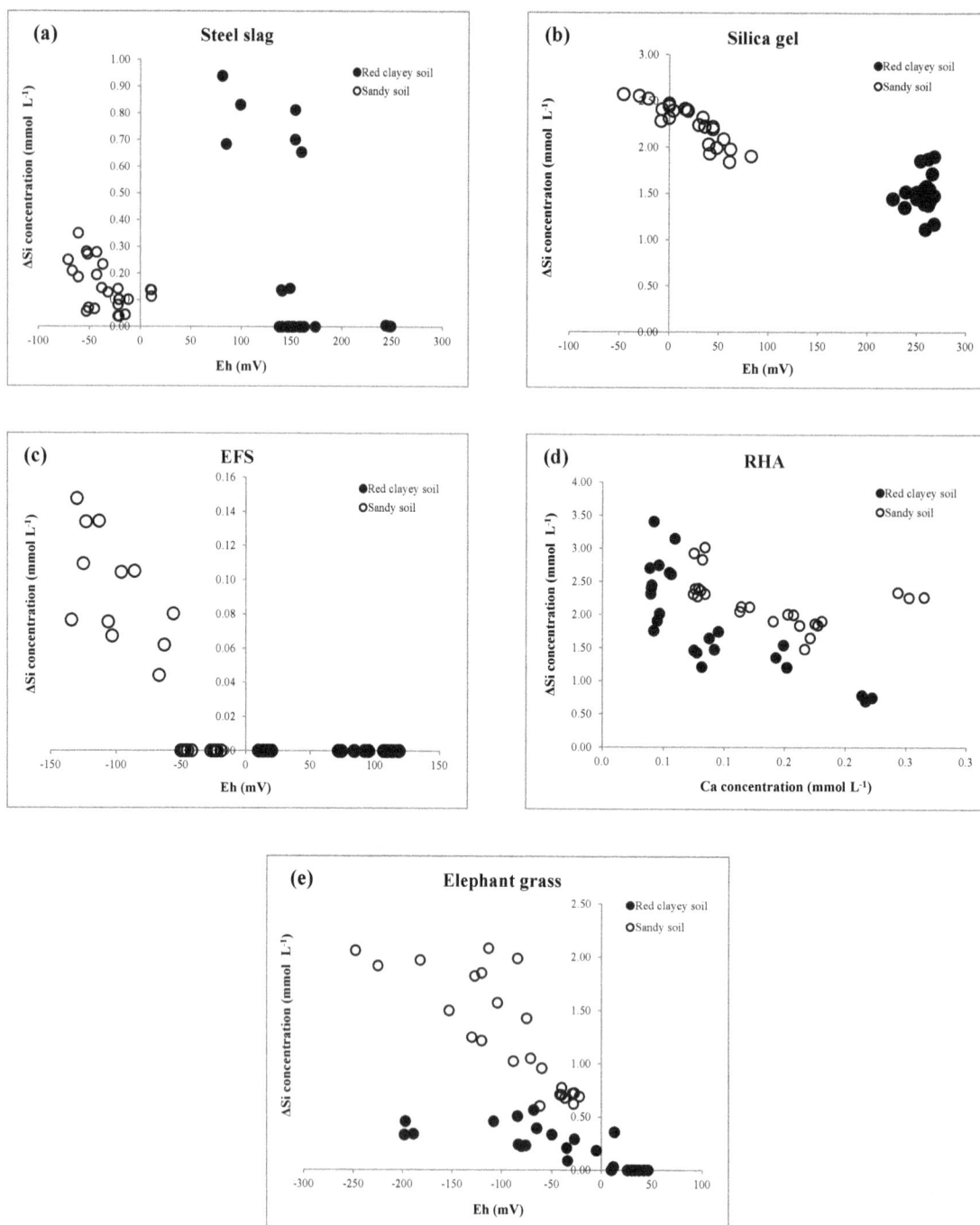

Figure 2. Correlations coefficient of Si concentration from material and other elements in soil solution

3.5.1 Steel Slag

Si release from steel slag in red clayey soil showed positive correlation with Fe release and negative one with Eh change in soil solution (Figure 2a). At conditions in which pH of soil solution was weak alkaline, Si was not affected by pH, but was negatively correlated with Eh. This indicated that Si release of the steel slag was basically controlled by dissolution of the slag with lowering Eh. Steel slag contains Si as calcium silicate form and the slag dissolution and Si release proceeded by following two steps, i) Ca and Mg are dissolved by ion exchange reaction with hydrogen ion in water (first step), ii) Si-O-Si and/or Al-O-Si chemical bonds are cleaved by hydrolysis (second step) (Kato & Owa, 1996). Kato and Owa (1996) also reported that lowering pH increase Si release and in contrast high Ca content in soil or soil solution decreased it. Beside, Liang et al. (2015) reported that low soil Eh at flooding condition, normally leads to an increase in available Si concentration. Lower Eh of sandy soil (Table 1) probably led to release of Si from the beginning of incubation. However, total amount of Si released from the steel slag was higher in red clayey soil, which was due to lower pH and lower exchangeable Ca content of the red clayey soil.

3.5.2 Silica Gel

Silica gel released highest concentration of Si among all the materials in both acid and weak alkaline conditions. We observed that Si release from silica gel in sandy soil solution condition was higher than in red clayey soil by 40% (Table 2). It seemed that high pH and low Eh (Figure 2b) in soil solution increased Si release from silica gel.

3.5.3 Fly Ash

Fly ash is made up of highly insoluble, glass-like particles, consisting of amorphous ferro-aluminosilicate and quartz (Haynes et al., 2013). The total amount of Si released from fly ash was higher in red clayey soil, which was due to lower pH in soil solution (Table 3).

3.5.4 Electric Furnace slag and Japanese Si Fertilizer

EFS consist of $CaSiO_3/MgSiO_2$ (Tubana & Heckman, 2015), where Si release was affected by reduction of Eh and pH in soil solution. EFS and JSF did not aparently release Si in red clayey soil solution (Table 2), which is possibly due to specific range of pH and relatively high Eh comparing with sandy soil condition. According to Meyer (1999), Si solubility is lowered at pH range of 6.5-7.5. Thus, soil pH, 7.2 and 6.9 for EFS and JSF (Table 3), respectively in red soil suppressed Si release from the materials.

3.5.5 Rice Straw Compost

RSC stably released Si in both soil conditions, although it was higher in sandy soil than in red clayey soil. Exchangeable Ca and Mg in red clayey soil might suppress Si release, but exchangeable Ca or Mg in sandy soil seemed not to influence the Si release. It was confirmed as an effective soil Si amendment.

3.5.6 Rice Husk-Biochar and Rice Husk-Ash

Silicon in rice husk was concentrated and increased in its availability by ashing (Table 1). The release of Si from RHA looked negatively affected by Ca (Figure 2d). Thus, we assume Si in RHA was changed in Ca bind form and Ca in soil or material itself might suppresse Si release. While Ca did not suppresse Si release from RHB. Eh was a possible factor influencing Si release from RHB as the release of Si was higher with low Eh in sandy soil (Table 3).

3.5.7 Cacao SB

The results of this study revealed that there was an inverse relationship between Ca and Si concentration in red clayey soil solution. Higher pH and exchangeable Ca content in sandy soil tended to reduce Si release from cacao SB. This is more likely because Ca binds Si in cacao SB which might hardly dissolve under higher pH and Ca condition. It seems there is a higher potential to release Si in acidic and low exchangeable Ca soil condition.

3.5.8 Elephant Grass and Media of Mushroom

Silicon release of elephant grass (Figure 2e) and MM was negatively correlated with Eh change. Elephant grass and MM had relatively high potential to reduce soil Eh but they also enhance release of Fe and Mn especially in red clayey soil (Table 2). Eh of soil control the solubility of Fe and Mn oxides (Patra & Neue, 2010). Solubilitzation of Fe and Mn possibly influence Si release from elephant grass and MM.

3.6 Prospectives of Local Materials as Silicon Fertilizers

Most of the land area in Indonesia are acidic soil, of which area is around 102,000,000 ha. The dominant acidic soil types are Ultisols and Oxisols, and some belong to Entisols, Inceptisols and Spodosols (Subagyo et al., 2000; Mulyani et al., 2009). In order to discuss the possibility to use examined materials in the present study for Si amendments in Indonesian paddy fields, we focused on the results found in red clayey soil as most of the land area in Indonesia are acidic soil. Overall Si materials, release of Si into red clayey soil solution were high for steel slag, cacao SB, RSC and also RHB as local Si materials. These can be candidates of Si amendments of paddy soil in Indonesia. The amount of Si release from EFS, elephant grass and MM were relatively low. Therefore, we assume these materials are not effective to improve paddy soil available Si in Indonesia. Although silica gel could be of course a good Si fertilizer as it was used in Japan, it is presently expensive for Indonesian local farmers.

In terms of availability of these materials in Indonesia, steel slag is promising. Rice straw is the biggest waste in Indonesia with the amount of 55 million ton per year (Setiarto, 2013). So far in Indonesia, there has been no research on Si content in rice straw and straw compost. Thus we cannot give recommendation which is better between rice straw and rice straw compost. Generally, agricultural waste in Indonesia is burned to accelerate land preparation. Based on our result that rice straw compost release high Si, thus farmers can sell the rice straw compost to increase income. Furthermore we suggest to make rice straw as compost.

Rice husk is also available throughout Indonesia. For the use of rice straw and husk, it is easy to collect and use in rice producing areas as farmers' groups exist. Besides, its function as Si amendment, *i.e.* Si release, can be easily improved by burning in paddy fields as the present study exhibited.

Cacao production for some regions in Indonesia such as Sumatera, Java and Sulawesi were 0.36; 0.40 and 0.46 ton ha^{-1} on 2013 (Tree Crop Estate Statistics of Indonesia, 2014). Cacao shell waste has not been optimally used in Indonesia (Murni et al., 2012), thus its agricutural use is recommendable. Furthermore, we may improve this material by charring to increase available Si in soil.

4. Conclusions

EFS, Fly ash and organic materials under submerged condition in paddy fields, improved Si availability. Besides, the addition of local materials such as steel slag reduces the soil Eh. Local materials such as steel slag, rice straw and husk, and cacao shell could be used as Si amendments in paddy fields in Indonesia. Additionally, other materials with relatively low Si release (*i.e.* elephant grass and MM) could be used to improve the availability Si in soil.

Acknowledgements

This work was partly supported by Japan Society for the Promotion of Science KAKENHI Grant Number 24405047. The first author gives thanks for the scholarships provided by Indonesian Agency for Agricultural Research and Development. Ministry of Agriculture Republic of Indonesia.

References

Anonymous. (n.d.). *Badan Meteorologi Klimatogi dan Geofisika (BMKG), 2000-2013.*

Badan Pusat Statistik (BPS). (2015). *Statistics of South Sumatera Province.*

Banten. (September 23, 2016). *Geographical conditions of Banten province.* Retrieved from http://www.bantenprov.go.id

Beckwith, R. S., & Reeve, R. (1963), Studies on soluble silica in soils. I. The sorption of silicic acid by soils and minerals. *Aust J. Soil Res., 1*, 157-68. In Liang, Y., Nicolic, M., Belanger, R., Gong, H., & Song, A. (2015). Silicon in agriculture. *From theory to practice* (p. 56). Springer Dordrecht Heidelberg New York London. http://dx.doi.org/10.1071/SR9630157

Bunker, B. C., Tallant, D. R., Headley, T. J., Turner, G. L., & Kirkpatrick, R. J. (1988). *Chemistry of glasses* (Vol. 29, p. 106). In Kouassi, S. S., Andji, J., Bonnet, J. P., & Rossignol, S. (2010). Dissolution of waste glasses in high alkaline solutions. *Ceramics-Silikaty, 54*(3), 235-240.

Cornelis, J. T., Delvaux, B., Georg, R. B., Lucas, Y., Ranger, J., & Opfergelt, S. (2011). Tracing the origin of dissolved silicon transferred from various soil-plant systems towards rivers: A review. *Biogeosciences, 8*, 89-112. http://dx.doi.org/10.5194/bg-8-89-2011

Datnoff, L. E., & Rodrigues, F. A. (2005). The Role of Silicon in Suppressing Rice Diseases. *APSnet Features.* http://dx.doi.org/10.1094/APSnetFeature-2005-0205

Dematte, J. L. I., Paggiaro, C. M., Beltrame, J. A., & Ribeiro, S. S. (2011). Uso de silicatos em cana-de-acucar. *Informacoesonomicas* (No. 133, pp. 7-12). In Anderson, C. M., & Lauricio, E. (2013). Silicon: Fertilization and nutrion in higher plants. *Amazonian Journal of Agricultural and Environmental Sciences.*

Environmental Protection Agency (EPA). (1993). *Standard for the use or disposal of sewage sludge* (40 CFR Part 257, p. 288).

FAO. (2010). *Agricultural commodity projections to 2010.*

Flint, E. P., & Wells, L. S. (1934). Study of the system CaO-SiO$_2$-H$_2$O at 30 C and of the reaction of water on the anhydrous calcium silicates. *Part of Bureau of Standards Journal of Research* (Vol. 12). U.S. Department of Commerce. http://dx.doi.org/10.6028/jres.012.060

Foy, C. D. (1992). Soil chemical factors limiting plant root growth. *Adv. Soil Sci., 19*, 97-149. In Datnoff, L. E., & Rodrigues, F. A. (2005). The Role of Silicon in Suppressing Rice Diseases. *APSnet Features.* http://dx.doi.org/10.1007/978-1-4612-2894-3_5

Gascho, G. J. (2001). Silicon sources for Agriculture. In L. E. Datnoff, G. H. Snyder, & G. H. Korndorfer (Eds.), *Silicon in Agriculture*. Elsevier Science, Amsterdam, Netherlands. http://dx.doi.org/10.1016/S0928-3420 (01)80016-1

Hansen, H. C. B., Wetche, T. P., & Rauland-Rasmussen, K. (1994). Stability constants for silicate adsorbed to ferrihydrite. *Clay Minerals, 29*, 341-450. http://dx.doi.org/10.1180/claymin.1994.029.3.05

Haynes, R. J. (2014). A contemporary overview of silicon availability in agricultural soils. *J. Plant Nutr. Soil Sci., 177*, 831-844. http://dx.doi.org/10.1002/jpln.201400202

Haynes, R. J., Belyaeva O. N., & Kingston, G. (2013). Evaluation of industrial wastes as sources of fertilizer silicon using chemical extractions and plant uptake. *J. Plant Nutr. Soil Sci., 176*, 238-248. http://dx.doi.org/10.1002/jpln.201200372

Husnain, W. T., Setyorini, D., Hermansah, Sato, K., & Masunaga, T. (2008). Silica availability in soils and river water in two watersheds on Java Island, Indonesia. *Soil Sci. Plant Nutr., 54*, 916-927. http://dx.doi.org/10.1111/j.1747-0765.2008.00313.x

Imaizumi, K., & Yoshida, S. (1958). Edaphological studies on silicon supplying power of paddy fields. *Bulletin Natl. Inst. Agric. Sci., Series B*(8), 261-304.

Jawa Barat. (September 23, 2016). Retrieved from http://www.jabarprov.go.id

Kashem, M. A., & Singh, B. R. (2001). Metal availability in contaminated soils: I. Effects of flooding and organic matter on changes in Eh, pH and solubility of Cd, Ni and Zn. *Nutrient Cycling in Agronoecosystems, 61*, 247-255. http://dx.doi.org/10.1023/A:1013762204510

Kato, N., & Owa, N. (1996). Dissolution of slags in water and calcium chloride solution: Effects of solution pH and calcium concentration on solubilities of slags. *Jpn. J. Soil Sci. Plant Nutr., 67*, 626-632. http://dx.doi.org/ 10.1080/00380768.1997.10414757

Kato, N., & Owa, N. (1997). Dissolution of slag fertilizers in a paddy soil and Si uptake by rice plant. *Soil Sci. Plant Nutr., 43*(2), 329-341.

Kendrick, J. K. (2006). Pedogenic silica accumulation. *Encyclopedia of soil science* (2nd ed., Vol. 2, p. 1251). Taylor and Francis.

Koyama, T., & Sutoh, M. (1987). Simultaneous multi element determination of soils, plant and animal samples by inductively coupled plasma emission spectrophotometry. *Soil Science & Plant Nutrition, 58*, 578-585.

Liang, Y., Nicolic, M., Belanger, R., Gong, H., & Song, A. (2015). Silicon in agriculture. *From theory to practice* (p. 45, 58, 230). Springer Dordrecht Heidelberg New York London. http://dx.doi.org/10.1007/978-94-017-9978-2

Ma, J. F., & Yamaji, N. (2006). Silicon uptake and accumulation in higher plants. *Trends in Plant Science, 11*(8). http://dx.doi.org/10.1016/j.tplants.2006.06.007

Makabe-Sasaki, S., Kakuda, K., Sasaki, Y., & Ando, H. (2013). Effect of slag silicate fertilizer on dissolved silicon in soil solution based on the chemical properties of Gleysols. *Soil Science and Plant Nutrition, 59*, 271-277. http://dx.doi.org/10.1080/00380768.2012.763022

Marafon, A. C., & Endres, L. (2013). Silicon: Fertilization and nutrition in higher plant. *Amazonian Journal of Agricultural and Environmental Science, Rev. Cienc. Agrar., 56*(4), 380-388. http://dx.doi.org/10.4322/ rca.2013.057

Marxen, A., Klotzbücher, T., Jahn, R., Kaiser, K., Nguyen, V. S., Schmidt, A., ... Vetterlein, D. (2016). Interaction between silicon cycling and straw decomposition in a silicon deficient rice production system. *Plant Soil, 398*, 153-163. http://dx.doi.org/10.1007/s11104-015-2645-8

Meyer, P. (1999). Behavior of silica in ion exchange and other systems. *Proc. International Water Conf. 64*.

Ministry of Energy and Mineral Resouces. (2014). *Outlook Energy Indonesia*. Ministry of Energy and Mineral Resouces, Republic of Indonesia.

Ministry of Industry Republic of Indonesia. (2014). *Profile of steel industry*.

Molina, F. V. (2014). Soil colloids. *Properties and ion binding* (p. 339). CRC Press. Taylor & Francis Group.

Mulyani, A., Rachman, A., & Dariah, A. (2009). Proliferation of acid land, potential and availability for agricultural development. *Book of Natural phosphate: Utilization of natural phosphate is used directly as a fertilizer source* (pp. 25-46). Indonesian Soil Research Institute, Bogor.

Murni, R., Akmal, & Okrisandi, Y. (2012). Utilization cocoa pods fermented with phanerochaete chrysosporium as forage substitution in goat feed. *Agrinak, 2*(1), 6-10.

Patra, P. K., & Neue, H. U. (2010). Dynamics of water soluble silica and silicon nutrition of rice in relation to changes in iron and phosphorus in soil solution to soil drying and reflooding. *Archived of Agronomy and Soil Science, 56*(6), 605-622. http://dx.doi.org/10.1080/03650340903192042

Pereira, H. S., Korndorfer, G. H., Moura, W. F., & Correa, G. F. (2003). Extractors of available silicon in slags and fertilizers. *Revista Brasileira de Ciencia do Solo, 27*, 265-274. In Haynes, R. J., Belyaeva, O. N., & Kingston, G. (2013). Evaluation of industrial wastes as sources of fertilizer silicon using chemical extractions and plant uptake. *J. Plant Nutr. Soil Sci., 176*, 238-248. http://dx.doi.org/10.1590/S0100-06832003000200007

Ponnamperuma, F. N. (1965). Dynamic aspects of flooded soil and nutrition of the rice plant. *The Mineral nutrition of rice plant* (pp. 295-328). Baltimore, MD: Johns Hopkins. In Tubana, B. S., & Heckman, J. R. (2015). *Silicon in soils and plants*. In F. A. Rodrigue, & L. E. Datnoff (Eds.), *Silicon and Plant diseasses*. Springer International Publishing Switzerland.

Ponnamperuma, F. N. (1972). The chemistry of submerged soils. *Advances in Agronomy, 24*, 45-51. http://dx.doi.org/10.1016/s0065-2113(08)60633-1

Rabovsky, J. (1995). Biogenic amourphous silica. *Scandinavian Journal of Work, Environ & Health, 21*(Suppl. 2), 108

Ramenzanianpour, A. A. (2014). *Cement replacement materials*. Springer Geochemistry/Mineralogy. http://dx.doi.org/10.1007/978-3-642-36721-2_2

Sahrawat, K. L. (2005). Fertility and organic matter in submerged rice soils. *Current Science, 88*(5).

Savant, N. K., Korndorfer, G. H., Datnoff, L. E., & Snyder, G. H. (1999). Silicon nutrition and sugarcane production: A review. *J. Plant Nutr. 22*(12), 1853-1903. In Haynes, R. J., Belyaeva, O. N., & Kingston, G. (2013). Evaluation of industrial wastes as sources of fertilizer silicon using chemical extractions and plant uptake. *J. Plant Nutr. Soil Sci., 176*, 238-248. http://dx.doi.org/10.1080/01904169909365761

Savant, N. K., Synder, G. H., & Datnoff, L. E. (1997). Silicon management and sustainable rice production. *Adv Agron, 58*, 151-199. In Tubana, B. S., & Heckman, J. R. (2015). Silicon in soil and plants. In F. A. Rodrigue, & L. E. Datnoff (Eds.), *Silicon and Plant diseasses*. Springer International Publishing Switzerland. http://dx.doi.org/10.1016/S0065-2113(08)60255-2

Setiarto, R. H. B. (2013). Prospects and potential of rice utilization of lignocellulose into compost, silage and biogas by microbial fermentation. *Jurnal Selulosa, 3*(2), 51-66.

Snyder, G. H., Matichenkov, V., & Datnoff, L. E. (2007). Silicon. In A. Barker, & D. Pilbeam (Eds.), *Handbook of Plant Nutrition* (Chapter 19, pp. 551-568). Taylor and Francis, Boca Raton, FL.

Subagyo, H., Nata, S., & Agus, B. S. (2000). Soil-agricultural land in Indonesia. *Book of Indonesian land resources and their Management* (pp. 21-66). Soil and Agro-climate Research Center, Bogor.

Sumida, H. (1992). Silicon supplying capacity of paddy soils and characteristics of silicon uptake by rice uptake in cool regions in Japan. *Bull. Tohoku. Agric. Exp. Stn, 85*, 1-46.

Tree Crop Estate Statistics of Indonesia. (2014). *Cocoa 2013-2015*. Directorate General of Estate Crops.

Tubana, B. S., & Heckman, J. R. (2015). Silicon in soils and plants. In F. A. Rodrigue, & L. E. Datnoff (Eds.), *Silicon and Plant diseasses*. Springer International Publishing Switzerland. http://dx.doi.org/10.1007/978-3-319-22930-0_2

Wei, C. F., Yang, J. H., Gao, M., Xie, D. T., Li, Q. Z., Li, H. L., … Xiang, T. C. (1997). Study on availability of silicon in paddy soils from purple soil. *J. Plant Nutr Fertil., 3*, 229-36. In Liang, Y., Nicolic, M., Belanger, R., Gong, H., & Song, A. (2015). Silicon in agriculture. *From theory to practice* (p. 58). Springer Dordrecht Heidelberg New York London.

Xiao, X., Chen, B., & Zhu, L. (2014). Transformation, morphology and dissolution of silicon and carbon in rice straw derived biochars under different pyrolytic temperatures. *Environ. Sci. Technol., 48*, 3411-3419. http://dx.doi.org/10.1021/es405676h

5

Effect of Storage Methods on Carbohydrate and Moisture of Cassava Planting Materials

Baraka B. Mdenye[1,2], Josiah M. Kinama[2], Florence M. Olubayo[2], Benjamin M. Kivuva[3] & James W. Muthomi[2]

[1] Masasi District Council, Masasi, Mtwara, Tanzania

[2] College of Agriculture and Veterinary Sciences, Faculty of Agriculture, University of Nairobi, Nairobi, Kenya

[3] Kenya Agricultural and Livestock Research Organization (KALRO), Kenya

Correspondence: Baraka B. Mdenye, Masasi District Council, P.O. Box 21, Masasi, Mtwara, Tanzania. E-mail: bamide2001@gmail.com

Abstract

Storage of cassava (*Manihot esculenta* Cruntz.) planting materials has been a challenge because of its properties of moisture and carbohydrates loss under storage. Two varieties of cassava cuttings 1 m long, stored for four months under four different storage methods in two locations Kabete and Kiboko. The storage methods were clamp under double shade (CUDS), horizontal under shade (HUS), vertical under shade (VUS) and the control horizontal under open ground (HOUG). In each storage method data loggers were installed to record temperature and RH. Percentage carbohydrate, moisture content (MC), 100% dry cuttings (DC) and cuttings dried to 25% or more of its stored length but not 100% were measured at intervals of 4 weeks. Data were subjected ANOVA and means separated using LSD. CUDS performed better than other storage methods in all parameters measured. The results showed cuttings stored under CUDS lost less moisture than those stored in HUOG. The moisture loss in CUDS was from 70.16%-56.69% while that of HUOG dropped from 70.16% to 27.26% within 8 weeks after storage. High rate of carbohydrate loss was observed in Kiboko than Kabete. Mean temperatures were 25 °C Kiboko and 22 °C Kabete. The results showed that temperature had effect on loss of carbohydrate. The results have proven that safe storage of cassava planting material is affected by plant related factors as well as environmental conditions.

Keywords: carbohydrate, cassava cuttings, cassava planting materials, moisture, storage methods

1. Introduction

Cassava contribute to food security and livelihood to majority of small scale farmers in semiarid areas. It is also a source of raw materials to more than 1000 microprocessors and traders around the world (Balagopalan, 2002). Cassava is a source of carbohydrate in Africa after maize and rice (Aerni, 2005). The ability of cassava to grow in marginal land as well as flexibility in harvest of tuber when needed make it the best crop of choice for most poor farmers. Worldwide cassava production increased from 163 MT in 1980 to 270 MT in 2013.A lot of effort has been put to increase cassava production to cater food, energy and animal feeds requirements in Africa but planting material has been a challenge to most farmers. Most farmers use planting materials from previous crop which, normally have diseases infection as well as low nutrients (Ogero et al., 2012).

Cassava planting materials require storage especially when climatic condition such floods, drought and low temperature or delayed land preparation and other factors. However, use of fresh planting materials is preferable than stored cuttings (Leihner, 1983; Lozano et al., 1977) as it has been observed that the longer the duration cassava planting materials are stored, deteriorate their sprouting. Sprouting ability depend on storage conditions (Oka et al., 1987) as well as other factors like temperature and moisture of field. Leihner (1986) reported that cassava stems lose carbohydrate reserves during storage mainly in form of total carbohydrate and reducing sugars. More lignified cuttings contain small amount of food reserved for shoots development during sprouting (Lozano et al., 1977). Despite good storage conditions, long storage durations bring about some losses in moisture, carbohydrates, and nutrients, which would partially account for reduced early vigour (Leihner, 1982). Cassava cuttings dehydrate when stored. The rate of moisture loss is high when the cuttings are stored in open air and exposed to sun (Leihner, 1982). Moisture loss on planting material are influenced by plant factors (level of lignification and moisture content at harvest) and environmental factors (radiation, humidity, temperature and wind speed) (Leihner, 1982).

When harvesting and planting are separated in time, a farmer can decide to leave some portion of crop as seed for next planting. But this can cause pests carryover and cause big loss to small scale farmers (Leihner, 1982). Also where land is unavailable, storage of planting materials is inevitable.

2. Materials and Methods

2.1 Description of Sites

The experiment was conducted in two sites namely University of Nairobi Kabete Campus and KARLO Kiboko. Kabete is situated about 15 km to the west of Nairobi city and lies at 1°15′S latitude and 36°44′E longitude and at altitude of 1930 m above sea level (masl) (Onyango et al., 2012). Kabete has a bimodal distribution of rainfall, with long rains from early March to late May and the short rains from October to December (Onyango et al., 2012). The mean annual temperature is 18 °C and total annual rainfall ranging between 700-1500 mm (Wasonga et al., 2015).

The second site was KARLO-Kiboko which lies within longitudes 37°.43212'E and latitudes 2°.12933'S, and 821.7 m above sea level in Makueni County, 187 km east of Nairobi, Kenya (Kivuva et al., 2015). The location receives between 545 mm and 629 mm of rainfall coming in two seasons. The long rains season is between April and May while the short rains season is between October and January. The mean annual temperature is 22.6 °C, where by the mean annual maximum temperature is 28.6 °C and mean annual minimum temperature is 16.5 °C (The Kenya Gazette, 2010).

2.2 Source of Cassava Stem Cuttings

The stem cuttings comprised varieties Karembo and KME4 were obtained from KARLO Thika. Planting and harvesting of planting materials was done at the same time as well as planting location to avoid the difference in accumulation of carbohydrate. The materials were selected on basis of the diseases free and high yielding of the varieties. KME4 has maturity of 8-10 months, fresh tuber yield is 38 t ha^{-1}, resistance to cassava mosaic virus and cassava brown streak. Karembo mature in 8 months and fresh tuber yield range from 50-70 t ha^{-1}, it has great tolerance to cassava mosaic virus and cassava brown streak virus (Kenfap Services Limited, 2013).

2.3 Experimental Design

The design was split plot in randomized complete block design (RCBD) (Petrenko, 2014). The main plot being storage methods (with 4 levels) sub plot being varieties (2 levels). The storage duration was in 5 terms (0, 4, 8, 12, and 16 weeks after storage (WAS)). Each main plot had 30 cuttings while the sub plot had 15 cassava cuttings from single variety each having 100 cm length with diameter range of 1.5 cm-3.4 cm. These 15 cuttings were tagged numbers 1-15 (Ravi & Suryakumari, 2005). Cassava cuttings were stored in four different storage methods namely; clamp under double shade (CUDS) (Plate 2), horizontal in open air under shade on the soil (HUS), vertical in open air under shade on soil (VUS) (Sales & Leihner, 1980; Plate 3) and horizontal under open ground with no shade on the soil (HUOG) (control or farmer's way of storing cassava cuttings when waiting transport or planting) (Plate 4). The shade was made by simple wooden poles and grass thatch (Plates 1 and 3). Three hitag 2 xsense loggers were installed to monitor temperature and relative humidity in clamp, under shade and under direct sun light. Two cuttings per storage method were sampled to form 6 cuttings. The 6 cuttings were cut into 20 cm cuttings after removing 10 cm from each end. The cuttings were mixed together and some were sampled for moisture and carbohydrate analysis.

Plate 1. (a) Simple shade and (b) different storage methods under shade

2.4 Carbohydrate Determination

In each storage method six samples of 20 cm each were taken to the laboratory for carbohydrate tests. The sample cuttings ware cut at the middle and both sides of the cut were grated to obtain composite sample.

Plate 2. Clamp under double shade storage method

Note. (a) clamp structure frame, (b) arranged cuttings for storage (c) clamp storage method covered by grass then 0.06 m^3 of soil.

Plate 3. Different storage methods

Note. (a) Vertical under shade with lower end of cassava touching the soil, (b) horizontal under shade, (c) cuttings stored horizontal in open ground under direct sun light.

Total carbohydrate in dry samples was estimated by the anthrone method which is a simple calorimetric method with relative insensitivity to interference from other cellular components (Clegg, 1956; Ravi & Suryakumari, 2005). 1 gram from sample were taken in duplicate and transferred to graduated 100 ml beaker. Then 10 ml of distilled water were added and then stirred thoroughly to dispense the sample.

From 10 ml of sample suspensions 13 ml of 52% perchloric acid were added in order to solubilize starch in samples (Rose et al., 1991). The suspensions were stirred for 20 minutes and then diluted to 100 ml. Then the suspensions were filtered to 250 ml flask and the solution were diluted to the mark to form a stock solution (Clegg, 1956). From the stock solution 10 ml was drawn and diluted to 100 ml. 1 ml of the diluted sample, standard sample and blank were pipetted into individual test tubes, then in each test tube 5 ml of anthrone reagent in concentrated sulphuric acid was added. The reactions in this process were, concentrated H_2SO_4 catalyses the dehydration of sugars to form furfural (from pentose's) or hydroxy methyl furfural (from hexoses). Adding anthrone into the sample give condensation product with bluish or green coloured.

The test tubes then transferred to boiling water for exactly 12 minutes then cooled to room temperature. From the test tubes the solutions were transferred to glass cuvettes to read absorbance at 630 nm wave length (Clegg, 1956).

The formula used to calculate the concentration of carbohydrate in the samples was:

$$\frac{\text{Absabance of sample}}{\text{Concentration of sample}} = \frac{\text{Absabance of standard}}{\text{Concentration of standard}} \tag{1}$$

2.5 Determination of Moisture Loss from Stored Cassava Stem Cuttings

The moisture content was determined by constant temperature oven method (ISTA, 2015). From the composite sample, 2 g of sample were measured in duplicate into moisture dishes and put to oven at 103 °C for more than 17 hrs to obtain constant weight. Calculations and expression of results for each replicate in three decimal places using the following formula (ISTA, 2015):

$$\frac{\text{Loss of weight} \times 100}{\text{Intinial weight}} = \frac{M_2 - M_3 \times 100}{M_2 - M_1} \tag{2}$$

Where, M_1 is weight (in grams) of empty container; M_2 is weight (in grams) of container + sample before drying; M_3 is weight in grams of container + dry sample.

2.6 Data Analysis

The data were subjected to analysis of variance (ANOVA) (Kroonenberg & van Eeuwijk, 1998; Cohen & Brooke, 2004; Cheng & Shao, 2006; Smith, 2006) to determine the difference between storage methods and varieties. GenStat 15 edition were used. Means were separated using least significant difference (LSD) at $p \leq 0.05$ (Kivua et al., 2015). The assumptions were that the population was normally distributed, samples were independent, variance of population was equal and group of samples was equal.

Complete model used was (Kroonenberg & van Eeuwijk, 1998; Smith, 2006):

$$Y_{ijkl} = \mu + \alpha_i + \beta_j + \gamma_k + (\alpha\beta)_{ij} + (\alpha\gamma)_{ik} + (\beta\gamma)_{jk} + (\alpha\beta\gamma)_{ijk} + \varepsilon_{ijkl} \tag{3}$$

Where, μ is the general mean of the population α_I, β_j, γ_k is mean of storage methods, mean of varieties and mean of duration of storage, as main effects while $(\alpha\beta)_{ij}$, $(\alpha\gamma)_{ik}$, $(\beta\gamma)_{jk}$ are corresponding two-way interaction effect and $(\alpha\beta\gamma)_{ijk}$ three-way interaction effect. ε_{ijkl} represent the expected error (Kroonenberg & van Eeuwijk, 1998). The computed data composed percentage dry cuttings more than 25% of its storage length but not 100% (%DC > 25% SL), Percentage dry cuttings 100% of its storage length (%DC), percentage moisture of stored cuttings (%MC) and carbohydrate content of stored cuttings sampled at specific duration according to Equations 2, 4, 5 and 6.

$$\%DC > 25\% \ SL = \frac{DC > 25 \ cm}{TNC} \times 100 \tag{4}$$

Where, %DC > 25% SL sampled cuttings with > 25 cm of its stored length dried but less than 100 cm; DC > 25 cm = sample cuttings with > 25 cm of its stored length dry; and TNC = total number of cuttings at a given time.

$$\%DC = \frac{DC}{TNC} \times 100 \tag{5}$$

Where, %DC = cuttings sample 100% of its stored length dried; DC = total dried cuttings samples; and TNC = total number of cuttings at a given time

$$\% \ Carbohydrate = \frac{25 \times b}{a \times w} \tag{6}$$

Where, b = absorbance of diluted sample; a = absorbance of dilute standard sample; and w = weight of sample (g).

3. Results

3.1 Weather Data during the Experiment Duration

The mean temperatures were 24.12 °C and 12.53 °C in Kiboko and Kabete respectively. Rainfall in Kiboko was negligible while in Kabete was around 5.67 mm. RH were around 82% and 65% in Kiboko and Kabete respectively. Data obtained from ICRISAT Kiboko and Kabete meteorological stations (Figure 1).

Data recorded by hitag 2 xsense data loggers (Table 1) were different from each storage methods. Minimum was recoded in CUDS and highest was under HUOG in both locations.

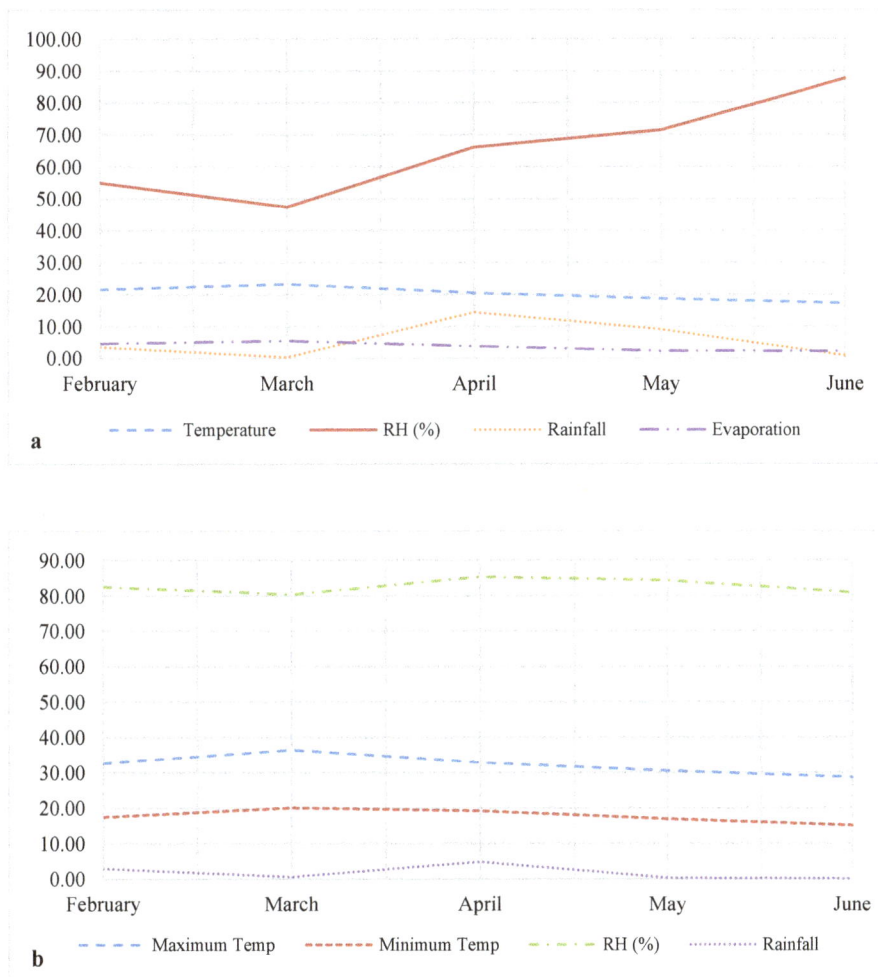

Figure 1. (a) Weather condition at Kiboko; (b) weather condition at Kabete in 2016

Table 1. Temperature and relative humidity in storage methods

Site	Storage method	Min T °C	Max T °C	Mean T °C	Min% RH	Max% RH	Mean% RH
Kabete	CUDS	12.25	24.5	18.78	38.00	100.00	72.05
Kabete	HUS &VUS	11.00	29.50	19.16	20.00	100.00	66.23
Kabete	HUOG	9.00	39.50	21.13	20.00	100.00	61.89
Kiboko	CUDS	17.50	32.75	24.99	20.00	92.00	60.13
Kiboko	HUS&VUS	14.50	40.00	25.42	20.00	100.00	56.66
Kiboko	HUOG	12.75	45.50	28.00	20.00	100.00	40.91

Note. T = temperature (°C), RH = relative humidity, Max = maximum, Min = minimum, CUDS = clamp under double shade, HUS = horizontal under shade, VUS = vertical under shade and HUOG = horizontal in under open ground.

3.2 The Percentage of Cuttings that Had Dried 25% of Its Storage Length but Less Than 100% (%DC > 25% SL)

The results showed highly significant difference ($p < 0.001$) among storage methods as well as varieties. The interaction between storage methods and varieties show significant difference ($p < 0.01$). Among storage methods, CUDS has the minimum increase of % DC > 25% SL, at 4 and 8 weeks after storage. Highest percentages (more than 50%) were found in horizontal under open ground with no shade on the soil only 4 weeks after storage.

CUDS performed better in all location and almost in all traits this can be explained by lowest mean temperature of 18.78 °C and 24.99 °C in Kabete and Kiboko respectively. Low mean temperature and medium RH reduce the desiccation of stored cuttings. The maximum temperature was recorded by temperature data logger's in horizontal under open ground on soil which was 45.50 °C in Kiboko. The maximum temperatures were recorded around 11:00 hours to 17 hours. Also the results shown that CUDS had high average of relative humidity of 72.05% and 60% in Kabete and Kiboko respectively while lowest mean relative humidity were recorded in HUOG 61.89% and 40.91% Kabete and Kiboko respectively. This justify why cuttings stored in HUOG lost more moisture than cuttings stored in CUDS. Low relative humidity means cuttings will loss water to surrounding. High evaporation of 5.62 mm was recorded at Kabete when average rainfall was 0.35 mm and also at the same time low RH was recorded. This mean that when there is low RH, cuttings loss more moisture content as evaporation become high. It also explains why cuttings stored in Kiboko dehydrated faster than that stored in Kabete.

The results also showed that the performance of HUS when the temperature was low was better than VUS. But under high temperature VUS lower end of cassava cuttings touching the soil did better than HUS. The mean temperature under shade was 19.16 °C and 25.42 °C, at Kabete and Kiboko respectively. While the maximum was 29.50 °C and 40.00 °C, Kabete and kiboko respectively. Relative humidity was 66.23% and 56.66% Kabete and Kiboko respectively. Thus Kabete site is good environment for cassava cuttings storage as compared to Kiboko because of low temperature and medium RH which is not triggering sprouting in storage or make stored cuttings to desiccate (Figure 2).

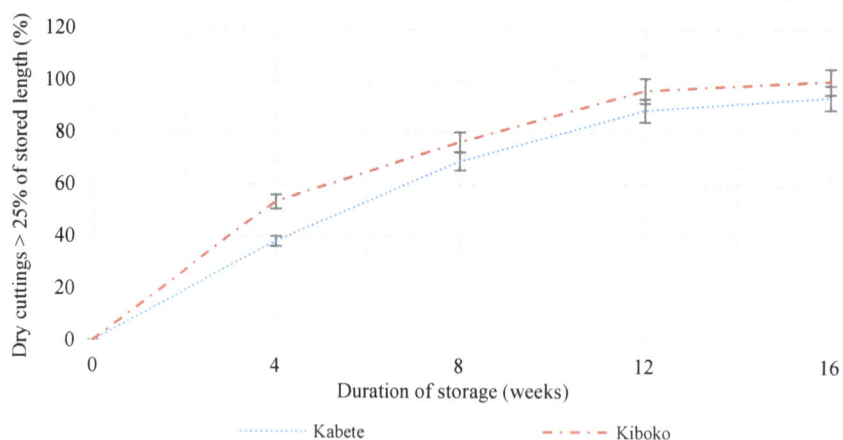

Figure 2. Rate of dehydration of cassava cuttings at Kabete and Kiboko for a duration of 16 weeks after storage

KME4 showed the best storability as compared to Karembo by having low% DC > 25% SL relative to that of Karembo.

3.3 Percentage Cuttings Dried 100% of Its Stored Length (%DC)

There was significant difference among storage methods at $p < 0.01$ as well as varieties (Figure 3). CUDS had lowest average dry cuttings of 34.40% as compared to other storage methods. The highest was under HUOG with 52.56% for whole period of storability test. Also KME4 had less dried cuttings 32.14% as compared to 47.15% of Karembo. Duration of storage was highly significant at $p < 0.001$ as results shown that the longer the storage percentage dried cuttings increased.

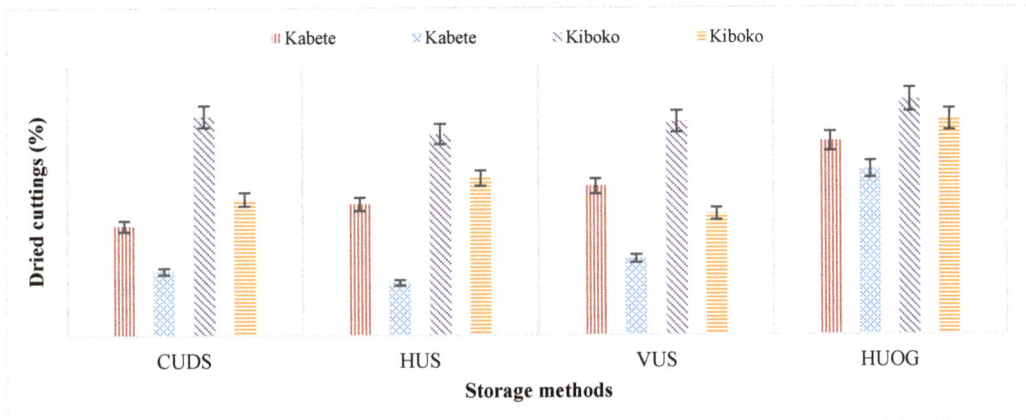

Figure 3. Effects of storage methods and variety on cassava cuttings drying in storage

Note. CUDS = clamp under double shade, HUS = horizontal under shade, VUS = vertical under shade and HUOG = horizontal in under open ground.

Storage was better in Kabete than in Kiboko due difference in RH, and temperature. Average relative humidity measured by Meteorology station Kabete was 60.04% and temperature was 23.54 °C as compared to average relative humidity measured by ICRISAT Kiboko field station of 83.91% and mean temperature of 33.18 °C. Variety KME4 performed better in both sites than Karembo. KME4 stored in Kiboko was performing better than Karembo stored in Kabete regardless of the difference in temperature and relative humidity.

3.4 Percentage Moisture of Stored Cuttings (MC%)

Moisture content of stored cuttings at different storage duration showed highly significant difference among sites ($p < 0.001$). The mean of moisture content of stored cuttings at 0 weeks of storage it was 70.16% both at Kabete and Kiboko. After 16 weeks of storage moisture content reduced to 14.23% at Kiboko and 39.70% at Kabete depending on methods of storage. Thus, the results shown that the rate of moisture content loss influenced by environmental conditions of particular location (Figure 4). The rate of dehydration in Kiboko was higher as compared to Kabete. Meaning that the moisture content of cuttings depends on weather conditions of locality of storage.

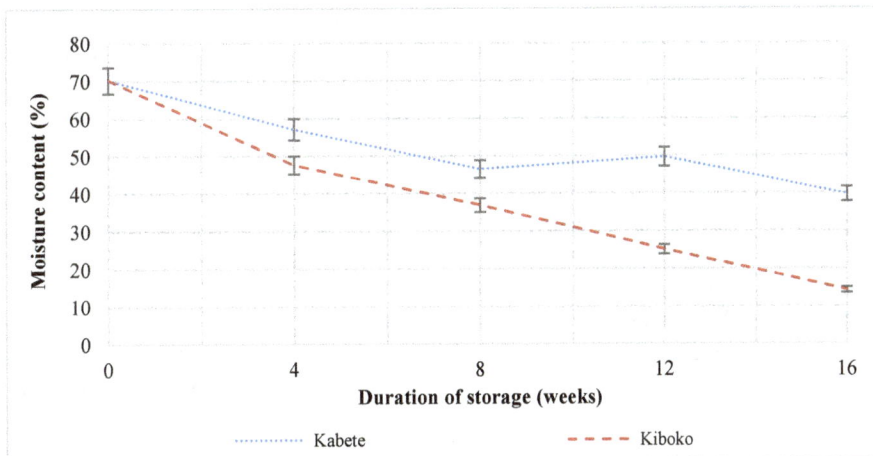

Figure 4. General percentage moisture content of cuttings with reference to duration of storage in two locations

Moisture reduction from stored cuttings in both Kabete and Kiboko site, dehydration of cuttings stored under CUDS was lower as compared to other storage methods while under HUOG dehydration of cuttings was highest. Thus, the best storage methods were CUDS followed by VUS and then HUS. Rate of dehydration of stored cuttings depends on varieties. Whereby, the rate of moisture loss of Karembo was higher compared to KME 4. CUDS performed better in all duration of storage from week 0 to week 16 after storage for stored cuttings.

3.5 Carbohydrate Content of Stored Cuttings

The amount of carbohydrate in cuttings during storage differed significantly among locations at $p < 0.001$. The results indicate that cuttings stored in Kiboko lost more carbohydrate than cuttings stored in Kabete (Figure 5). Duration of storage showed significant difference between weeks of storage at $p > 0.001$. The results showed that the more farmer store cuttings for long duration the more cuttings consume carbohydrate for maintenance. Also the results showed highly significant difference among varieties at $p < 0.001$. This mean the consumption of carbohydrate during storage differ with varieties. The interaction between duration of storage and variety was highly significant at $p < 0.001$ meaning that if the farmer store cassava planting material for a long time, it will lose more carbohydrate.

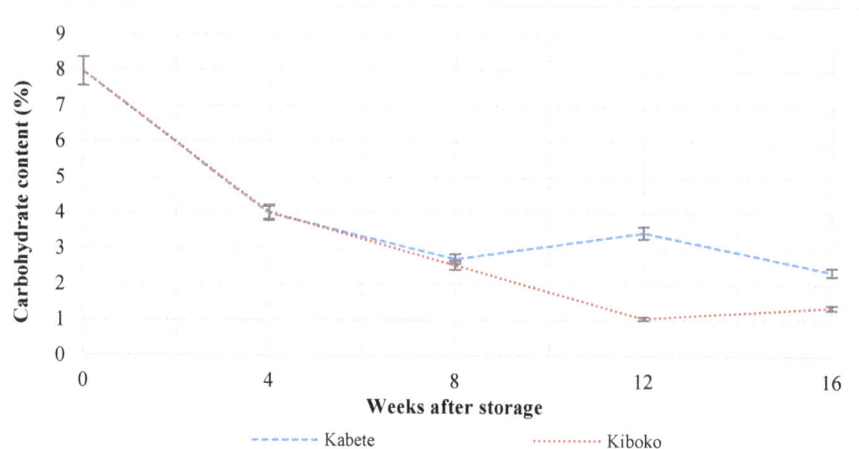

Figure 5. Average carbohydrate content reduction of cuttings with reference to duration of storage

4. Discussion

Cassava is vegetatively propagated and its cuttings are expensive as compared to true seed of other crops like maize and its establishment depends on quality planting materials. If 25% of original length of cassava dried during storage, then only 75% of stored cuttings is available for planting. Difference among the storage methods was observed in results as highly significant in stem dehydration and moisture loss. The carbohydrate loss was non-significant at p < 5% between storage methods but location of storage as well as varieties were significant among treatments. The rate of carbohydrate consumption in stored cuttings differed with location of experiment, Kiboko being hotter than Kabete had higher respiration rate than Kabete.

CUDS was shown to lose less moisture content than other methods. At 16 weeks after storage, the cuttings had an average of 53.54% moisture content as compared to control which had an average of 38.83%. This result may be attributed to difference in temperature and relative humidity among storage methods. Ratanawaraha et al. (2000) argued that storage under shade is better than under full sunlight. Direct sunlight containing heat energy which will accelerate the rate of moisture loss. The maximum temperature of 45 °C and 39.5 °C was recorded in the control at Kiboko and Kabete respectively. From the results it showed some increase in carbohydrate as well as moisture content of the stored cuttings at 12 WAS. This can be attributed by rainfall in March to April. The rainfall triggered some stored cuttings to sprout and form leaves which were contributing to photosynthesis. But general trend was decreasing in moisture and carbohydrate of stored cuttings. Similar results were obtained by Leihner, (2002) who showed that storing planting material under inadequate condition can cause cassava stake to loss 70% of sprouting ability if they were stored for 15 days at 24 °C. The difference in moisture loss is determined by cultivar, plant related factor such as degree of lignification at harvesting time and length of stored cuttings as well as environmental factors such as temperature, RH, radiation and air. Ravi and Suryakumari (2005) in their experiment of the novel technique to increase the shelf life of cassava planting materials they found similar factors affecting rate of moisture loss. The cuttings stored using HUOG in both sites lost moisture from 70.16% to 27.26% in just 8 weeks of storage probably due to be exposed full solar radiation and wind (Kinama et al., 2005). Kinama found that when soil doesn't have cover it loses more moisture through evaporation than when it contains cover crop or mulch. When there is high evaporation than rainfall results to high loss of moisture content of planting materials. It also explains why cuttings stored in Kiboko dehydrated faster than that stored in Kabete because evaporation in Kiboko is high than in Kabete. Pilbeam, Daamen, and Simmonds (1994) showed that evaporation in Kiboko is around 131.2 mm-224.9 mm from year 1991-1992. When stored cuttings are in open ground without cover or shade will lose more moisture content than when under shade and covered like in CUDS. According to Leihner (1982) cassava stored planting materials lose faster moisture in shorter stems than in long stems as the loss of moisture was recorded to start from two cut ends of stem cuttings increasing to the middle of the stem. This was resulting to some cuttings loss its viability from 25% of its length to 100%. It was recorded that only at 4 WAS the range loss of cuttings moisture were 31.11% to 58.33% depending on method of storage and location of storage. The loss of moisture was significantly different between varieties Karembo dehydrated more than KME4 stored in the same environment. Pérez et al. (2011) argued that cassava variety have difference capacity withstanding storage duration from harvesting to planting. This difference influences crop establishment and yield. The range of moisture content after 16 weeks of storage was 56.16%-43.27% for KME4 and 49.09%-34.30% for Karembo depending on site of experiment. Kiboko is a hot area with mean temperature of 25 °C and Kabete of 22 °C. This can explain the difference of moisture loss among the sites. In clamp storage in Kiboko the mean temperature recorded was almost similar (25.22 °C and 25.82 °C) perhaps due to hot air movement across the ventilation of clamp which was meant to reduce the RH in storage to control storage sprouting. This can explain the difference in performance of the clamp under double shade between Kabete and Kiboko. Thus Kabete site is good environment for cassava cuttings storage as compared to Kiboko because of low temperature and medium RH which is not triggering sprouting in storage or make stored cuttings to desiccate.

The results showed high significant differences among location and variety in carbohydrate loss of stored planting materials. The results showed significant loss during first 4 weeks after storage then after the rate of loss decreased. Ravi and Suryakumari (2005) found that carbohydrate content of stored cuttings decreases significantly in one month after storage. There after the change in content was less as compared to first four weeks. This rate can be due to reason during four weeks' moisture content of cuttings still high so even metabolic activities will be high. But other scenario can be due to stress of wounding of stored cuttings hence plant will be struggling to heal the wound caused in harvesting. The carbohydrate observed at 0 week of storage was 9.21% and 6.68% for Karembo and KME4 respectively. According to Kozlowsk (1991) carbohydrate consumption during storage can be due to maintenance respiration to keep planting material alive. Leihner (2002) reported that physiological deterioration of cassava planting materials linked with two main factors which are respiration and dehydration. He further said the respiration will be accelerated when cuttings are stored in hot environment than being stored in dry and cool environment. This can explain why the decrease in carbohydrate in Kiboko site was high as compared to Kabete. But also the results showed us that cuttings stored in HUOG lost less carbohydrate than other methods, it can be due to the fact that when cuttings lose high amount of moisture the maintenance respiration also will reduce or stop. Oka et al. (1987) found that respiration rate of stored cuttings increases soon after harvest of planting materials then after decrease before increase at slow rate again. This indicate that, variety and storage methods should be considered when a farmer wants to store the cassava cuttings with reference to a certain duration.

5. Conclusion

According to results obtained from this study storability of cassava depend on cultivar and environmental factors especially relative humidity, temperature, wind and radiation. It's better if cassava planting materials will be stored under shade and provide cover to insulate from high temperature and direct radiation. Where possible prolonged storage should be avoided since it contributes to carbohydrate loss during storage which has impact in crop establishment and vigour. High temperature has influence in carbohydrate loss as it increases respiration rate of stored cuttings Long term storage of cassava cuttings is possible under CUDS methods with temperature less than 20 °C.

Acknowledgements

The authors appreciate Alliance for a Green Revolution in Africa (AGRA) for funding the study through the University of Nairobi. The opinions expressed herein are those of the author(s) and do not necessarily reflect the views of AGRA.

Reference

Aerni, P. (2006). Mobilizing Science and Technology for Development: The Case of the Cassava Biotechnology Network (CBN). *AgBioForum, 9*(1), 1-14.

Cheng, B., & Shao, J. (2006). Testing Treatment Effects in Two-Way Linear Models: Additive or Full Model? *The Indian Journal of Statistics, 68*(2000), 392-408.

Clegg, K. M. (1956). The application of the anthrone reagent to the estimation of starch in cereals. *J. Sci. Food Agric., 40*(7). http://dx.doi.org/10.1002/jsfa.2740070108

Food and Agriculture Organization of the United Nations (FAO). (2013). *Save and grow: Cassava*. Italy, Rome.

ISTA. (2015). *International Rules for Seed Testing* (Vol. 2015, Full Issue i-19-8(276)).

Kenfap Services Limited (KSL). (2013). Re-introduction and commercialization of cassava for improved livelihoods through whole value chain model. *Ganze cassava value chain report*. Nairobi, Kenya.

Kinama, J. M., Stigter, C. J., Ong, C. K., Ng'anga, J. K., & Gichuki, F. N. (2005). Evaporation from soils below sparse crops in contour hedgerow agroforestry in semi-arid Kenya. *Agricultural and Forest Meteorology 130*(3), 149-162. http://dx.doi.org/10.1016/j.agrformet.2005.03.007

Kivuva, B. M., Sibiya, J., Githiri, S. M., & Yencho, G. C. (2015). Screening sweetpotato genotypes for tolerance to drought stress. *Field Crops Research, 171*, 11-22. http://dx.doi.org/10.1016/j.fcr.2014.10.018

Kozlowski, T. T., Kramer, P. J., & Stephen, G. (1991). *The physiological ecology of woody plants*. Academic Press, San Diego. http://dx.doi.org/10.1093/treephys/8.2.213

Kroonenberg, P. M., & van Eeuwijk, F. A. (2016). Multiplicative Models for Interaction in Three-Way ANOVA, with Applications to Plant Breeding. *International Biometric Society Stable, 54*(4), 1315-1333.

Leihner, D. (2002). Agronomy and cropping system. In R. J. Hillocks, J. M. Thresh, & A. C. Ballots (Eds.), *Cassava: Biology, production and utilization* (pp. 91-113). CABI International. http://dx.doi.org/10.1079/97 80851995243.0091

Leihner, D. E. (1982). Current practices in the production of cassava planting material. In J. H. Cock (Ed.), *Global workshop on root and tuber crops propagation* (pp. 41-46). Proceedings of a Regional Workshop held in Cali, Colombia, September, 13-16.

Leihner, D. E. (1984). Storage effects on planting material and subsequent growth and root yield of cassava (*Manihot esculenta* Crantz). *Proceedings of the sixth symposium of the international society for tropical root crops* (pp. 257-266). Lima, Peru, February 20-25, 1983.

Leihner, D. E. (1986). Storage and regeneration of cassava (*Manihot esculenta* Crantz) planting material. *Proceedings of a global workshop on the propagation of root and tuber crops* (pp. 131-138). September 12-16, 1983. Centro international de agriculture tropical (CIAT), Cali, Colombia.

Lozano, J. C., Toro, J. C., Castro, A., & Bellotti, A. C. (1977). Production of planting materials. *Cassava information centre*. Centro Internacional de Agricultura Tropical (ClIAT). Cali, Colombia S.A.

Nassar, N. M. A., Abreu, L. F. A., Teodoro, D. A. P., & Graciano-Ribeiro, D. (2010). Drought tolerant stem anatomy characteristics in *Manihot esculenta* (Euphorbiaceae) and a wild relative. *Genet. Mol. Res., 9*(2), 1023-103. http://dx.doi.org/10.4238/vol9-2gmr800

Ogero, K. O., Mburugu, G. N., Mwangi, M., Ombori, O., & Ngugi, M. (2012). *In vitro* Micropropagation of Cassava through Low Cost Tissue Culture. *Asian Journal of Agricultural Science, 4*(3), 205-209.

Oka, M., Sinthuprama, S., & Limsila, J. (1987). Variations in some characteristics of cassava stems during storage prior to taking cuttings. *Japan Agricultural Research Quarterly, 21*(3).

Onyango, C. M., Harbinson, J., Imungi, J. K., Onwonga, R. N., & Kooten, O. (2012). Effect of nitrogen source, crop maturity stage and storage conditions on phenolics and oxalate contents in vegetable amaranth (*Amaranthus hypochondriacus*). *Journal of Agricultural Sciences, 4*(7), 219-230. http://dx.doi.org/10.5539/ jas.v4n7p219

Otoo, J. (1996). *Rapid Multiplication of Cassava: IITA Research Guide 51*. IITA feature. Retrieved February 23, 2015, from http://www.iita.org/cms/details/trn_mat/IRG51/irg51.htm

Pérez, J. C., Ceballos, H., Calle, F., Morante, N., & Lenis, J. I. (2011). Cassava genetic improvement. In R. H. Howeler (Eds.), *The cassava handbook, A reference manual based on the Asian regional cassava training course held in Thailand.*

Pilbeam, C. J., Daamen, C. C., & Simmonds, L. P. (1994). Analysis of water budgets in semi-arid lands from soil water records. *Expl. Agric., 31*, 131-149. http://dx.doi.org/10.1017/S0014479700025229

Pooja, N. S., & Padmaja, G. (2015). Effect of single and sequential cellulolytic enzyme cocktail on the fermentable sugar yield from pre-treated agricultural residues of cassava. *American Journal of Biomass and Bioenergy, 4*(1), 39-53. http://dx.doi.org/10.7726/ajbb.2015.1004

Ratanawaraha, C., Senanarong, N., & Suriyapan, P. A. (2000). Review of Cassava in Asia: Evidence from Thailand and Vietnam. *Status of cassava in Thailand: Implications for future research and development.* Rome, Italy.

Ravi, V., & Suryakumari, S. (2005). Novel technique to increase the shelf life of cassava stem stored for propagation. *Adv. Hort. Sci. (Italy), 19*, 123-129. Retrieved from http://www.jstor.org/stable/42882403

Rose, R., Rose, C. L., Omi, S. K., Forry, K. R., Durall, D. M., & Bigg, W. L. (1991). Starch determination by perchloric acid vs enzymes: Evaluating the accuracy and precision of six colorimetric methods. *Journal of Agricultural and Food Chemistry, 39*, 2-11. http://dx.doi.org/10.1021/jf00001a001

Sales, A. M., & Leihner, D. E. (1980). Influence of period and conditions of storage on growth and yield of cassava. In E. J. Weber, J. C. Toro, & M. Graham (Eds.), *Cassava cultural practices.* Proceeding, Salvador, Bahia, Brazil, March 18-21.

Setter, T. L., Duque, L., & Alves, A. (2008). Drought tolerance mechanisms in cassava. *Cassava: Meeting the challenges of the new millennium* (p. 148). Scientific meeting of the global cassava partnership. Ghent: IPBO.

Van Eeuwijk, F. A., & Kroonenberg, P. M. (1998). Multiplicative Models for Interaction in Three-Way ANOVA, with Applications to Plant Breeding. *Biometrics, 54*, 1315-1333. http://dx.doi.org/10.2307/2533660

Wasonga, D. O., Ambuko, J. L., Chemining'wa, G. N., Odeny, D. A., & Crampton, B. G. (2015). Morphological characterization and selection of spider plant (*Cleome Gynandra*) Accessions from Kenya and South Africa. *Asian Journal of Agricultural Sciences, 7*(4), 36-44. http://dx.doi.org/10.19026/ajas.7.2198

Zhu, W., Lestander, T. A., Orberg, H., Wei, M., Hedman, B., Ren, J., ... Xiong, S. (2015). Cassava stems: a new resource to increase food and fuel production. *GCB Bioenergy, 7*, 72-83. http://dx.doi.org/10.1111/gcbb.12112

6

The Content of Sulphur in the Soil and Plant from Park Areas Exposed to Traffic Pollution

Hanna Jaworska[1], Anetta Siwik-Ziomek[2] & Katarzyna Matuszczak[1]

[1] Department of Soil Science and Soil Protection, UTP University of Science and Technology, Bydgoszcz, Poland

[2] Department of Biochemistry, UTP University of Science and Technology, Bydgoszcz, Poland

Correspondence: Hanna Jaworska, Department of Soil Science and Soil Protection, UTP University of Science and Technology, Bernardyńska 6 St., Bydgoszcz, Poland. E-mail: hjawor@utp.edu.pl

Abstract

Sulphur occurs in many environmental compounds. Source of this element may be natural as also anthropogenic origin, for example related with the development of road traffic. The aim of this study was to evaluate the impact of traffic on the content of total and sulphate sulphur in forest soils and plant material. The selected physicochemical properties of soils were determined: soil texture by laser diffraction method, soil pH by potentiometric method, total organic carbon (TOC) by Tiurin method. The content of total and sulphate sulphur in research material was determined by Bardsley-Lancaster method modified by COMN-IUNG. All analyses were performed in three replicates and the verification of the results was based on the certified material Till-3. Statistical analysis of the results were performed in Statistica 12.0 for Windows Pl software. Examined research material was characterized by medium, high and anthropogenic origin content of total and sulphate sulphur. Undertaken studies showed that the traffic could have an adverse influence on the content of sulphur in soils and plant material.

Keywords: total sulphur, sulphate sulphur, forest soils, pine bark, pine needle, traffic road

1. Intoduction

Landscape parks are unique places in terms of species richness, diversity of landscape and variety of geomorphic forms (Helmuth & Dudek, 2002). Through insight into many advantages, these areas are under legal protection (Regulation No. 92, item. 880). Myślęcinek Forest Park for Culture and Leisure, constituting an area of research, is the biggest city park in Poland. These site is located near to urban areas and the surface takes 830 ha (Dynarz & Wiśniewski, 1997).

Sulphur occurs in many environmental components e.g. air, water, soil. It is an element widely occurring in ecosystems. Sources of sulphur in environment may be natural as also anthropogenic origin. The biggest threat with regard to the amount of sulphur emissions to the environment comes from human activities i.e. exploitation of natural resources, the development of road traffic, storage of waste and many others (Sherer, 2001). Road traffic is the source of substances in environmental components (Pallvani & Harrison, 2013). The content of selected compounds are directly proportional to vehicle speed. Higher velocity causes the increased emissions (Duong & Byeong-Kyu, 2011). Pollution from traffic contributes to gradual degradation process of soil and vegetation cover in the areas located about 500 m from road (Sławiński, Gołąbek, & Senderak, 2014). Dust fall and gas concentrations have major impact on the calculated percentage of deterioration environmental quality of the forest taking into account the distance of forest surface from the emission source. This is because it undoubtedly determines the amount of precipitation and the concentration of gases in base: the farther the distance, the smaller fall of the dust and gas concentrations (K. Sporek & M. Sporek, 2007). Furthermore, fuel combustion processes are the main source of sulphur in environment (Bąbelewska, 2013). Sulphur compounds may be affected directly on change the soil pH (Jaggi & Freedman, 1992). Lower soil pH can cause launching other compounds for example heavy metals like aluminium, nickel, mercury (Karczewska & Kabała, 2010). Therefore, pollutant emissions such as sulphur must be monitored to provide proper quality of environmental components and prevent of their degradation (Franco et al., 2013).

The increased concentration of sulphate sulphur ($S\text{-}SO_4^{-2}$), which has the highly significance as a phytotoxic factor, contribute to changes in a given ecosystem (Jakubus, 2006). Sulphate sulphur which is a source of sulphur for flora is considered as a measure of the availability of this element in the soil (Filipek-Mazur, Lepiarczyk, & Tabak, 2013). Sulphate sulphur constitute from 3-50% of the total content of this element in the soil. This is an unstable form, which provides an easy leaching and also an easy consumption by plants (Motowicka-Terelak & Terelak, 1998; Jakubus, 2006). Furthermore, high content of sulphur may cause the disturbance of photosynthesis process and plant growth (Eguagie et al., 2015). Increased content of sulphate sulphur in the soil, as well as the content of trace elements, is not only a significant phytotoxic factor but it can also be considered as an indicator of human pressure, contributing to the dying forests (Jakubus, 2006). Exists four degrees of content assessment of total and sulphate sulphur, from low content to anthropogenic to increase (Kabata-Pendias, 2010).

Pinus sylvestris L. is a species characterized by high variability, which is associated with the wide prevalence. Pine bark and needles are extremely sensitive bioindicators of pollution of natural environment (Chrzan, 2012, Robles et al., 2003). Pine needles remain on the plant for several seasons and under the accumulation of contaminants followed by accumulation these toxins in to them (Dmuchowski & Bytnerowicz, 1995). Tree bark is a tissue commonly used in many environmental research as an bioindicator because of its ability for long-term accumulation of pollution both dusts and gases (Grodzińska, 1977). Assessment of sulphur content in soils is necessary in terms of impact of this element for mobility of heavy metals and deterioration of chemical properties of soil especially launch of aluminium and losses of magnesium (Motowicka-Terlak & Dudka, 1991). The content of sulphur in Polish soils takes average values from 500 to 5000 $mg{\cdot}100\ g^{-1}$ (Gorlach & Mazur, 2001).

The aim of conducted study was to assess the content of total and sulphate sulphur in forest soils and bark and three-years increments of needles of *Pinus sylvestris* L. from park areas exposed to traffic pollution. Due to the proximity of high-traffic road the area must be monitored.

2. Materials and Methods

2.1 Materials

Research material was collected along the exit road of Bydgoszcz with high-traffic from Myślęcinek Forest Park for Culture and Leisure (Poland) (Figure 1). Bydgoszcz is located in the temperate transitional climate zone (Bąk & Łabędzki, 2014). Soil material included 26 soil samples collected from 13 research points located 75 m away from road, and the distance between points was 50 m. Samples were taken from two depths: 0-20 cm (surface samples) and 20-40 cm (subsurface samples) during three years of investigation.

Plant material included 13 bark samples and 13 three-years increments of needles of Pinus sylvestris L. collected from 13 research points located in the same way as the soil research points (Table 1) during one years of investigation.

Figure 1. Location of research points

Table 1. Three-years increments of pine needles

Sample	Repeating	Year		
		1	2	3
1	(1)			
	(2)			
2	(1)			
	(2)			
3	(1)			
	(2)			
4	(1)			
	(2)			
5	(1)			
	(2)			
6	(1)			
	(2)			
7	(1)			
	(2)			
8	(1)			
	(2)			
9	(1)			
	(2)			
10	(1)			
	(2)			

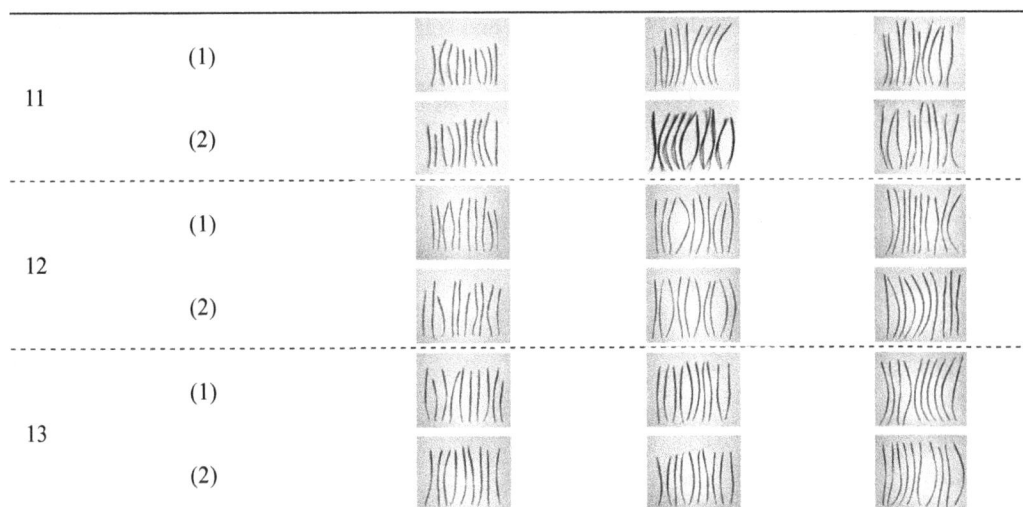

11	(1)			
	(2)			
12	(1)			
	(2)			
13	(1)			
	(2)			

2.2 Methods

Following analysis were conducted on soil samples: pH by the potentiometric method in H_2O (in the ratio 1:2.5) and in 1 M KCl (in the ratio 1:2.5), total organic carbon (TOC) by Tiurin method in the solution of dichromate (VI) potassium and soil texture by laser diffraction method using a Mastersizer 2000. The content of total sulphur and sulphate sulphur in the soil was determined by Bardsley-Lancaster method (Bardsley & Lancaster, 1960) modified by COMN-IUNG. All analysis were performed in three replicates and the verification of the results was based on the certified material Till-3.

Following analysis were conducted on pine bark samples: the content of total sulphur and sulphate sulphur were determined by Bardsley-Lancaster method (Bardsley & Lancaster, 1960) modified by COMN-IUNG.

Following analyses were conducted on pine three-years increments needles samples: the content of total sulphur and sulphate sulphur was determined by Bardsley-Lancaster method (Bardsley & Lancaster, 1960) modified by COMN-IUNG. Furthermore morphological analysis of pine needles was performed with the use of computer software radixNova 1.0 (Stypczyńska et al., 2012). This analysis allows to obtain a length, width and surfaces of the pine needles at each increment. Calculations were performed using this software are carried out with the use of non-statistical methods. They consist in on a detailed analysis of the course of pixel images or scans forming the axes of the various parts of the plant.

Pearson correlation coefficient ($p < 0.05$) was calculated between the content of total and sulphate sulphur in all kinds of samples and their measured properties. Statistical analysis were performed in the Statistica 12.0 for Windows Pl software.

3. Results

3.1 Soil Properties

The analysis of texture allowed to classify the investigated soils to 2 texture classes: sand and loamy sand (USDA, 2012). Content of organic carbon ranged from 12.8 to 48.3 $g \cdot kg^{-1}$ in surface samples and from 4.3 to 19.9 $g \cdot kg^{-1}$ in subsurface samples. In the analysed samples pH_{H2O} ranged from 4.6 to 7.1, while pH_{KCl} from 3.8 to 6.8 (Table 2).

Table 2. Physicochemical properties and the content of total and sulphate sulphur in soil samples (mean for three years of investigation)

Soil sample and depth (cm)		pH $_{H2O}$	pH $_{KCl}$	TOC (g·kg^{-1})	Clay (%)	S total (mg·100 g^{-1})	S-SO$_4^{2-}$ (mg·100 g^{-1})
1	0-20	5.7	4.9	48.3	0	53.1	-
	20-40	5.2	4.4	11.2	1	23.5	3.7
2	0-20	5.0	4.1	20.6	1	23.4	2.1
	20-40	4.7	4.0	19.9	1	24.3	2.8
3	0-20	5.2	4.1	14.1	1	25.1	3.5
	20-40	5.2	4.6	4.3	1	23.8	2.8
4	0-20	5.1	4.2	30.4	0	27.4	-
	20-40	4.9	4.2	7.7	0	22.3	2.6
5	0-20	5.2	4.3	17.2	1	24.8	2.3
	20-40	5.2	4.4	11.1	1	23.6	3.0
6	0-20	4.9	4.0	27.9	1	25.4	2.9
	20-40	5.0	4.4	6.7	1	23.9	3.6
7	0-20	6.3	5.6	32.9	1	26.4	2.4
	20-40	5.5	4.6	10.9	1	22.2	2.7
8	0-20	5.2	4.4	16.3	1	22.4	1.3
	20-40	5.6	4.7	11.5	2	24.4	2.3
9	0-20	5.0	4.2	18.0	3	25.6	1.8
	20-40	4.9	4.3	14.4	3	19.3	2.1
10	0-20	4.6	3.8	32.3	1	27.4	2.2
	20-40	4.7	4.0	17.9	1	23.3	2.6
11	0-20	7.1	6.8	18.9	1	26.5	1.9
	20-40	5.4	4.4	18.9	1	25.4	1.1
12	0-20	6.2	5.8	13.4	1	25.4	3.8
	20-40	6.0	5.5	13.3	2	26.6	1.8
13	0-20	4.9	4.1	12.8	1	26.1	1.8
	20-40	5.2	4.2	7.4	1	27.2	1.5

Note. "-" lack of samples.

3.2 Analysis of the Sulphur Content

Total content of analysed elements have taken the following values in soil samples: total sulphur from 22.4 mg·100 g^{-1} to 53.1 mg·100 g^{-1} in surface samples and from 19.3 to 27.2 in subsurface samples, sulphate sulphur from 1.3 to 3.8 mg·100 g^{-1} in surface samples and from 1.5 to 3.7 mg·100 g^{-1} in subsurface samples (Table 2).

Total content of analysed elements have taken the following values in pine bark samples: total sulphur from 39.2 mg·100 g^{-1} to 84.4 mg·100 g^{-1} and sulphate sulphur from 10.5 to 157.8 mg·100 g^{-1} (Table 3).

Total content of analysed elements have taken the following values in pine needles samples: 1-years increments of needles total sulphur from 78.3 mg·100 g^{-1} to 341.9 mg·100 g^{-1} and sulphate sulphur from 1.2 to 4.9 mg·100 g^{-1}, 2-years increments of needles total sulphur from 87.1 to 415.0 mg·100 g^{-1} and sulphate sulphur from 1.5 to 5.3 mg·100 g^{-1}, 3-years increments of needles total sulphur from 159.2 to 461.5 mg·100 g^{-1} and sulphate sulphur from 0.8 to 5.7 mg·100 g^{-1} (Table 4).

Table 3. The content of total and sulphate sulphur in pine bark

Sample	S total (mg·100 g^{-1})	S-SO$_4^{2-}$ (mg·100 g^{-1})
1	39.2	41.7
2	62.4	97.4
3	46.6	46.9
4	39.3	157.8
5	64.4	124.9
6	73.1	119.6
7	49.8	113.7
8	51.1	10.7
9	80.5	14.5
10	84.4	56.3
11	57.6	20.1
12	53.4	10.5
13	55.1	92.9

3.3 Morphological Analysis of Pine Needles

Morphological analysis of three-years increments of needles of *Pinus sylvestris* L. allows to determine length, width and surface at each increment.

The 1-year increments of pine needles have taken the following values: length from 4.38 to 5.93 cm, width from 0.06 to 0.09 cm and surface from 0.25 to 0.52 cm. The 2-year increments of pine needles have taken the following values: length from 5.45 to 7.21 cm, width from 0.07 to 0.09 cm and surface from 0.40 to 0.63 cm. The 3-year increments of pine needles have taken the following values: length from 6.75 to 7.95 cm, width from 0.08 to 0.12 cm and surface from 0.56 to 0.97 cm (Table 4).

Table 4. Morphological analysis and the content of total and sulphate sulphur in pine needles

Sample	Year	Length (cm)	Width (cm)	Surface (cm^2)	S total (mg·100 g^{-1})	S-SO$_4^{2-}$ (mg·100 g^{-1})
1	1	4.38	0.08	0.33	152.3	4.9
	2	6.39	0.08	0.51	185.9	5.3
	3	7.41	0.09	0.67	237.2	5.7
2	1	5.93	0.07	0.40	206.6	3.4
	2	6.11	0.09	0.52	114.9	4.9
	3	7.16	0.09	0.62	228.5	5.2
3	1	4.17	0.08	0.33	216.5	3.5
	2	7.21	0.07	0.53	266.6	4.2
	3	7.25	0.08	0.57	192.0	3.8
4	1	5.37	0.06	0.33	78.3	3.4
	2	6.08	0.08	0.51	87.1	1.5
	3	7.11	0.09	0.63	213.1	0.8
5	1	5.51	0.07	0.37	341.9	1.2
	2	6.61	0.08	0.55	415.0	3.9
	3	7.81	0.09	0.73	461.5	4.2
6	1	4.82	0.09	0.43	215.5	3.0
	2	6.04	0.09	0.56	272.8	3.9
	3	7.95	0.12	0.97	297.4	4.4
7	1	5.74	0.08	0.48	208.3	3.6
	2	6.90	0.09	0.59	242.5	3.2
	3	7.71	0.08	0.63	257.2	3.9
8	1	4.90	0.09	0.44	194.0	3.4
	2	7.16	0.09	0.63	198.4	4.1
	3	7.23	0.10	0.72	209.4	4.2
9	1	5.33	0.09	0.52	158.9	2.3
	2	6.42	0.10	0.59	184.5	2.9
	3	7.49	0.11	0.79	202.0	3.4
10	1	5.72	0.07	0.38	146.3	3.7
	2	6.03	0.07	0.40	146.2	3.8
	3	6.86	0.09	0.64	159.2	3.8
11	1	4.73	0.08	0.36	117.6	2.9
	2	5.45	0.09	0.47	182.3	3.2
	3	7.21	0.10	0.68	189.3	3.8
12	1	4.02	0.06	0.25	120.7	1.8
	2	6.81	0.08	0.55	159.3	2.8
	3	7.74	0.08	0.65	175.5	2.9
13	1	5.44	0.06	0.34	113.6	3.4
	2	5.83	0.07	0.43	116.4	3.7
	3	6.75	0.08	0.56	169.8	4.3

3.4 Statistical Analysis

Calculated correlation coefficients confirmed the significant relations: in soil samples between the content of total sulphur and the content of sulphate sulphur (r = -0.532 at p < 0.05) (Figure 2); in pine bark between the content of total sulphur and two properties of soil samples: [H+] in pH $_{H2O}$ (r = -0.441 at p < 0.05) (Figure 3a) and the content of clay (r = 0.534 at p < 0.05) (Figure 3b); in pine needles between the content of total sulphur and the surface of needles (r = 0.434 at p < 0.05) (Figure 4).

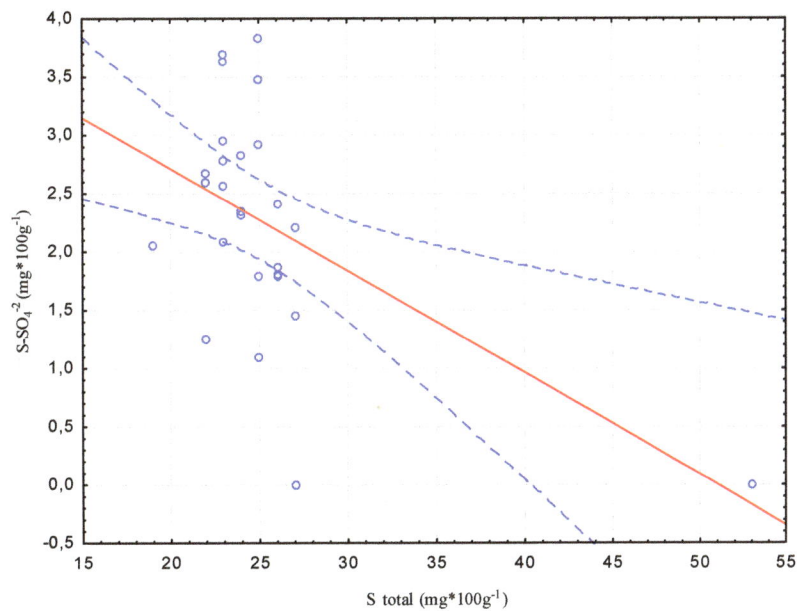

Figure 2. Correlation between the content of total sulphur and sulphate sulphur in soil samples

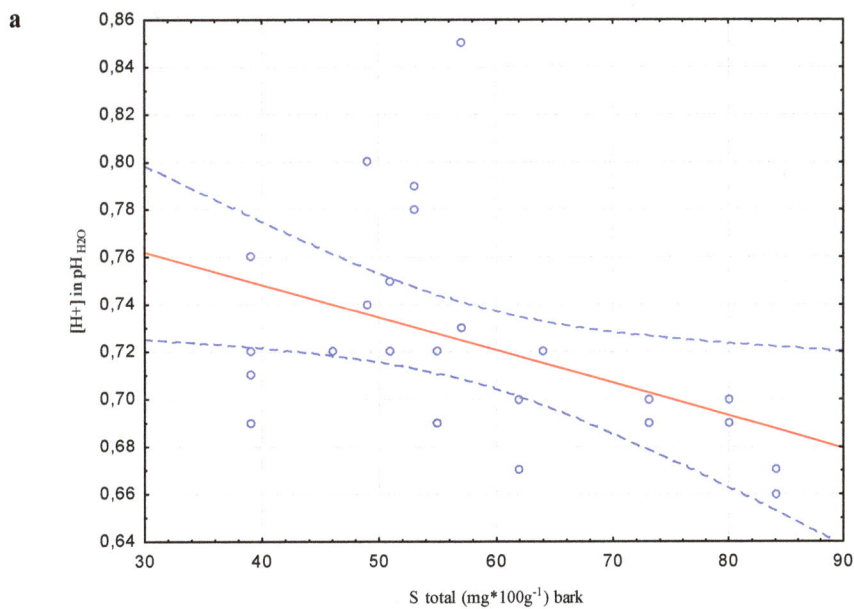

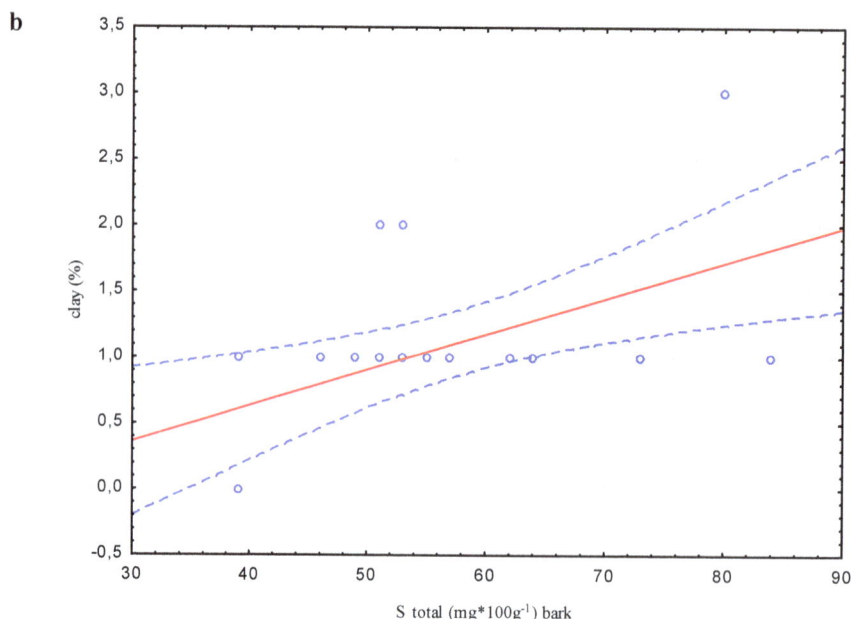

Figure 3. Correlation between [H+] in pH $_{H2O}$ (a) and clay (b) and the total content of sulphur in pine bark

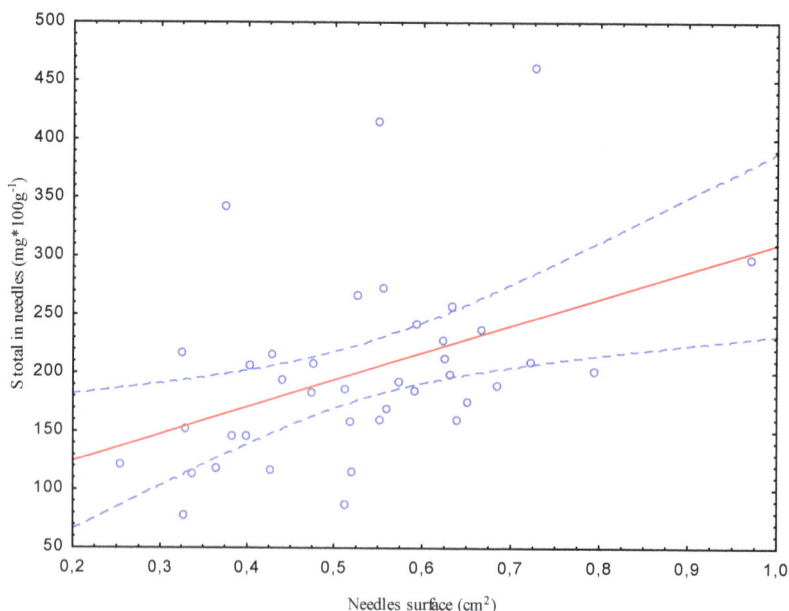

Figure 4. Correlation between the total content of sulphur in needles and their surface

4. Discussion

The content of total sulphur in examined soil sample ranged from 19.3 to 53.1 mg·100 g^{-1} and sulphate sulphur ranged from 1.3 to 3.7 mg·100 g^{-1}. Examined soils were characterized by medium and high content of total sulphur and also by medium, high and anthropogenic origin content of sulphate sulphur. According to the guidelines developed by Kabata-Pendias (2010) examined soil may be qualify as a soil with medium and high content of total sulphur and as a soil with medium, high and anthropogenic origin content of sulphate sulphur. Therefore, that sulphate sulphur is one of the forms in soil especially contributing to acidification of soils, monitoring the level of this compound is such an important factor. The content of total sulphur in examined pine barks ranged from 39.2 to 84.4 mg·100 g^{-1} and sulphate sulphur ranged from 10.5 to 157.8 mg·100 g^{-1}. Examined pine needles were characterized by a content of total sulphur in ranged from 78.3 to 461.5 and sulphate sulphur from 0.8 to 5.7 mg·100 g^{-1}. Referring the results obtained from the conducted research to the current state of knowledge can be stated that the total content of sulphur in plant materials was on a medium level and sulphate sulphur on low. This indicates a slight anthropogenic impact on the local flora. Despite this, due to the proximity of routes with high traffic occurs need for constant monitoring of the studied area.

Similar research were conducted by Licznar and Licznar (2005) in city park in Wroclaw (Poland). The mean content of total sulphur in soil from Szczytnicki Park (Wroclaw) was 116.0 mg·100 g^{-1}. Research conducted by Kalambasa and Godlewska (2010) showed that mean content of total sulphur in loamy sands was 6.5 mg·100 g^{-1}. In soil fertilised with macroelements mean content of sulphate sulphur ranged from 82.2 to 125.0 mg·100 g^{-1} (Siwik-Ziomek, Lemanowicz, & Koper, 2016). Another research reported that mean content of sulphate sulphur in forest soil ranged from 5.5 to 10.5 mg·100 g^{-1} (Koper, Piotrowska, & Siwik-Ziomek, 2008). Research conducted by Szopka et al. (2011) showed that soils from Karkonosze National Park were characterized by high volatility and the content of sulphate sulphur ranged from 200.0 to 36,000.0 mg·100 g^{-1}. Kulczycki and Spiak (2004) in the arable layer (0-20 cm) of eighty soils from south-west Poland, reported that the participation of the sulphate sulphur was about 700,000.0 mg·100 g^{-1}. Other studies indicated that the content of sulphate sulphur in soils of road Siedlce city ranged from 150,000.0 to 1,700,000.0 mg·100 g^{-1} (Kalembasa & Godlewska, 2005). Boratyński et al. (1975) reported that in methodological research of soil samples content of S-SO$_4^{-2}$ ranged from 0.3 to 0.5 mg·100 g^{-1}. According to research conducted by Terelak and Motowicka-Terelak (2000) the content of sulphate sulphur in Polish soils ranged from 0.01 to 50.0 mg·100 g^{-1}. More than 55% of soils are areas with low content of S-SO$_4^{-2}$. Surface of areas with medium and high content of sulphate sulphur constitutes appropriately: 25.1 and 13.1%, and anthropogenically contaminated by sulphur only 3.7% (Terelak, 2005).

Close to the Kola smelters content of total sulphur in pine bark was a twice as high as the other side of border of the Finland (Poikolainen, 1997). Research conducted by Manninen et al. (1997) showed that the content of sulphate sulphur in pine needles takes values from 0.6 to 1.2 mg·100 g^{-1}. In pine needles from northernmost Europe the content of sulphate sulphur ranged from 7,410.0 to 20,170.0 mg·100 g^{-1} (Raitio, Tuovinen, & Anttila, 1995). Received results were very high and probably it was connected with SO$_2$ emissions into the atmosphere in the surrounding areas. In areas highly industrialized the content of total sulphur in pine needles ranged from 6,000.0 to 10,000.0 mg·100 g^{-1}, while in areas where there is no pollution of air content of total sulphur ranged from 300.0 to 1,200.0 mg·100 g^{-1} (Grodzińska, 1977). Research conducted by Staszewski et al. (2008) reported that the content of sulphate sulphur in 1-year increments of pine needles from Pieniny National Park was 700.0 mg·100 g^{-1}. Other studies on the pine needles from this area showed that the content of total sulphur ranged from 500.0 to 999.0 mg·100 g^{-1} (Panek & Szczepańska, 2005). Assessment of sulphur content in soils is necessary in terms of impact of this element for mobility of heavy metals and deterioration of chemical properties of soil especially launch of aluminium and losses of magnesium (Motowicka-Terlak & Dudka, 1991). The content of sulphur in Polish soils takes average values from 500 to 5000 mg·100 g^{-1} (Gorlach & Mazur, 2001).

To sum up, undertaken studies shows that the traffic have slight influence on the content of total and sulphate sulphur in soils and plant material. So far, conducted only few similar studies, relating to the assessment of environmental pollution by sulphur compounds as a result of traffic road. Many other studies were primarily aimed to assess the various industries on the content these elements in environmental components. Therefore undertaken research complementing the state of knowledge.

References

Bąbelewska, A. (2013). The impact of sulphur dioxide from Częstochowa Agglomeration on acidity degree of *Pinus sylvestris* L. bark of "Zielona Góra" and "Sokole Góry" Nature Reserves. *Natural Environment Monitoring, 14,* 69-77.

Bąk, B., & Łabędzki, L. (2014). Prediction of precipitation deficit and excess in Bydgoszcz Region in view of predicted climate change. *Journal of Water and Land Development, 23,* 11-19. http://dx.doi.org/10.1515/jwld-2014-0025

Bardsley, C. E., & Lancaster, J. D. (1960). Determination of reserve sulphur and soluble sulphates in soil. *Soil Science Society of America, Proceedings, 240,* 265-268. http://dx.doi.org/10.2136/sssaj1960.0361599500 2400040015x

Boratyński, K., Grom, A., & Ziętecka, M. (1975). Research on the content of sulphur in soil. Part I. Mothodological research on determination of sulphate sulphur in soil. *Soil Science Annual, 26*(3), 121-139.

Chrzan, A. (2015). Necrotic bark of common pine (*Pinus sylvestris* L.) as a bioindicator of environmental quality. *Environmental Science and Pollution Research, 22*(2), 1066-1071. http://dx.doi.org/10.1007/s11356-014-3355-0

Council Directive 92/880 of 16 April 2004 on nature protection.

Dmuchowski, W., & Bytnerowicz, A. (1995). Monitoring environmental pollution in Poland by chemical analysis of Scots pine (*Pinus sylvestris* L.) needles. *Environmental Pollution, 87*, 87-104.

Duong, T. T., & Byeong-Kyu, L. (2011). Determining contamination level of heavy metals in road dust from busy traffic areas with different characteristics. *Journal of Environmental Management, 92*(3), 554-562. http://dx.doi.org/10.1016/j.jenvman.2010.09.010

Dynarz, R., & Wiśniewski, H. (1996). *Natural and educational values of Myślęcinek Forest Park for Culture and Leisure.* University Publisher, Bydgoszcz, Poland.

Eguagie, M. O., Aiwansoba, R. P., Omofomwan, K. O., & Oyanoghafo, O. O. (2016). Impact of simulated acid rain on the growth, yield and plant component of Abelmoschus caillei. *Journal of Advances in Biology & Biotechnology, 6*(1), 1-6. http://dx.doi.org/10.9734/JABB/2016/24804

Filipek-Mazur, B., Lepiarczyk, A., & Tabak, M. (2013). Nitrogen and sulphur fertilization on yielding and zinc content in seeds of winter rape 'Baldur' cultivar. *Ecological Chemistry and Engineering, 20*(11), 1359-1368. http://dx.doi.org/10.2428/ecea.2013.20(11)123

Franco, V., Kousoulidou, M., Muntean, M., Ntziachristos, L., Hausberger, S., & Dilara, P. (2013). Road vehicle emission factors development: A review. *Atmospheric Environment, 70*, 84-97. http://dx.doi.org/10.1016/j.atmosenv.2013.01.006

Gorlach, E., & Mazur, T. (2001). *Soil chemistry* (pp. 84-104). Warsaw: Polish Scientific Publishers.

Grodzińska, K. (1977). Acidity of tree bark as a bioindicator of forest pollution in southern Poland. *Water, air and Soil Pollution, 8*(1), 3-7. http://dx.doi.org/10.1007/BF00156718

Helmuth, J. L., & Dudek, C. L. (2002). Traveller information en route needs and available services for visitors to national parks. *Transportation Research Board 81st Annual Meeting, 24*. USA: Washington.

Jaggi, B., & Freedman, M. (1992). An examination of the impact of pollution performance on economic and market performance: Pulp and paper firms. *Journal of Business Finance & Accounting, 19*(5), 697-713. http://dx.doi.org/10.1111/j.1468-5957.1992.tb00652.x

Jakubus, M. (2006). Sulphur in environment. *University Publisher Poznan University of Life, 61*.

Kabata-Pendias, A. (2010). *Biogeochemistry of trace elements.* Scientific Publishing PWN, Warsaw.

Kalembasa, S., & Godlewska, A. (2005). The content of total and sulphate sulphur in soils of road Siedlce city. *Ecological Engineering, 12*, 31-32.

Kalembasa, S., & Godlewska, A. (2010). Total sulphur and its fractions as well as activity of arylsulphtase in soil depending on waste organic materials and liming. *Environment Protection Engineering, 36*(1), 5-11.

Karczewska, A., & Kabała, C. (2010). The soils polluted with heavy metals and arsenic in Lower Silesia – The need and methods of reclamation. *Scientific Journal of Wrocław University of Environmental and Life Science. Series of Agronomy, 576*, 59-79.

Koper, J., Piotrowska, A., & Siwik-Ziomek, A. (2008). Activity of dehydrogenases, invertase and rhodanase in forest rusty soil in the vicinity of 'Anwil' nitrogen plant in Wloclawek. *Ecological Chemistry and Engineering, 15*(3), 237-243.

Kulczycki, G., & Spiak, Z. (2004). Content of total and sulphate sulphur in soils from south-west Poland. *IUNG, 1*(18), 75-81.

Licznar, S. E., & Licznar, M. (2005). The impact of urban agglomeration of Wroclaw on levels of humus soil Szczytnicki Park. *Soil Science Annual, 56*(1-2), 113-118.

Mannine, S., Huttunen, S., & Kontio, M. (1997). Accumulation of sulphur in and on scots pine needles in the subarctic. *Water, Air, and Soil Pollution, 95*(1), 147-164. http://dx.doi.org/10.1023/A:1026449222526

Motowicka-Terelak, T., & Terelak, H. (1998). *Sulphur in Polish soils – State and the dangerous.* Library of Environmental Monitoring, Warsaw.

Motowicka-Terlak, T., & Dudka, S. (1991). Chemical degradation of soils contaminated with sulphur and its impact on crops. *IUNG*, 1-95.

Pallavi, P., & Harrison, R. M. (2013). Estimation of the contribution of road traffic emissions to particulate matter concentrations from field measurements: A review. *Atmospheric Environment, 77*, 78-97. http://dx.doi.org/10.1016/j.atmosenv.2013.04.028

Panek, E., & Szczepańska, M. (2005). Trace metals and sulphur in selected plant species in the Małe Pieniny Mts. *Mineral Resources Management, 21*(1), 89-109.

Poikolainen, J. (1997). Sulphur and heavy metal concentrations in Scots pine bark in northern Finland and the Kola peninsula. *Water, Air, and Soil Pollution, 93*(1), 395-408. http://dx.doi.org/10.1007/BF02404769

Raitio, H., Tuovinen, J. P., & Anttila, P. (1995). Relation between sulphur concentrations in the Scots pine needles and the air in northernmost Europe. *Water, Air, and Soil Pollution, 85*(3), 1361-1366. http://dx.doi.org/10.1007/BF00477171

Robles, C., Greff, S., Pasqualini, V., Garzino, S., Bousquet-Mélou, A., Fernandez, C., ... Bonin, G. (2003). Phenols and flavonoids in aleppo pine needles as bioindicators of air pollution. *Soil Science Society, 32*(6), 2265-2271. http://dx.doi.org/10.2134/jeq2003.2265

Scherer, H. W. (2001). Sulphur in crop production—Invited paper. *European Journal of Agronomy, 14*(2), 81-111. http://dx.doi.org/10.1016/S1161-0301(00)00082-4

Siwik-Ziomek, A., Lemanowicz, J., & Koper, J. (2016). Sulphur and phosphorus content as well as the activity of hydrolases in soil fertilised with macroelements. *Journal of Elementology, 21*(3), 847-858. http://dx.doi.org/10.5601/jelem.2015.20.3.982

Sławiński, J., Gołąbek, E., & Senderak, G. (2014). Influence of transport pollution on soil and cultivated vegetation of the wayside. *Ecological Engineering, 40*, 137-144.

Sporek, K., & Sporek, M. (2007). Bioaccumulation of sulphur and calcium in scots pine (*Pinus sylvestris* L.) bark in the Załęczański Landscape Park. *Proceedings of ECOpole, 1*(1/2), 245-248.

Staszewski, T., Kubiesa, P., Łusik, W., Uziębło, A. K., & Szdzuj, J. (2008). Monitoring of changes in space forest in Pieniny National Park. *Pieniny Moutains - Nature and Human, 10*, 3-9.

Stypczyńska, Z., Dziamski, A., Schmidt, J., & Jendrzejczak, E. (2012). Reaction of lawn grasses cultivars of genus Festuca on water deficits and the sod regeneration level based on morphometric root experiments. *Acta Scientiarum Polonorum. Agricultura, 11*(3), 85-94.

Szopka, K., Karczewska, A., Kabala, C., & Kulczyk, K. (2011). Sulphate sulphur in forest soil from Karkonoski National Park. *Environmental Protection and Natural Resources, 50*, 61-70.

Terelak, H. (2005). Heavy metals and sulphur in soils of farmland of Poland. *Journal of Ecology and Health, 9*(5), 259-264.

Terelak, H., & Motowicka-Terelak, T. (2000). The heavy metals and sulphur status of agricultural soils in Poland. *Soil Quality, Sustainable Agriculture and Environmental Security in Central and Eastern Europe, 69*, 37-47. http://dx.doi.org/10.1007/978-94-011-4181-9_3

USDA (United States Department of Agriculture). (2012). *Soil Survey Manual*. Scientific Publishers Journal Department.

7

Effects of Varied Nitrogen Supply and Irrigation Methods on Distribution and Dynamics of Soil NO$_3$-N during Maize Season

Dongliang Qi[1,2] & Tiantian Hu[1,2]

[1] College of Water Resources and Architectural Engineering, Northwest A&F University, Yangling, China

[2] Key Laboratory of Agricultural Soil and Water Engineering in Arid and Semiarid Areas of Ministry of Education, Northwest A&F University, Yangling, China

Correspondence: Tiantian Hu, College of Water Resources and Architectural Engineering, Northwest A&F University, Weihui Road 23, Yangling 712100, China. E-mail: hutiant@nwsuaf.edu.cn

The research is financed by the Special Fund for Agro-scientific Research in the Public Interest (201503124), National Natural Science Fund of China (51079124) and the National High Technology Research and Development Program of China (2011AA100504).

Abstract

A field experiment was carried out to investigate the effects of different supply methods of nitrogen (N) fertilizer and irrigation on the spatial distribution and dynamics of soil NO$_3$-N for maize (*Zea mays* L.) grown in northwest China in 2012 and 2014. In 2012, there were three irrigation methods: alternate furrow irrigation (AI), fixed furrow irrigation (FI) and conventional furrow irrigation (CI). Three N supply methods: alternate N supply (AN), fixed N supply (FN) and conventional N supply (CN), were applied at each irrigation method. In 2014, the fixed treatments were excluded. Soil NO$_3$-N in horizontal direction was measured to 100 cm soil profile. For 2012, at filling stage, compared to CI, AI increased soil NO$_3$-N concentration under the plant by 4.5 to 7.4% in 0-40 cm soil profile and decreased that by 9.9 to 14.4% in 40-80 cm for three N supply methods. NO$_3$-N concentration between two sides of the ridge was comparable for AN and CN coupled with AI or CI. When compared to CI, AI reduced soil NO$_3$-N concentration in 60-100 cm by 4.8 to 8.7% from 12 collars stage to maturity over different positions when coupled with CN. Soil residual NO$_3$-N at maturity in 0-100 cm was the lowest in AI coupled with CN or AN. The 2014 experiment verified the above results. Therefore, alternate furrow irrigation coupled with conventional or alternate N supply brought an optimum spatial distribution of soil NO$_3$-N during maize season, resulting in little soil residual NO$_3$-N at maturity.

Keywords: soil NO$_3$-N distribution, soil NO$_3$-N dynamics, nitrogen supply method, irrigation method, soil residual NO$_3$-N, *Zea mays*

1. Introduction

Declining freshwater resources have stimulated research into developing novel irrigation strategies to increase crop water use efficiency (Morison, Baker, Mullineaux, & Davies, 2008). Partial root-zone irrigation (PRI) is a new strategy of deficit irrigation. PRI can be applied in two ways: alternate PRI and fixed PRI. In alternate PRI, half of the root zone is irrigated while the other half is dried, and then the previously well-watered side of the root system is allowed to dry while the previously dried side is fully irrigated (Kang, Zhang, Liang, Hu, & Cai, 1997). However, in fixed PRI, a fixed half of the root zone is always irrigated while the other half is always dried. Alternate PRI is considered a water-saving irrigation technique and is being intensively studied on field crops (Kang, Liang, Hu, & Zhang, 1998; Tang, Li, & Zhang, 2010; Shahnazari, Liu, Anderson, Jacobesen, & Jensen, 2007).

Soil nitrate (NO$_3$-N) is the dominated nitrogen (N) form in dryland soil. Its sustainable supply in root zone has predominant effect on crop yield and N fertilizer use. In a large part, low N use efficiency is attributed to potential NO$_3$-N leaching and residual NO$_3$-N at harvest. Studies have demonstrated that residual NO$_3$-N (Tarkalson, Payero, Ensley, & Shaprio, 2006; Wang, F. Li, Zhang, G. Li, & Vance, 2012) as well as N leaching (Tafteh & Sepaskhah, 2012; Zotarelli et al., 2009) increased as N fertilizer application rates were higher. It is fully understood that N and water have a huge interaction and irrigation significantly affects N status in soil. Excess application of water resulted in NO_3-N leaching (Ju, Liu, Zhang, & Roelcke, 2004; Gheysari, Mirlatifi, Homaee, Asddi, & Hoogenboom, 2009). Furthermore, compared to conventional furrow irrigation (CI) and fertilization, separation of N fertilizer and water with alternate PRI increased NO_3-N in the upper soil layers (0-60 cm) by 30 to 60%, while reduced that in the deeper soil layers (60-200 cm) by 13 to 33% (Han et al., 2014). Placement of N fertilizer in the ridge rather than the furrow could lower N leaching in fixed PRI or CI (Benjamin, Havis, Ahuja, & Alonso, 1994; Benjamin, Poter, Duke, & Alnoso, 1997), so did placement of N fertilizer in non-irrigated rather than irrigated furrow in fixed PRI (Zhu et al., 2013). These suggest that potential residual and leaching of NO_3-N in soil not only depends on the rates of N and water, but also on N and water placement in soil profile.

The position of NO_3-N in soil profile, determining soil N leaching and residual, varies largely during the growing season. Besides N source and the application rate, distribution and dynamics of NO_3-N in soil depends on the volume and placement of irrigation water in the active root zone. Because NO_3-N tended to accumulate towards the boundary of the wetted volume (Bar-Yosef & Sheikholslami, 1976; Li, Zhang, & Ren, 2003), the use of irrigation strategies that limit the wetted volume in the root zone may change NO_3-N distribution and improve N use efficiency (NUE) (Singandhupe, Rao, Patil, & Brahmanand, 2003). In furrow irrigation, a symmetrical distribution of NO_3-N across the ridge was found in both alternate PRI and CI, but NO_3-N concentration in the non-irrigated furrow was higher than that in the irrigated one under fixed PRI (Liu, Zhang, Yang, Wang, & Li, 2011). Alternate PRI could promote upward movement of NO_3-N within a deeper soil layer when compared to CI (Wang et al., 2014). Placement of N fertilizer to the non-irrigated rather than irrigated furrow under alternate PRI was favored for maintaining distribution of NO_3-N more at the upper soil profile (0-40 cm) for a longer time (Xing, Wang, L. Li, & S. Li, 2003). Nevertheless, these researches on the distribution and dynamics of NO_3-N under PRI were mainly conducted in a pot (Wang et al., 2014) or a semi-arid area (Xing et al., 2003; Liu et al., 2011). Moreover, the effect of N application patterns on changing NO_3-N under PRI has not been considered.

Thus, this study was conducted for maize production in an arid area of northwest China to investigate how distribution and dynamics of NO_3-N is influenced by different N supply and irrigation methods.

2. Materials and Methods

2.1 Experimental Site

A field study was conducted during the 2012 and the 2014 growing seasons at Wuwei Experimental Station for Efficient Use of Crop Water, Ministry of Agriculture, northwest China (latitude 37°52′20″N, longitude 102°50′50″E, altitude 1581 m). The site is in a typical continental temperate climate zone with mean annual precipitation of 164.4 mm, mean annual evapotranspiration of 2000 mm. Mean annual sunshine duration is over 3000 h and mean annual temperature is 8.8 °C. The groundwater level is consistently 25-30 m below the soil surface. The soil is a light sandy loam. Total precipitation in the growing season was 129 mm in 2012 and 174 mm in 2014 (Figure 1). In the top soil layer (0-40 cm), organic matter is 15.90 g kg^{-1}, total N is 0.85 g kg^{-1}, total phosphorus is 0.93 g kg^{-1}, available phosphorus is 6.22 mg kg^{-1}, and available potassium is 236.24 mg kg^{-1}.

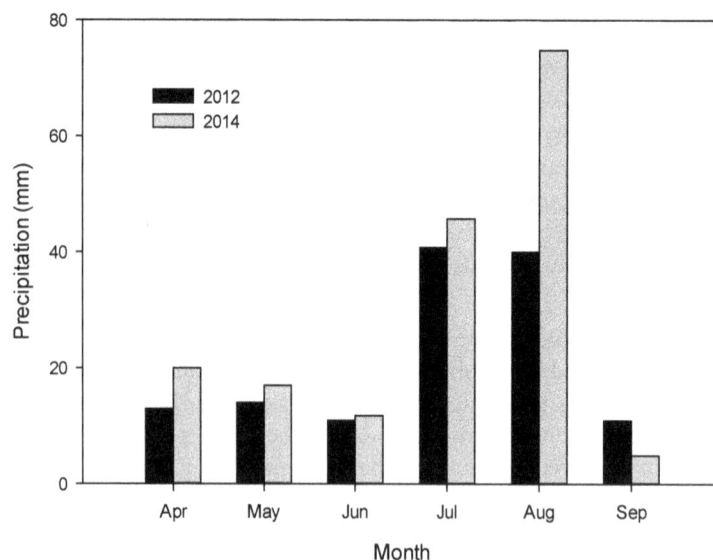

Figure 1. Monthly precipitation during the growth period at the experimental site in 2012 and 2014

2.2 Crop Management

Furrow irrigation was adopted in the field experiment. A trapezoid fracture surface was established for furrows and ridges. Furrows were 30 cm in depth and 20 cm in width at bottom. Ridges were 20 cm and 35 cm in width at top and bottom, respectively. This resulted in a ridge spacing of 55 cm. All experimental ridges were built in a west-east direction. Superphosphate fertilizer was applied at 77 kg ha^{-1} one day before furrows were established. Ridges were then covered using plastic film. Each plot was 24 m^2 (4 m × 6 m) in 2012 and (4 m × 8 m) 32 m^2 in 2014. Seven ridges were established for each plot in each year. Grain maize, cultivar 'Golden northwest No. 22' (*Zea mays* L.) were sown in the ridges at a density of 73000 plants ha^{-1} on April 19 and 20 in 2012 and 2014, respectively. Crop was harvested on September 20 and 22 in 2012 and 2014, respectively.

2.3 Experimental Design

The experiment factors were irrigation method and N fertilizer supply method. In 2012, irrigation methods included conventional furrow irrigation (CI), alternate furrow irrigation (AI) and fixed furrow irrigation (FI). CI means that all furrows were irrigated for every irrigation event. AI means that one of the two neighboring furrows was alternately irrigated during consecutive watering. FI means that irrigation was fixed to one of the two neighboring furrows. N supply methods included conventional N supply (CN), alternate N supply (AN) and fixed N supply (FN). CN means that N fertilizer was applied to all furrows. AN means that N fertilizer was alternately applied to one of the neighboring two furrows in consecutive fertilization. FN means that N fertilizer was fixed to one of every two furrows. This experimental plan yielded 9 treatments, *i.e.* CIAN, CICN, CIFN, AIAN, AICN, AIFN, FIAN, FICN and FIFN. In addition, FIFN was conducted in two ways, named FIFNS (application of N fertilizer to the irrigated furrow under FI) and FIFND (application of N fertilizer to the non-irrigated furrow under FI), respectively.

In 2014, based on the results of 2012, fixed treatments (FI and FN) were excluded, and only the AIAN, AICN, CIAN and CICN treatments were conducted. All treatments were arranged in a randomized block design with three replicates in two years.

Twice as much water and/or N was applied to the irrigated/fertilized furrow in AI/AN and FI/FN as that to the furrow in CI/CN treatment, so that the total amount of water and/or N was the same for all treatments. Urea was applied at a rate of 200 kg N ha^{-1} to the center of the furrows in 5 cm deep, which is the optimum N rate for maize production in local area (Yang & Su, 2009). N fertilizer application included basal application (50%) and topdressing at 12 collars (25%) and tasseling (25%). According to Zhang, Ma, and E (2007), irrigation was applied after planting and at the 6 collars, 12 collars, tasseling and filling stages of maize (75 mm per time), respectively. The growth stage of maize was determined according to Ritchie and Hanway (1982). The irrigation water was supplied by a pipe with a diameter of 55 mm, and the amount of water applied was measured with a water meter installed at the discharging end of the pipe. Irrigation and N fertilizer topdressing was conducted on the same day. The position details of localized irrigation and N application are described in Table 1.

Table 1. Time and position of localized irrigation and nitrogen (N) fertilization

Items	Seeding	6 collars	12 collars	Tasseling	Filling
DAP of	3(2012)	45	84	98	119
irrigation (d)	3(2014)	45	82	96	115
Location of AI	Both furrows	South furrow	North furrow	South furrow	North furrow
Location of FI	Both furrows	South furrow	South furrow	South furrow	South furrow
Location of CI	Both furrows	Both furrows	Both furrows	Both furrows	Both furrows
DAP of N	-1(2012)	/	84	98	/
application (d)	-1(2014)	/	82	96	/
Location of AN	South furrow	/	North furrow	South furrow	/
Location of FN	South furrow	/	South furrow	South furrow	/
	North furrow [a]	/	North furrow [a]	North furrow [a]	/
Location of CN	Both furrows	/	Both furrows	Both furrows	/

Note. "/" represents no treatment. DAP, days after planting; AI, alternate furrow irrigation; CI, conventional furrow irrigation; FI, fixed furrow irrigation; AN, alternate nitrogen supply; CN, conventional nitrogen supply; FN, fixed nitrogen supply. Fixed N/water treatments were only conducted in 2012. [a] represents N supplied only for FIFND treatment. FIFNS and FIFND treatments denote application of N fertilizer to the irrigated (south) and non-irrigated (north) furrow under FI, respectively.

2.4 Soil Sampling and Measurement

Soil samples for NO_3-N measurement were taken from each plot before experiment and at 6 collars, 12 collars, tasseling, filling and maturity growth stages, which corresponds to -2, 44, 82, 97, 117 and 152 days after planting (DAP) in 2012, and -2, 43, 80, 94, 113 and 149 DAP in 2014, respectively. These sampling dates were earlier than corresponds fertilization/irrigation dates. Three plants in the middle row in each plot were randomly chosen for soil sampling position during 6 collars to maturity. Before sampling, shoots were cut near the soil surface. A hand-driven auger with 7 cm diameter was used for sampling. The sampling was collected to 100 cm depth from three positions around one plant. The three positions were: (1) directly over the crown of the plant; (2) south and (3) north side of the plant. For position (2) and (3), sampling sites were positioned one quarter of row spacing directly opposite the crown. The core was sectioned into 20 cm depths and a 10-15 g sample was used for soil NO_3-N concentration determination.

NO_3-N was determined using a Flow Solution IV Analyzer (FSIV, O.I. Analytical, U.S.A.) after extraction with a 1:5 ratio (w/w) soil: 1 mol L^{-1} KCl solution (Bao, 2000). The amount of NO_3-N (kg N ha^{-1}) stored in 0-100 cm soil profile at harvest was calculated according to the equation modified by Emteryd (1989):

$$Y_i = T_i \times BD_i \times [NO_3]_i \times 0.1 \tag{1}$$

Where, T_i is the thickness of soil layer in cm; BD_i is the bulk density in g cm^{-3}, 1.32, 1.40, 1.55, 1.58 and 1.60 g cm^{-3} for soil layer of 0-20, 20-40, 40-60, 60-80, 80-100 cm, respectively (Liu, Li, Pan, Qu, & Du, 2009); $[NO_3]_i$ is the soil NO_3-N concentration in mg kg^{-1}, which was the mean NO_3-N concentration across the different positions, and 0.1 is the conversion coefficient.

2.5 Statistical Aanalysis

To compare NO_3-N concentration and residual NO_3-N amount among treatments, analysis of variance (ANOVA) was performed using the general linear model-univariate procedure from SPSS 12.0 software. FIFND treatment was excluded in analysis of ANOVA in 2012, and the means were compared for any significant differences among all treatments using the Duncan's multiple range tests at the significant level of $P = 0.05$. Since treatment amount differed among years, ANOVA was performed separately in two years.

3. Results

3.1 Spatial Distribution of Soil NO₃-N

3.1 Spatial Distribution of Soil NO_3-N

In consideration that the last N fertilization application was at tasseling (VT) stage of maize. Filling (R_2), which is the next sampling time to follow VT, is one of the pivotal stages for N uptake by maize (Tsai, Huber, Glover, & Wareen, 1984), R_2 was chosen to characterize spatial distribution of NO_3-N.

In 2012, irrigation method, N supply method, and their interaction had a significant impact on NO_3-N concentration at three positions of the plant in 0-40 cm soil profile except for the interaction in 20-40 cm under the plant (UP). Among them, at the marked significant level, N supply method influenced NO_3-N concentration at north of the plant (NP) and south of the plant (SP). Moreover, the significant impact was also observed for N supply method at NP and SP in 40-80 cm, and irrigation method at NP in 40-60 cm and SP in 60-80 cm (Table 2).

In 2014, a little different result occurred for all the four treatments (Table 2). The irrigation method had a significant impact on NO_3-N concentration at three positions of the plant in 0-40 cm soil profile, and the significant impact of irrigation method extended to 40-100 cm for NP and SP (excluded NP in 80-100 cm). A significant impact on NO_3-N concentration by N supply method was only observed for NP in 40-60 cm and SP in 20-40 cm (Table 2).

Table 2. Variance analysis for soil NO_3-N concentration among different treatments at filling stage (R_2) in 2012 and 2014

Year	Position of sampling	Factors	Soil depth (cm)				
			0-20	20-40	40-60	60-80	80-100
2012	North of plant	IM	*	*	*	NS	NS
		NSM	**	**	*	*	NS
		IM×NSM	**	*	NS	NS	NS
	South of plant	IM	*	*	NS	*	NS
		NSM	**	**	*	*	NS
		IM×NSM	**	*	NS	NS	NS
	Under the plant	IM	*	*	NS	NS	NS
		NSM	*	*	NS	NS	NS
		IM×NSM	*	NS	NS	NS	NS
2014	North of plant	IM	*	*	*	*	NS
		NSM	NS	NS	*	NS	NS
		IM×NSM	NS	NS	NS	NS	NS
	South of plant	IM	*	*	*	*	*
		NSM	NS	*	NS	NS	NS
		IM×NSM	NS	NS	NS	NS	NS
	Under the plant	IM	*	*	NS	NS	NS
		NSM	NS	NS	NS	NS	NS
		IM×NSM	NS	NS	NS	NS	NS

Note. *, ** means significant at the 0.05 and 0.01 levels, respectively. IM and NSM represent irrigation and nitrogen supply method, respectively. FIFND treatment was excluded in the variance analysis in 2012, symbol of FIFND treatment is shown in Table 1.

In 2012 (Figure 2), in 0-40 cm soil depth, an approximate symmetrical distribution of NO_3-N across the ridge was observed for AIAN, AICN, CICN and CIAN treatments (Figures 2a, 2b, 2d and 2e). Concentration of NO_3-N under SP (42 cm at the horizontal direction) was observed for FN treatments in three irrigation methods (Figures 2c, 2f and 2i) except for FIFND treatment (Figure 2j), in which NO_3-N concentration under SP was 1.4 to 1.8 times higher than that under NP (14 cm at the horizontal direction). This reversed in FICN and FIFND treatments and the disparity was larger (Figures 2h and 2j), *i.e.*, NO_3-N concentration under SP was much lower than that under NP. In 40-80 cm soil depth, the situation was a little different: under SP, concentration of NO_3-N was observed for FI treatments in three N supply methods (Figures 2g, 2h and 2i) except for FIFND treatment (Figure 2j). Compared to CI, under UP (28 cm at the horizontal direction), AI increased NO_3-N concentration in 0-40 cm soil depth by 4.5 to 7.4 % ($P > 0.05$) and reduced that in 40-80 cm soil depth by 9.9 to 14.4% ($P < 0.05$) among different N supply methods (Figures 2a, 2b, 2c, 2d, 2e and 2f). In 80-100 cm soil depth, NO_3-N concentration was comparable among all treatments (about 10 mg kg^{-1}). In 2014 (Figure 3), for AIAN, AICN, CICN and CIAN treatments, spatial distribution of NO_3-N was similar to that in 2012. Overall, alternate furrow irrigation coupled with conventional N supply or alternate N supply not only benefited the uniform distribution of NO_3-N in the horizontal direction, but also inhibited downward movement of NO_3-N in the soil profile.

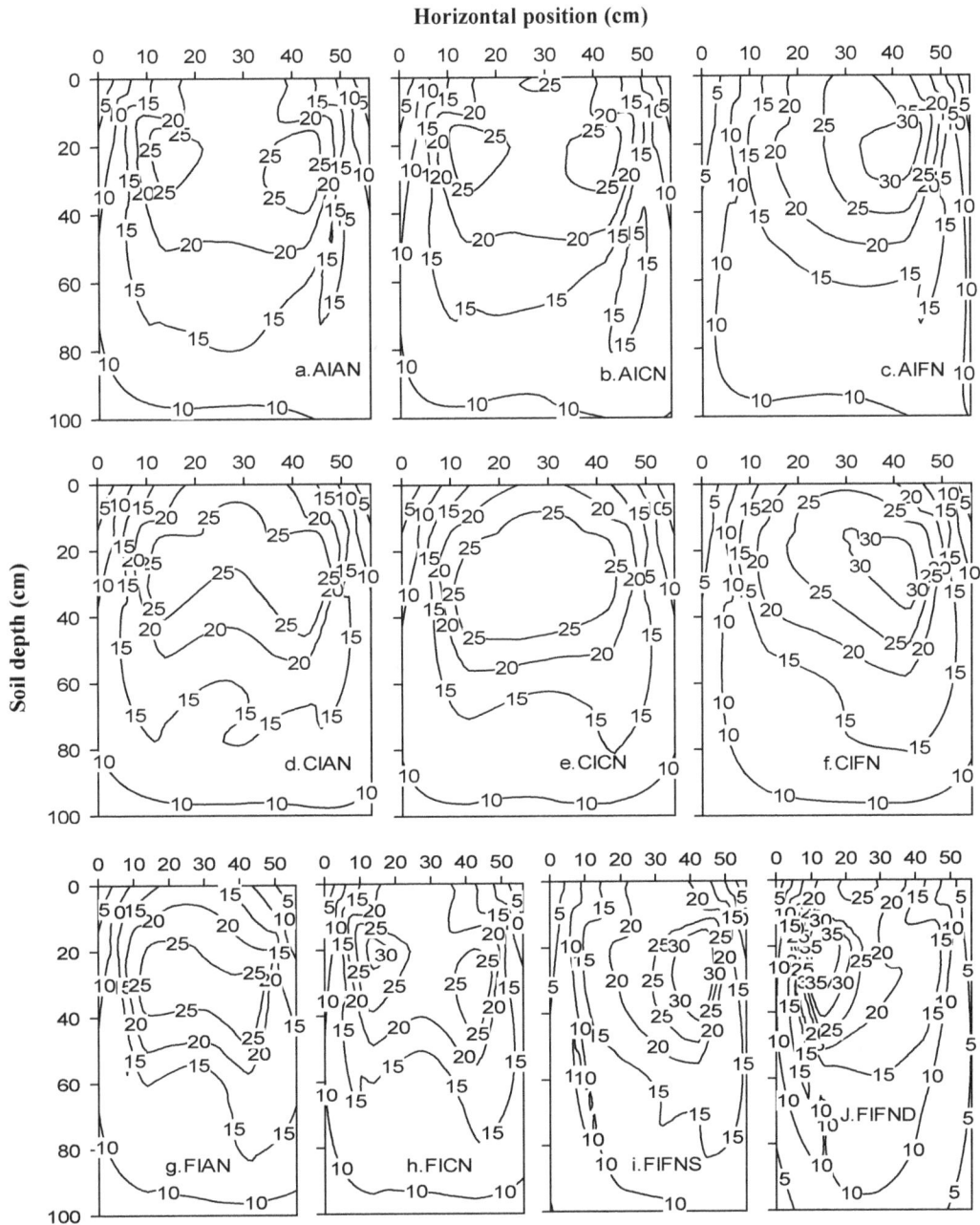

Figure 2. Spatial distribution of soil NO_3-N (mg kg^{-1}) for different treatments at filling (R_2) in 2012

Note. 14, 28 and 42 cm at the horizontal axis represents north, under and south of the plant, respectively.

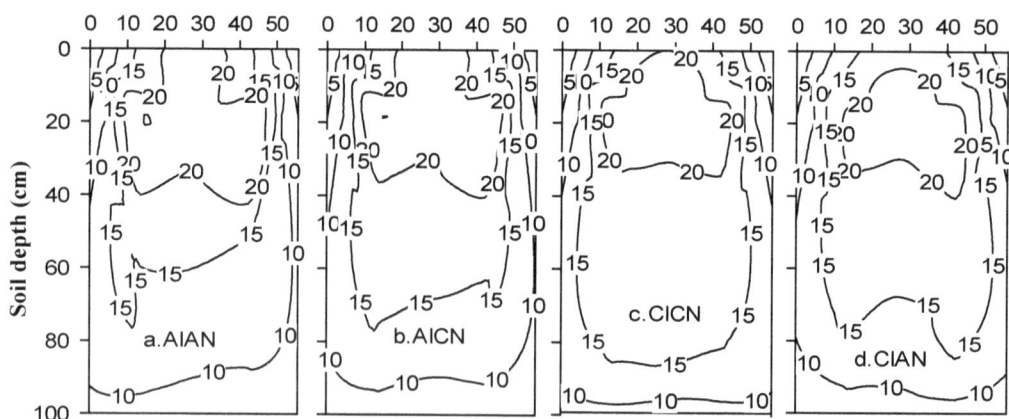

Figure 3. Spatial distribution of soil NO_3-N (mg kg^{-1}) for different treatments at filling (R_2) in 2014

Note. 14, 28 and 42 cm at the horizontal axis represents north, under and south of the plant, respectively.

3.2 Dynamics of Soil NO₃-N in Different Soil Layers

The NO_3-N dynamics of AICN, CICN, FIFNS and FIFND treatments was shown to reflect several trends of all treatments and avoid too many lines to be unclear at the meantime. Data was pooled if no significant difference was observed among either or both adjacent soil layers and different sampling positions. Over the different sampling positions and soil layers, a similar dynamics of NO_3-N in the whole growing period occurred in all treatments. Specifically, a significantly increase trend of NO_3-N concentration was observed from planting to 6 collars (V_6), slightly decrease from V_6 to 12 collars (V_{12}), then significantly increase from V_{12} to V_T, and significantly decrease from V_T to maturity (R_6) for all treatments (Figures 4 and 5). Apparently, it displayed a type of 'M' for this changing process.

Detailed patterns of changing soil NO_3-N, however, differed among treatments (Figures 4 and 5). In 2012 (Figure 4), after basal N fertilizer was applied, the increase of NO_3-N concentration in 0-100 cm soil profile in SP from planting to V_6 was greater in FIFNS while smaller in FIFND treatment than those of the other treatments ($P < 0.05$). The opposite was true for that UP in 0-40 cm and NP in 0-60 cm. From V_6 to V_{12}, compared to the other treatments, the decrease of NO_3-N concentration in 0-60 cm was greater in FIFNS treatment of SP and FIFND treatment of NP ($P < 0.05$). After N fertilizer was topdressing at V_{12}, the increase of NO_3-N concentration in 0-40 cm in SP and NP from V_{12} to VT showed similarities compared to that from planting to V_6. However, after N fertilizer was topdressing at V_T, no increase of NO_3-N concentration in 0-100 cm was observed at R_2 (Figure 4). Moreover, compared to the other treatments, the decrease of NO_3-N concentration in 0-40 cm from V_T to R_6, was smaller in FIFND treatment of SP and FIFNS treatment of NP ($P < 0.05$). The decrease of NO_3-N concentration in 40-60 cm from V_T to R_2 was greater in AICN treatment of SP and FIFND treatment of NP than those of the other treatments ($P < 0.05$). The decrease of NO_3-N concentration in 40-60 cm in NP from R_2 to R_6 was greater in CICN treatment while that in 60-100 cm in SP from R_2 to R_6 was smaller in FIFND treatment than those of the other treatments ($P < 0.05$). In addition, compared to CICN treatment, AICN treatment reduced the NO_3-N concentration by 4.8 to 8.7% in 60-100 cm from V_{12} to R_6 over the different positions.

In 2014 (Figure 5), compared to CICN treatment, the decrease of NO_3-N concentration in 0-100 cm soil profile from VT to R_2 was significantly increased by AICN treatment ($P < 0.05$). Compared to CICN treatment, AICN treatment reduced the NO_3-N concentration by 5.1 to 9.6% in 60-100 cm from V_{12} to R_6 over the different positions. Moreover, the dynamics of NO_3-N concentration of AIAN treatment from V_T to R_6 was comparable to that of AICN in two years (data not shown).

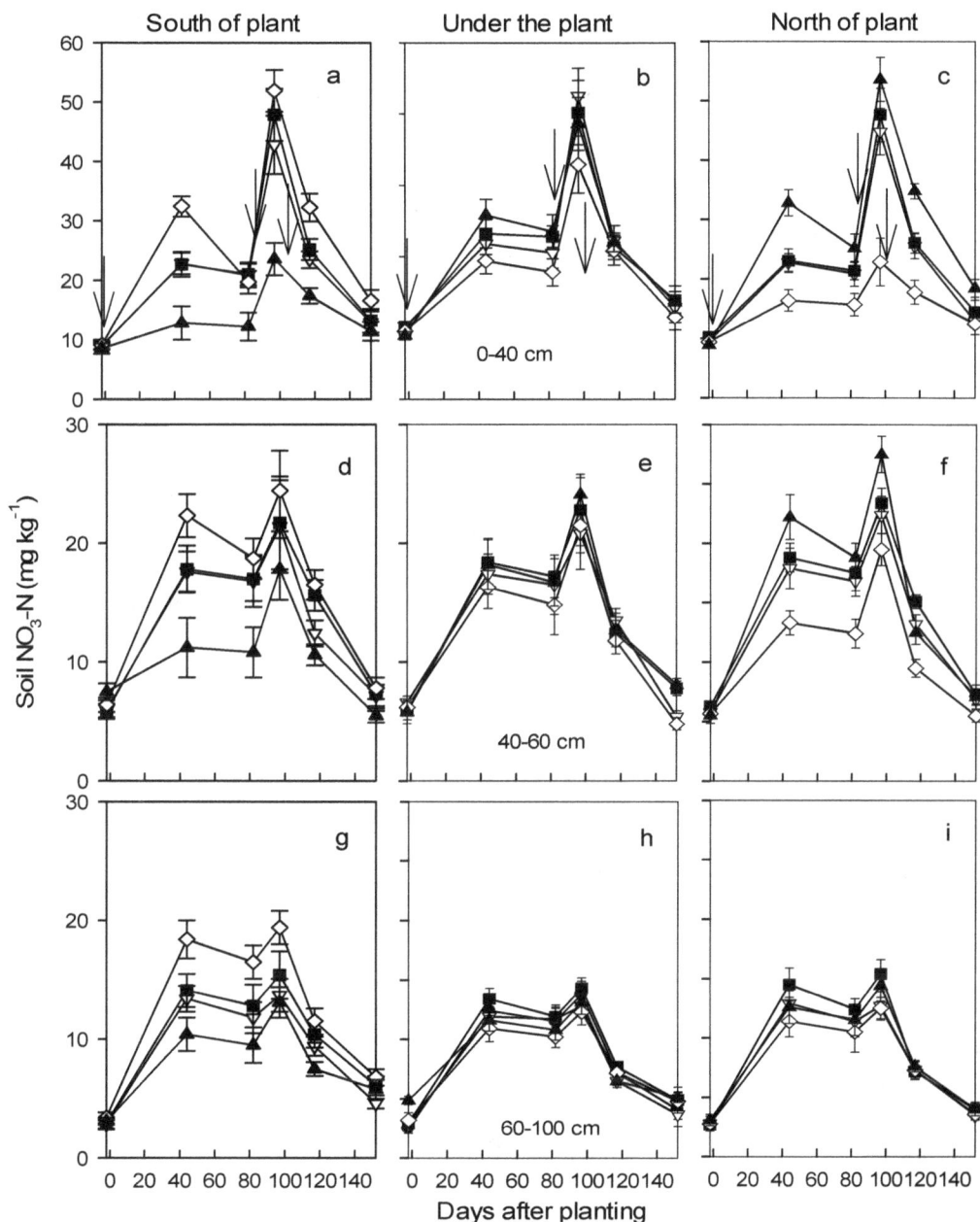

Figure 4. Dynamics of soil NO_3-N concentration following days after planting at different positions in 0-100 cm soil layer for AICN, CICN, FIFNS and FIFND treatments in 2012

Note. Symbol of ■, ▲, ▽ and ◇ denotes the treatment of CICN, FIFND, AICN and FIFNS, respectively. Arrows represents application of N fertilizer. Days after planting of -2, 44, 82, 97,117 and 152 d represents before planting and development stage of maize at 6 collars (V_6), 12 collars (V_{12}), tasseling (VT), filling (R_2) and maturity (R_6), respectively.

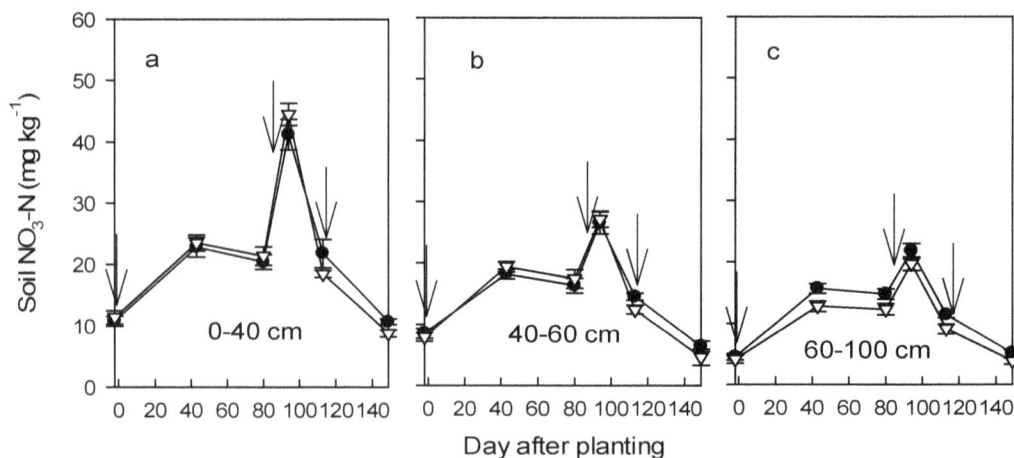

Figure 5. Dynamics of soil NO_3-N concentration following days after planting in 0-100 cm soil layer for CICN and AICN treatments in 2014

Note. Symbol of ● and ▽ denotes the treatment of CICN and AICN, respectively. Data are combined across different positions. Arrows represents application of N fertilizer. Days after planting of -2, 43, 80, 94, 113 and 149 d represents before planting and development stage of maize at 6 collars (V_6), 12 collars (V_{12}), tasseling (VT), filling (R_2) and maturity (R_6), respectively.

3.3 Residual Soil NO_3-N

In 2012 (Table 3), in 0-40 cm soil profile, FIFND had the largest residual NO_3-N while AIAN and AICN the smallest among the treatments. In 40-60 cm, residual NO_3-N of both FIFNS and CIFN was significantly higher than that of the other treatments. In 60-80 cm, residual NO_3-N was the smallest in FIFND treatment. In 80-100 cm, FIFNS treatment had the largest residual NO_3-N among the treatments. In 0-100 cm soil profile, compared to CI, residual NO_3-N was significantly increased by FI while reduced by AI in any N supply method. Compared to CN, residual NO_3-N of AN was comparable while that of FN increased significantly irrespective of irrigation method. The largest residual NO_3-N was found in FIFND treatment, followed by FIFNS treatment, and the smallest in AIAN and AICN treatments (Table 3).

In 2014 (Table 3), for each individual soil layer as well as 0-100 cm soil profile, for both AI and CI, residual NO_3-N had no response to N supply method. On the contrary, compared to CI, residual NO_3-N was significantly decreased by AI in either N supply method (excluded in 40-60 cm). These results indicated that alternate furrow irrigation coupled with conventional or alternate N supply was useful to reduce residual NO_3-N in 0-100 cm soil profile at maturity.

Table 3. Soil residual NO_3-N (kg ha^{-1}) in 0-100 cm soil profile at maturity (R_6) in 2012 and 2014

Year	Treatment	Soil depth (cm)					
		0-20	20-40	40-60	60-80	80-100	0-100
2012	AIAN	29.2±2.1c	28.4±1.8c	15.6±0.7b	12.1±0.6b	10.1±0.3b	95.3±2.9e
	AICN	28.6±2.2c	27.2±2.1c	14.8±1.3b	11.3±0.8b	9.7±0.5b	91.5±3.4e
	AIFN	31.3±3.4b	30.6±2.7b	16.6±1.1b	13.5±1.0b	11.7±0.8b	103.8±4.8d
	CIAN	31.3±2.8b	30.9±2.8b	15.5±1.0b	12.8±1.3b	11.7±0.8b	103.1±4.2d
	CICN	30.7±2.7b	30.5±3.0b	16.6±0.9b	12.9±0.7b	11.5±0.4b	102.2±2.4d
	CIFN	34.6±3.6ab	33.1±3.5b	18.5±1.4a	14.9±1.2a	11.3±0.9b	112.4±3.5c
	FIAN	32.7±3.8b	31.5±2.8b	16.2±1.1b	15.5±1.8a	11.3±0.8b	107.2±3.4cd
	FICN	32.6±3.5b	31.6±1.8b	14.6±0.8b	15.0±0.5a	12.5±1.0b	106.3±4.5cd
	FIFNS	34.8±4.1ab	32.5±3.2b	18.9±1.3a	16.6±1.3a	13.7±1.3a	116.5±4.8b
	FIFND	44.5±4.8a	42.8±4.4a	14.6±0.7b	10.6±0.9c	9.8±0.4b	123.3±6.1a
2014	AIAN	23.2±2.3b	21.4±1.5b	12.4±0.7a	10.0±0.5b	8.5±0.3b	75.5±3.5b
	AICN	22.6±1.8b	20.2±2.7b	11.5±0.5a	9.6±0.6b	8.6±0.8b	73.5±4.1b
	CIAN	25.4±2.2a	23.4±1.5a	12.8±1.1a	11.3±0.5a	10.1±0.4a	83.0±5.0a
	CICN	27.8±3.0a	22.5±2.1a	13.4±0.8a	11.8±0.6a	10.5±0.6a	86.0±4.7a

Note. Values are means± standard error (n = 3). Different letters in the same column within same year indicate significant difference ($P < 0.05$).

4. Discussion

Dynamics of NO_3-N concentration in soil profile displayed a type of 'M' during the growing season, with two peak points at 6 collars and tasseling of maize stage, respectively. In addition, the decrease of NO_3-N concentration from 6 collars to 12 collars was slight but sharp from tasseling to filling (Figures 4 and 5).This was obviously related to the changes of N application and crop N requirement. The first soil sampling was earlier than the first N application while all of them were ahead of the sowing, resulting in the lowest and first peak point of NO_3-N concentration at the beginning of the growing season and at 6 collars, respectively. Crop N uptake increased gradually with maize growth going, leading soil NO_3-N decreasing from 6 collars. Soil NO_3-N reached its second peak point at tasseling after the second N application at 12 collars. Though 25% of 200 kg N ha^{-1} has been supplied at tasseling, from tasseling to filling when maize plant was in its reproductive period, the N demand was dramatically increased (Hirel, Gouis, Ney, & Gallais, 2007) and much more than that from 6 collars to 12 collars (Chikowo, Mapfumo, Nyamugafata, Nynamadzawo, & Giller, 2003), resulting in a sharp decrease of NO_3-N concentration from tasseling.

However, an enhanced decrease of NO_3-N concentration from tasseling to filling was observed for south of the plant (SP) of AICN treatment and north of the plant (NP) of FIFND treatment in 40-60 cm soil profile (Figure 4d, f). For AICN treatment, on one side, AI could enhance root growth compared to CI (Liang, Zhang, & Wang, 1996; Mingo, Theobald, Bacon, Davis, & Dodd, 2004), and the effect was amplified by its coupling with CN (Qi, Hu, Wu, & Niu, 2015). When crop N requirement was increased from tasseling to filling (Hirel et al., 2007), N supply from 0-40 cm may not be enough, thus stimulating the relatively developed roots under AICN (Qi, Hu, Wu, & Niu, 2015) to enhance the N absorption from 40-60 cm. On the other side, NO_3-N in 40-60 cm could move upward as a function of mass flow due to more consumption of NO_3-N in 0-40 cm (Song & Li, 2005). In addition, the irrigated position for AI was located in SP at tasseling and the relatively adequate water supply might enhance depletion of NO_3-N at SP (Li et al., 2009). For FIFND treatment, root growth at NP (the N fertilizer supplied side) in 0-40 cm may be inhibited because of high N concentration (Tian, Chen, & Liu, 2008), which might induce more roots to extend to deeper soil layers. Thus, NO_3-N in 40-60 cm under NP might become an ideal N source for absorption when plants are in dire need of N nutrition. This compares well with the findings that the enhanced decrease of NO_3-N in local area was closely related to the improved root growth (Wang, de Kroon, & Smits, 2007). These suggested that dynamics of soil NO_3-N were related to the supply of N and water, crop N requirement and the root growth.

In comparison to CI, residual NO_3-N in 0-100 cm soil profile was significantly reduced by AI under three N supply methods (Table 3). This was sustained by the results that AI plants absorbed more N than CI (data not shown). These results are in line with the conclusion that AI irrigation practice had better N-fertilizer recovery with minimal mineral N left in the soil for maize (Kirda et al., 2005) and potato (Shahnazari et al., 2008). Shahnazari et al. (2008) suggested that reduced residual NO_3-N under AI can be ascribed to both enhanced N uptake by the plants and increased N losses to the air via denitrification. However, contrary results for residual NO_3-N were reported by Tan et al. (2005) and Liu et al. (2011). They insisted that compared to CI, leaching of NO_3-N was reduced due to lateral flow increased by AI (Pan & Kang, 2000; Zhou et al., 2008). As a result, increased residual NO_3-N was found in AI rather than CI. Differences in N fertilizer rate, irrigation volume, climate condition, soil water characteristic and their interaction might contribute to the difference. Indeed, this needs to be further investigated.

Obvious concentration of soil NO_3-N was observed in the N applied side for both FIFND and FIFNS treatments, and the difference was that the concentration was higher in FIFND and deeper in FIFNS (Figures 2i and 2j). This could be explained as follows: For FIFND treatment in which N fertilizer was supplied with separation from water, as soil moisture content determines the soil N availability and its transport to the roots (Hu, Li, & Zhang, 2009), N absorption from the non-irrigated side of FI might be reduced by severe water deficit. For FIFNS treatment in which N fertilizer was supplied with irrigation water, soil water moved mainly in vertical direction under FI (Data not shown), which brought about downward movement of NO_3-N thus being far from the root system. Moreover, residual NO_3-N of both FIFND and FIFNS at maturity was higher than that of the other treatments (Table 3). Skinner et al. (1999) demonstrated that the prerequisite for FI to increase N uptake and reduce residual NO_3-N was a climate condition that allowed adequate root development within its non-irrigated furrow. However, this experimental area is prone to large evapotranspiration (about 2000 mm in average), and received only 129 mm of precipitation during the maize growing season in 2012 (Figure 1). Meanwhile, our previous work showed that FI obviously reduced maize root growth in the same area (Qi et al., 2015). Moreover, N uptake by FIFND and FIFNS plants was significantly decreased compared to the other treatments (data not shown). Thus, neither FIFND nor FIFNS ought to be recommended for management of N and water in the arid area.

It cannot be ignored that N supply method, irrigation method and their interaction had a significant impact on NO_3-N concentration in 0-40 cm soil profile in 2012 but not in 2014 (excluded irrigation method) (Table 2). This was mainly ascribed to exclude fixed treatments in 2014, where NO_3-N distribution in 0-40 cm was comparable for AI or CI coupled with CN and AN (Figure 3). Moreover, compared to CI, AI increased NO_3-N content under the plant in 0-40 cm and decreased that in 40-80 cm for three N supply methods (Figures 2 and 3). This might be related to the differences in change of soil moisture between AI and CI. For AI, soil moisture content between the two parts of the root systems risen and fallen alternatively, while for CI, the soil moisture risen/fallen meantime during 6 collars to maturity (Data not shown). Thus, AI could keep more NO_3-N in the upper soil layer through stimulated lateral flow and reduced deep percolation (Pan & Kang, 2000; Zhou, Kang, Li, & Zhang, 2008).

5. Conclusions

At filling stage, under the same method of N supply, alternate furrow irrigation enhanced soil NO_3-N concentration under the plant in 0-40 cm soil profile and reduced that in 40-80 cm thanks to its increased lateral irrigation water flow. Conventional and alternate N supply coupled with alternate or conventional furrow irrigation brought a relatively uniform distribution of soil NO_3-N across the plant rows; while fixed N supply enhanced NO_3-N concentration under the N supplied side in 0-80 cm soil profile for three irrigation methods. Compared to conventional furrow irrigation, alternate furrow irrigation reduced soil NO_3-N concentration in 60-100 cm soil profile from 12 collars to maturity when coupled with alternate or conventional N supply, resulting in decreased soil residual NO_3-N in 0-100 cm. This might be related to reduce downward movement of NO_3-N and enhanced N uptake by plant under alternate furrow irrigation. Therefore, spatial distribution of soil NO_3-N during maize season was optimum and soil residual NO_3-N was minimal at maturity under alternate furrow irrigation as long as conventional or alternate N supply method are used.

References

Bao, S. (2000). *Soil and agricultural chemistry analysis*. Edition III, Beijing: Chinese Agriculture Press.

Bar-Yosef, B., & Sheikholslami, M. R. (1976). Distribution of water and ions in soils irrigated and fertilized from a trickle source. *Soil Science Society of American Journal, 40*, 575-582. http://dx.doi.org/10.2136/ sssaj1976.03615995004000040033x

Benjamin, J. G., Havis, H. R., Ahuja, L. R., & Alonso, C. V. (1994). Leaching and water flow patterns in every-furrow irrigation. *Soil Science Society of American Journal, 58*, 1511-1517. http://dx.doi.org/10.2136/ sssaj1994.03615995005800050034x

Benjamin, J. G., Porter, L. K., Duke, H. R., & Alonso, C. V. (1997). Corn growth and nitrogen uptake with furrow irrigation and fertilizer bands. *Agronomy Journal, 89*, 609-612. http://dx.doi.org/10.2134/agronj1997.00021962008900040012x

Chikowo, R., Mapfumo, P., Nyamugafata, P., Nynamadzawo, G., & Giller, K. E. (2003). Nitrate-N dynamics following improved fallows and maize root development in a Zimbawean sandy clay loam. *Agroforestry Systems, 59*, 187-195. http://dx.doi.org/10.1023/B:AGFO.0000005219.07409.a0

Emteryd, O. (1989). *Chemical and physical analysis of inorganic nutrients in plant, soil, water and air.* Sweden: Stencil No 10 Umea.

Gheysari, M., Mirlatifi, M. S., Homaee, M., Asadi, E. M., & Hoogenboom, G. (2009). Nitrate leaching in a silage field under different irrigation and nitrogen fertilizer rates. *Agricultural Water Management, 96*, 946-954. http://dx.doi.org/10.1016/j.agwat.2009.01.005

Han, K., Yang, Y., Zhou, C., Shangguan Y., Zhang, L., Li, N., & Wang, L. (2014). Management of furrow irrigation and nitrogen application on summer maize. *Agronomy Journal, 106*(4), 1402-1410. http://dx.doi.org/10.2134/agronj13.0367

Hirel, B., Gouis, J. L., Ney, B., & Gallais, A. (2007). The challenge of improving nitrogen use efficiency in crop plants: Towards a more central role for genetic variability and quantitative genetics within integrated approaches. *Journal of Experimental Botany, 58*, 2369-2387. http://dx.doi.org/10.1093/jxb/erm097

Hu, T., Kang, S., Li, F., & Zhang, J. (2009). Effects of partial root-zone irrigation on the nitrogen absorption and utilization of maize. *Agricultural Water Management, 96*, 208-214. http://dx.doi.org/10.1016/j.agwat.2008.07.011

Ju, X., Liu, X., Zhang, F., & Roelcke, M. (2004). Nitrogen fertilization, soil nitrate accumulation and policy recommendations in several agricultural regions of China. *American Biology Teachnology, 33*, 278-283. http://dx.doi.org/10.1579/0044-7447-33.6.300

Kang, S., Liang, Z., Hu, W., & Zhang, J. (1998). Water use efficiency of controlled alternate irrigation on root-divided maize plants. *Agricultural Water Management, 38*, 69-76. http://dx.doi.org/10.1016/S0378-3774(98)00048-1

Kang, S., Zhang, J., Liang, Z., Hu, X., & Cai, H. (1997). The controlled alternative irrigation - A new approach for water saving regulation in farmland. *Agricultural Research Arid Area, 15*(1), 1-6.

Kirda, C., Topcu, S., Kaman, H., Ulger, A. C., Yazici, A., Cetin, M., & Derici, M. R. (2005). Grain yield response and N-fertilizer recovery of maize under deficit irrigation. *Field Crops Research, 93*, 132-141. http://dx.doi.org/10.1016/j.fcr.2004.09.015

Li, J., Zhang, J., & Ren, L. (2003). Water and nitrogen distribution as affected by fertigation of ammonium nitrate from a point source. *Irrigation Science, 22*, 19-30. http://dx.doi.org/10.1007/s00271-003-0064-8

Liang, J., Zhang, J., & Wong, M. H. (1996). Effects of air-filled soil porosity and aeration on the initiation and growth of secondary roots of maize (*Zea mays*). *Plant Soil, 186*, 245-254. http://dx.doi.org/10.1007/BF02415520

Liu, X., Zhang, F., Yang, Q., Wang, J., & Li, Z. (2011). Transfer of mineral nitrogen in maize root zone soil under different irrigation methods. *Chinese Journal of Eco-Agriculture, 19*, 540-547. http://dx.doi.org/10.3724/SP.J.1011.2011.00540

Liu, Y., Li, Y., Pan, T., Qu, L., & Du, Z. (2009). Study on effects of different irrigation treatments on evapotranspiration and yield in spring maize. *Arid Research in Arid Area, 27*, 67-72.

Mingo, D. M., Theobald, J. C., Bacon, M. A., Davies, W. J., & Dodd, I. C. (2004). Biomass allocation in tomato (*Lycopersicon esculentum*) plants grown under partial root-zone drying: enhancement of root growth. *Function Plant Biology, 31*, 971-978. http://dx.doi.org/10.1071/FP04020

Morison, J. I. L., Baker, N. R., Mullineaux, P. M., & Davies, W. J. (2008). Improving water use in crop production. *Philosiphical Transactions of the Royal Society, 363*, 639-658. http://dx.doi.org/10.1098/rstb.2007.2175

Pan, Y., & Kang, S. (2000): Irrigation water infiltration in furrows and crop water use of alternate furrow irrigation. *Transactions CASE, 16*, 39-43.

Qi, D., Hu, T., Wu, X., & Niu, X. (2015). Rational irrigation and nitrogen supply methods improving root growth and yield of maize. *Transactions CASE, 31*, 144-149.

Ritchie, S. W., & Hanway, J. J. (1982). How a corn plant develops. *Special Report 48*. Review Iowa State University Cooperation Extensive Service, American, IA.

Shahnazari, A., Ahmadi, S. H., Laerke, P. E., Liu, F. L., Plauborg, F., Jacobsen, S. E., … Andersen, M. N. (2008). Nitrogen dynamics in the soil plant system under deficit and partial root-zone drying irrigation strategies in potatoes. *European Journal of Agronomy, 28*, 65-73. http://dx.doi.org/10.1016/j.eja.2007.05.003

Shahnazari, A., Liu, F., Andersen, M. N., Jacobsen, S., & Jensen, R. C. (2007). Effects of partial root-zone drying on yield, tuber size and water use efficiency in potato under field conditions. *Field Crops Research, 100*, 117-124. http://dx.doi.org/10.1016/j.fcr.2006.05.010

Singandhupe, R. B., Rao, G. G. S. N., Patil, N. G., & Brahmanand, P. S. (2003). Fertigation studies and irrigation scheduling in drip irrigation system in tomato crop (*Lycopersicum esculentum* L.). *European Journal of Agronomy, 19*, 327-340. http://dx.doi.org/10.1016/S1161-0301(02)00077-1

Skinner, R. H., Hanson, J. D., & Benjamin, J. G. (1999). Nitrogen uptake and partitioning under alternate and every furrow irrigation. *Plant Soil, 210*, 11-20. http://dx.doi.org/10.1023/A:1004695301778

Song, H., & Li, S. (2005). Effects of root uptake function and soil water on NO_3^--N and NO_4^+-N distribution. *Scientia Agricultural Science, 38*, 96-101.

Tafteh, A., & Sepaskhah. (2012). Yield and nitrogen in maize field under different nitrogen rates and partial root drying irrigation. *International Journal of Plant Production, 6*, 94-113. Retrieved from http://ijpp.gau.ac.ir/article_672_101.html

Tan, J., Wang, L., & Li, S. (2005). Movement and utilization of water and nutrient under different irrigation patterns. *Plant Nutrition Fertilizer Society, 11*, 442-448.

Tang, L., Li, Y., & Zhang, J. (2010). Partial root zone irrigation increase water use efficiency, maintains yield and enhance economic profit of cotton in arid area. *Agricultural Water Management, 97*, 1527-1533. http://dx.doi.org/10.1016/j.agwat.2010.05.006

Tarkalson, D. D., Payero, J. O., Ensley, S. M., & Shapiro, C. A. (2006). Nitrogen accumulation and movement under deficit irrigation in soil receiving cattle manure and commercial fertilizer. *Agricultural Water Management, 85*, 201-210. http://dx.doi.org/10.1016/j.agwat.2006.04.005

Tian, Q., Chen, F., & Liu, J. (2008). Inhibition of maize root growth by high nitrate supply is correlated with reduced IAA levels in roots. *Journal of Plant Physiology, 165*, 942-951. http://dx.doi.org/10.1016/j.jplph.2007.02.011

Tsai, C. Y., Huber, D. M., Glover, D. V., & Warren, H. L. (1984). Relationship of N deposition to grain yield and N responses of three maize hybrids. *Crop Science, 24*, 277-281. http://dx.doi.org/10.2135/cropsci1984.001 1183X002400020016x

Wang, C., Zhu, P., Shu, L., Zhu, J., Yu, H., Zhan, Y., & Yuan, M. (2014). Effects of alternate partial root irrigation and nitrogen forms on utilization and movement of nitrate in soil. *Transactions CASE, 30*, 92-101.

Wang, L., de Kroon, H., & Smits, A. J. M. (2007). Combined effects of partial root zone drying and patchy fertilizer placement on nutrient acquisition and growth of oilseed. *Plant Soil, 295*, 207-216. http://dx.doi.org/10.1007/s11104-007-9276-7

Wang, Q., Li, F., Zhang, E., Li, G., & Vance, M. (2012). The effects of irrigation and nitrogen application rates on yield of spring wheat (longfu-920), and water use efficiency and nitrate nitrogen accumulation in soil. *Australian Journal of Crop Science, 6*, 662-672. Retrieved from http://search.informit.com.au/document Summary;dn=362475432090775;res=IELHSS

Xing, W., Wang, L., Li, L., & Li, S. (2003). Effect of water-fertilized spatial coupling on corn in semiarid area ∏ dynamics distribution of water and available nitrogen in soil. *Soil, 35*, 242-247.

Yang, R., & Su, Y. (2009). Effects of nitrogen fertilization and irrigation rate on grain yield, nitrate accumulation and nitrogen balance on sandy farmland in the marginal oasis in the middle of Heihe River basin. *Acta Ecology Sinica, 28*, 1460-1469.

Zhang, L., Ma, Z., & E, S. (2007). Effects of ridge cultivation with plastic film Mulching furrow irrigation on yield and water use efficiency of seed corn. *Acta Agricultural Since, 16*, 83-86.

Zhou, Q., Kang, S., Li, F., & Zhang, L. (2008). Comparison of dynamic and static APRI-models to simulate soil water dynamics in a vineyard over the growing season under alternate partial root-zone drip irrigation. *Agricultural Water Management, 95,* 767-775. http://dx.doi.org/10.1016/j.agwat.2008.01.018

Zhu, P., Wu, S., Shu, L., Liu, F., Zhu, K., & Guo, J. (2013). Coupling effects of water and nitrogen application with partial root zone irrigation on growth of cucumber and nitrate distribution in soil. *Bulletin Soil Water Conversation, 33,* 6-16.

Zotarelli, L., Dukes, M. D., Scholberg, M. S. J., Munoz-Carpena, R., & Icerman, J. (2009). Tomato nitrogen accumulation and fertilizer use efficiency on a sandy soil, as affected by nitrate rate and irrigation scheduling. *Agricultural Water Management, 96,* 1247-1258. http://dx.doi.org/10.1016/j.agwat.2008.06.007

Abbreviations

AI, alternate furrow irrigation; FI, fixed furrow irrigation; CI, conventional furrow irrigation; AN, alternate nitrogen supply; FN, fixed nitrogen supply; CN, conventional nitrogen supply; UP, under the plant; SP, south of the plant; NP, north of the plant; V_6, V_{12}, VT, R_2 and R_6 represents 6 collars, 12 collars, tasseling, filling and maturity of maize development stage, respectively.

8

Nodulation, Nutrient Uptake and Yield of Common Bean Inoculated with *Rhizobia* and *Trichoderma* in an Acid Soil

Alice M. Mweetwa[1], Gwen Chilombo[1] & Brian M. Gondwe[1]

[1] Department of Soil Science, School of Agricultural Sciences, University of Zambia, Lusaka, Zambia

Correspondence: Alice M. Mweetwa, Department of Soil Sciences, School of Agricultural Sciences, University of Zambia, Box 32379, Lusaka, Zambia. E-mail: alicemweetwa@yahoo.com

Abstract

Common bean is an important source of protein, fat, carbohydrate, vitamins and minerals for both human beings and livestock. In Zambia, common bean is produced mostly by smallholder farmers whose current yields fall far short of the potential 2 t/ha due to various challenges. Among the biophysical constraints, poor soil fertility and acidity pose the greatest challenges. This study investigated the individual and dual inoculation of *Rhizobia* and *Trichoderma* to common bean in a phosphorus deficient, acid soil. Soil in which the common bean was grown was characterized for selected chemical properties before planting and at harvest using standard laboratory procedures. Soils were amended with nitrogen and phosphorus fertilizer at the rates of 100 kg N and 80 kg P_2O_5 per ha; 1 g/kg of seed of *Trichoderma harzianum*; 100 g/kg of seed of *Rhizobium tropici*; or a combination of both at the recommended rates at planting. Nodulation and nodule effectiveness were determined at 51 days after planting. Nitrogen and phosphorus accumulation in the above ground biomass, biomass and grain yields were determined at maturity. To determine differences among soil amendments, data were analysed using Analysis of Variance and Least Significant Difference at 95% confidence limit. Relationships among parameters were determined using correlation analysis. The results showed that amending soils with inorganic N at high rates can depress nodulation even in the presence of high levels of inorganic phosphorus. Inoculating common bean with *Rhizobia* and *Trichoderma* either singly or in combination increases nodule number and effectiveness per plant but may not result in higher nitrogen and phosphorus accumulation, or in an increase in subsequent biomass or grain yields. The low phosphorus accumulation in *Trichoderma* inoculated plants observed in this study needs to be investigated further by studying the extent of colonization and the accompanying changes in root volume.

Keywords: *Rhizobia*, *Trichoderma*, co-inoculation, common bean, nitrogen, phosphorus

1. Introduction

Common bean (*Phaseolus vulgaris* L.) is the most important food grain legume in the world (Beebe, 2009; Nakitto, Muyona, & Nakimbungwe, 2015). It is an important source of protein, fat, carbohydrates, vitamins and minerals for humans and livestock (Beebe, Gonzalez, & Rengifo, 2000). Protein content in most varieties is high and averages 25%. The crop is rich in the amino acids tryptophan and lysine, which are not typically found in the staple maize crop. Common bean therefore, provides a good alternative to animal protein to meet minimal daily requirements. In addition, common bean provides other health benefits that include weight control, reduced proneness to diabetes, colon cancer and heart problems (Nyau, 2014).

Current crop production in Zambia covers over 85,000 hectares of land mostly in the Northern, North western and Eastern provinces of Zambia (CSO, 2012). These provinces cover parts of agro-ecological regions II and III of the country, with average seasonal rainfall between 800 and 1000, and more than 1000 mm, respectively. In Zambia, the crop is grown mostly at subsistence and small-scale levels, with only a small portion being grown by commercial farmers. While the potential yield of most varieties is 2 t/ha, the average yield is only 0.58 t/ha (Hamazakaza et al., 2014). Constraints to common bean production have been cited as being the use of unimproved varieties, and attack by pests and diseases. Pests and diseases alone account for 25-50% of the total yield losses (Mweetwa, 2011). In addition, poor soil fertility prevalent in the common bean production areas limits its productivity. Specifically, the leaching of bases in high rainfall areas (agro-ecological region III) results in an accumulation of hydrogen and aluminium ions in the soils. The resulting acidity has both direct and indirect effects on plant growth; common bean grows well in soils with pH ranging from 6.5 to 7.5 (Izquierdo,

1990). On the other hand, common bean is particularly sensitive and highly susceptible to physical and chemical environmental stresses compared to other legumes (Kabahuma, 2013) making the crop very nutrient demanding (da Silva et al., 2014).

In Zambia, most small-scale producers of common bean do not typically use chemical fertilizers to enhance crop productivity. While chemical fertilizers are readily available and may provide immediate answers to current limiting nutrient levels in the soil, they tend to be expensive and in the long-term, cause soil acidification and pollution of soils and water. This has resulted in the promotion of more sustainable agricultural practices such as the use of organic fertilizers to address poor soil fertility. Organic fertilizers have the advantage of improving soil physical properties, acting as substrates for beneficial soil microorganisms, buffering soil reaction, controlling erosion and as well as controlling plant parasitic nematodes and fungi (Gupta, 2011; Panda, 2011) in addition to supplying plant nutrients to the soil. The challenge usually of using organic fertilizers, such as animal and green manures, is that they tend to be bulky and of variable quality, and nutrient release may also be slow. This, coupled with competition for alternative uses on most small-scale farms, limits their use as single solutions to limiting soil nutrients.

As a legume, common bean forms symbioses with a broad range of *Rhizobia* that biologically fix nitrogen in the nodules. Several factors influence nitrogen fixation in legumes; these include soil nutrient conditions, temperature and moisture (Kabahuma, 2013). In biological nitrogen fixation, phosphorus is critical for nodule formation and functioning as well as for specific nitrogenase activity (Kouas, Labidi, Debez, & Abdelly, 2005). In Zambia, regions of common bean production tend to be deficient in phosphorus and high in acidity, this limits biological nitrogen fixation and productivity of the crop. In addition, common bean crop productivity is often limited by nitrogen deficiency despite its ability to form symbiotic relationships with nitrogen fixing *Rhizobia* (Rodino, Santalla, De Ron, & Drevon, 2005); unlike other legumes, common bean is a poor nitrogen fixer; this is attributable to some genetic factors and its sensitivity to physical and environmental stresses (Yadegari & Rahmani, 2010).

The use of bio-fertilizers such as *Rhizobia* and other plant-growth promoting bacteria or fungi have been suggested to result in increased growth and subsequent yield in many crop species including legumes (Stajkovic et al., 2011). *Rhizobia* inoculation alone to seeds or application to the soil at planting has shown varied results in terms of stand count after emergence, stand count at harvest, biomass yield, number of nodules per plant, biological nitrogen fixation and grain yield of common bean (Musandu & Joshua, 2001; Mweetwa et al., 2014a). On the other hand, co-application of *Rhizobia* with other microorganisms such as *Pseudomonas flourescens*, *Bacillus* and *Azospirillum lipoferum* has been reported to result in improved dry weight, phosphorus uptake, nodulation, nitrogen fixation, enzyme activity, protein content, yield and yield components (El-Katany, 2010; Stajkovic et al., 2011; Yadegari & Rahmani, 2010; Samavat, Samavat, Mafakheri, & Shakour, 2012). The benefits of co-inoculating *Rhizobia* with other microorganisms are derived from the ability of these microorganisms to make available to the symbiosome essential nutrients such as phosphorus, to enhance water uptake by the host plant, to produce plant growth regulators and to inhibit some soil-borne plant pathogens through anti-fungal activity and siderophore production (Sharma, Jadega, Kataria, Anamika, & Kumar, 2014; Tatarani, Dichio, & Xiloyannis, 2012; Yadegari & Rahmani, 2010; Izquierdo, 1990).

Trichoderma species are filamentous fungi of the ascomycota known to exist in many soil types of the World (Tatarani et al., 2012). Many species of the *Trichoderma* genus are known to control various plant pathogens through a number of chemiotropic mycoparasitic interactions (Sharma et al., 2014). In addition, *Trichoderma* have been reported to enhance uptake of nutrients and water through stimulated production of lateral roots, and to promote plant growth and biomass yield through the stimulation of the production of several natural plant growth hormones (Tatarani et al., 2012). Dual inoculation of *Trichoderma* and *Rhizobia* to legumes has been reported with conflicting results in terms of the nature of the interactions between the two organisms, and their combined effect on nodulation, biomass production and grain yield. Inhibition of *Rhizobia* growth by *Trichoderma in vitro* has been attributed to competition for nutrients, antibiosis and lysis (Jayaraj & Ramabadran, 1999); on the other hand, compatibility, even synergism has been reported by others.

In view of the potential benefits to biological nitrogen fixation by enhanced phosphorus uptake through *Trichoderma*-plant interactions, a study was undertaken to investigate dual inoculation of *Rhizobia* and *Trichoderma harzianum* to common bean grown in a phosphorus deficient acid soil. This paper reports the effect of singly or co-inoculating *Rhizobia* and *Trichoderma harzanium* on nodulation, nutrient uptake, biomass production and grain yield of common bean under greenhouse conditions.

2. Materials and Methods

2.1 Site Description, Soil Collection and Characterization

Greenhouse experiments were conducted at the University of Zambia, 15°24′ South and 28°19′ East. Soils used in the study were collected from the Liempe University of Zambia Farm (15°24′ South and 28°28′ East). This site was specifically selected for soil collection because soils in that location are known to be acid and because the land had been left fallow for a number of seasons.

Soils were collected to a depth of 20 cm from ten (10) random spots and mixed thoroughly to make a composite sample. Soil intended for chemical characterization was air-dried and passed through a 2-mm sieve. Soil was then characterized for reaction (pH) and exchangeable bases. Soil reaction was determined in 0.01 M Calcium chloride using a soil: solution ratio of 1:2.5 using glass calomel combination electrode connected to a pH meter. The ammonium acetate method (Thomas, 1982) was used to determine exchangeable bases of the soil. These soil characteristics were also measured at harvest to determine the effects of the different soil amendments.

2.2 Fertilizer and Inoculation Treatments and Experimental Setup

The study was conducted as a greenhouse experiment during the 2013/14 cropping season. Common bean seeds of the common variety Lyambai were planted in pots with the following fertilizer or inoculation treatments: unamended controls; nitrogen and phosphorus fertilizer application to the soil at planting (100 kg N/ha and 80 kg P_2O_5/ha, respectively); 1 g/kg seed of *Trichoderma harzanium* (Eco-T); 100 g/kg seed of *Rhizobium tropici* (Biofix); a combination of *Trichoderma harzanium* and *Rhizobium tropici* as one application. Microbial inoculants were applied to the seeds just before planting. Each treatment was replicated four times and arranged in a completely randomized design (CRD). Four seeds were sown in each pot (5 kg soil) and thinned to one plant at 14 days after emergence. In order to determine both nodulation and yield of common bean due to these amendments, two sets of experiments were setup exactly as described above.

2.3 Determination of Nodulation and Nodule Effectiveness

At 51 days after planting, plant stems were cut just above the soil level. The remaining below ground biomass was rinsed under running water to remove soil particles from the roots. Sieves were placed at the bottom of the biomass while rinsing to capture any nodules that could have detached during the processing. Nodules from each plant were physically counted from the roots to determine number per plant. In addition, nodule effectiveness was determined by selecting four equal sized nodules; nodules were sliced open with a sharp surgical razor to expose their centre. Nodule colour was then scored as follows: pink and red nodules were scored as effective, while white, brown and/ or green nodules were scored as ineffective. Nodule effectiveness was then calculated as a percent.

2.3 Determination of Nitrogen and Phosphorus Uptake

At maturity, plants were cut just above the soil level. The harvested above ground biomass was dried and weighed and oven dried at 60 °C for 72 hours. Using the modified Kjedahl and Dry Ashing methods, tissue nitrogen and phosphorus were determined (Doll & Lucas, 1975; Kalra & Maynard, 1991).

2.4 Determination of above Ground Biomass and Grain Yield

Above ground biomass yield was determined by cutting the plant stems just above the soil, oven drying at 60 °C to constant weight and weighing;while the grain yield was determined by weighing the shelled pods and as well as the actual seed after shelling.

2.5 Data Management and Statistical Analyses

Effects of different amendments were compared using analysis of variance (ANOVA). Means were compared at 95% confidence using Least Significance Difference (LSD). Other relationships between parameters were made using correlation analysis in Excel Windows software.

3. Results

3.1 Effect of N and P Fertilizer, Rhizobium tropici and Trichoderma harzianum on Number of Nodules per Plant and Nodule Effectiveness in Common Bean

The number of nodules across amendments ranged from 0 to 228 per plant and was significantly affected by the type of amendment. The application of inorganic nitrogen and phosphorus did not yield any nodules, while inoculating with *Rhizobia* and *Trichoderma* resulted in more nodules per plant than the unamended plants. The number of nodules per plant was not significantly different between the single *Rhizobia* and *Trichoderma* treatments with 197 and 189 nodules per plant, respectively (Figure 1). The most number of nodules was observed in co-inoculated plants (228 nodules per plant).

Nodule effectiveness ranged from 25 to 88% and was highest in plants co-inoculated with *Rhizobia* and *Trichoderma* (Figure 1). Single inoculations with *Trichoderma* or *Rhizobia* did not significantly increase nodule effectiveness compared to the unamended controls. In summary, co-inoculating common bean seeds at planting with *Trichoderma* and *Rhizobia* gave the greatest response of both nodule number per plant and percent effectiveness, while single inoculations only significantly increased nodule number per plant and not nodule effectiveness.

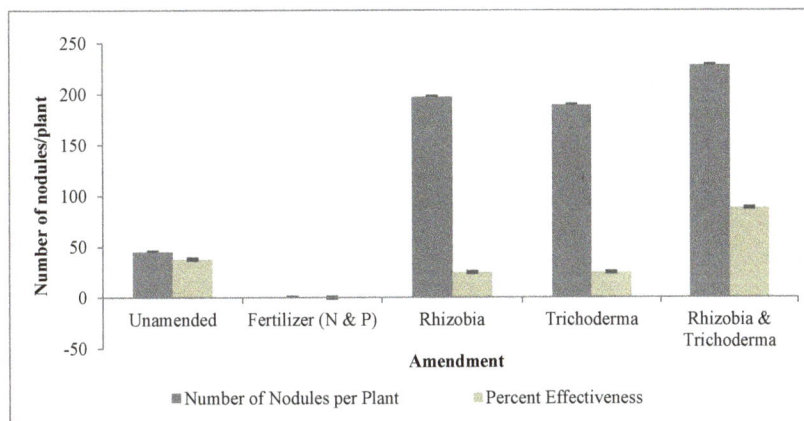

Figure 1. Number of nodules and nodule effectiveness of common bean plants in soils amended inorganic fertilizer, *Rhizobia* and *Trichoderma*

3.2 Nitrogen and Phosphorus Accumulation in Common Beans

Applying inorganic nitrogen and phosphorus resulted in the most nitrogen and phosphorus accumulation in plants at harvest than in the control plants and those treated with *Trichoderma*, *Rhizobia*, and *Rhizobia* and *Trichoderma*. Among the bio-fertilized, *Trichoderma* inoculated plants accumulated the least phosphorus, but did not differ significantly from the unamended control plants, plants inoculated with *Rhizobia*, and *Rhizobia* and *Trichoderma* (Table 1). Compared to the controls, amending soils with inorganic fertilizers resulted in approximately 11 fold increase in P accumulation, while *Rhizobia*, and *Rhizobia* and *Trichoderma* inoculation resulted in approximately 3 fold increases in P accumulation. While there was an increase in the total nitrogen in plants singly or co-inoculated with *Rhizobia* and *Trichoderma*, this increase was not significantly different from the control (Table 1).

Table 1. Nitrogen and phosphorus accumulation in common bean plants supplied with inorganic N and P, and inoculated with *Rhizobia* and *Trichoderma* at planting

Amendment	Total Nitrogen (mg/plant)	Total Phosphorus (mg/plant)
Unamended	90 b	8.89 c
Fertilizer (N & P)	780 a	110 a
Rhizobia	160 b	30 b
Trichoderma	170 b	15.7 bc
Rhizobia & *Trichoderma*	140 b	26.8 b
LSD	**192**	**14.9**
CV%	**2.13**	**2.15**
p-value	*0.034*	*0.0001*

There was a strong and positive relationship between N and P accumulation in plants ($r^2 = 0.972$) as well as between number of nodules per plant and N accumulation ($r^2 = 0.714$) as shown in Figure 2.

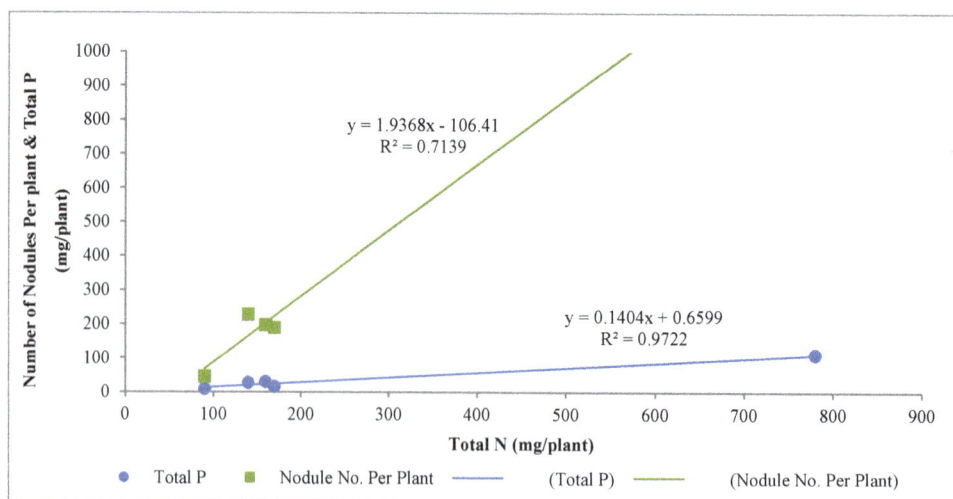

Figure 2. Relationship between Total P at harvest, number of nodules per plant at 51days after sowing and nitrogen accumulation in plants at harvest

3.3 Biomass and Grain Yield of Common Bean Due to N and P Fertilizer, Rhizobia and Trichoderma Application

Applying inorganic nitrogen and phosphorus to the soil resulted in the highest biomass and grain yields compared to the unamended and bio-fertilized common bean crops (Figure 3). While grain yield increased by 314, 163 and 277%, respectively from the unamended controls with *Rhizobia*, *Trichoderma*, and *Rhizobia* and *Trichoderma* application, respectively, these increases were not statistically significant. On the other hand, biomass yield increased significantly from the control by 747, 187, 23 and 230% with fertilizer, *Rhizobia*, *Trichoderma*, and *Rhizobia* and *Trichoderma* application, respectively. Inoculation with *Trichoderma* alone did not result in any significant increase in biomass yield from the unamended control (Figure 3).

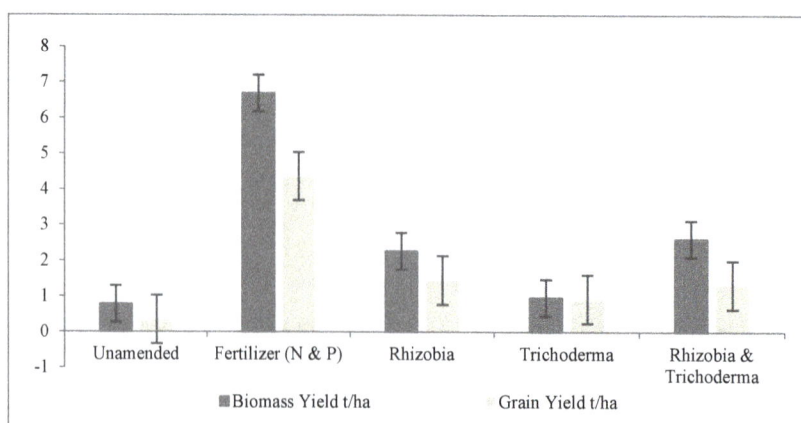

Figure 3. Biomass and grain yield (t/ha) of common bean supplied with inorganic N and P, and inoculated with *Rhizobia* and *Trichoderma*

Grain yield was positively correlated with the amounts of accumulated phosphorus and nitrogen by the plant with r^2 values of 1 and 0.97, respectively. On the other hand, there was a positive but weak relationship between grain yield and number of nodules per plant ($r^2 = 0.21$).

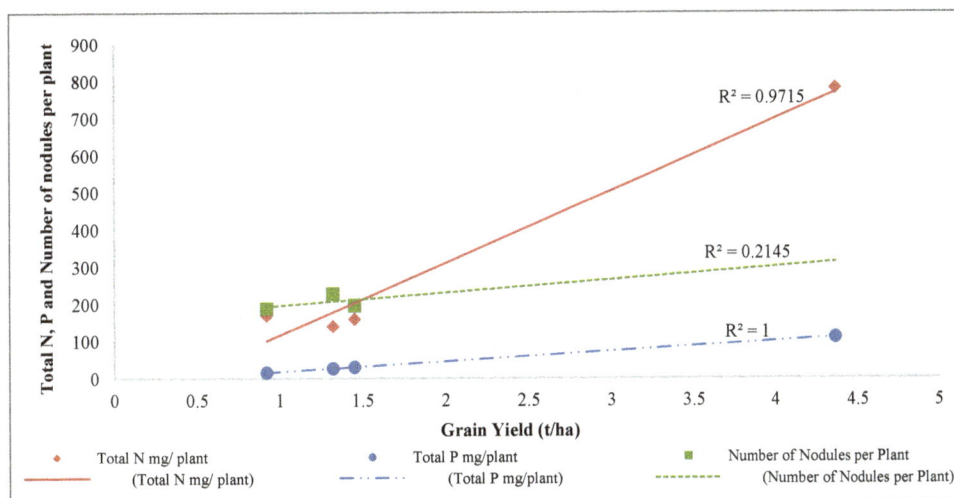

Figure 4. Relationship between Total N & P (mg/kg), number of nodules per plant, and grain yield (t/ha)

3.4 Changes in Soil Reaction and Exchangeable Bases Due to N and P Fertilizer, Rhizobia and Trichoderma Application

Soil pH declined from the initial 5.8 to 5.35, 5.11, 4.39, 5.19 and 4.18 in the unamended soils and soils amended with inorganic fertilizer, *Rhizobia*, *Trichoderma*, and *Rhizobia* and *Trichoderma* combined. The pH and Mg were lowest in soils to which *Rhizobia* was added either as a single inoculant or in combination with *Trichoderma*. The addition of *Rhizobia* appeared to depress pH and magnesium levels.

Calcium levels in soils amended with fertilizer, *Trichoderma*, and *Rhizobia* and *Trichoderma* were comparable to the levels at the start of the experiment (no differences). Unamended soils and soils to which *Rhizobia* was added alone had lower levels of calcium compared to the initial calcium levels. Amending soils with fertilizer resulted in the lowest levels of potassium at harvest.

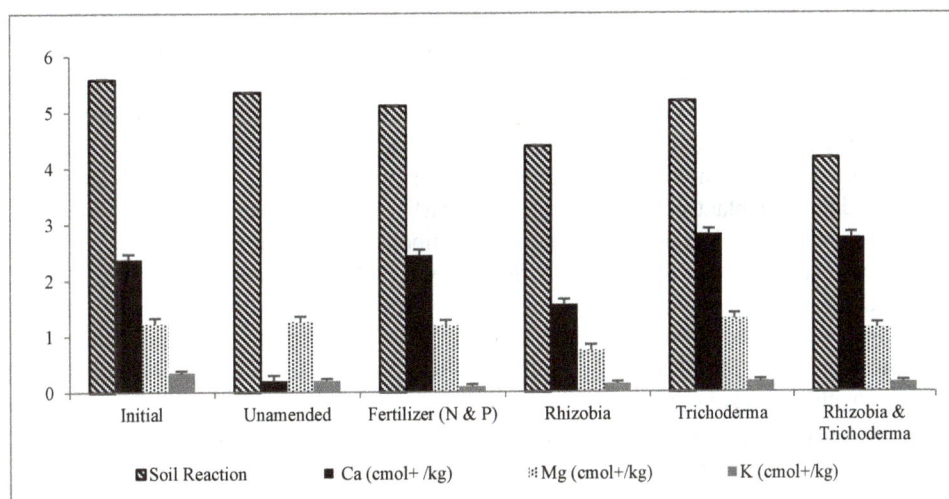

Figure 5. Changes in selected soil chemical properties due to to N and P fertilizer, *Rhizobia* and *Trichoderma* application

4. Discussion

The ability of legumes to form nodules has been attributed to many soil and biological factors that include levels of mineralizable N, levels of available P, soil reaction in the form of pH, type and vigour of legume, *Rhizobia* populations and their effectiveness in infecting and nodulating the host (Mohammadi, Shohrabi, Gholamreza, Khalesro, & Majidi, 2012).

Phosphorus plays a significant role in legume nodulation through its ability to enhance root development and proliferation thereby, affording the *Rhizobia* more sites for infection and initiation of nodule formation. For this reason, addition of phosphorus to the soil has been reported to result in enhanced nodulation of common bean in particular and other legumes, in general. Attar, Blavet, Selim, Abdelhamid, and Drevon (2012) reported higher nodule numbers in plants fertilized with phosphorus; in which case fertilization with 45 and 90 kg P_2O_5 per ha resulted in 31 and 37 nodules per plant, respectively compared to 20 in unfertilized plants. In addition, Muthamia, Kimani, Chemining'wa, and Esilaba (2015) also reported that phosphorus nutrition has a strong influence on the nodulation of common bean, with unfertilized plants having much lower numbers of nodules even in the presence of microbial inoculants. Adding phosphorus to the soil in other forms such as manure, has also been shown to have a positive effect on nodulation (Otieno et al., 1997). Our study on the other hand, reports a complete absence of nodules with the application of 80 kg P_2O_5/ha. We attribute the lack of nodulation to the addition of high levels of inorganic nitrogen at planting (100 kg N/ha). Application of nitrogen in high amounts to common bean at sowing has previously been reported to reduce nodule number, nodule fresh weight and nitrogenase activity (Izquierdo, 1990; Muller & Pereira, 1995; Otieno et al., 2007). Inorganic sources of nitrogen in the form of nitrate and ammonium are preferred by both the *Rhizobia* and legume hosts to biologically fixed dinitrogen which is energetically costly to acquire. Therefore, when levels of soil nitrogen are sufficient, nodulation is inhibited and existing nodules can degrade (Otieno et al., 1997). On the other hand, small starter doses of inorganic nitrogen have been shown to stimulate nodule initiation (Musandu & Joshua, 2001).

Because of the role that phosphorus plays in nodulation, number of nodules is severely restricted when it is limiting (Kouas et al., 2005). While low soil P may limit the numbers of nodules initiated, it does not directly influence functional characteristics nor restrict nitrogenase enzyme activity (Kouas et al., 2005). It must, however, still be noted that phosphorus plays a major role in determining the energy budget of the symbiosome in the form of ATP.

The current study reports significantly increased numbers of nodules with *Rhizobia* alone, *Trichoderma* alone, and *Rhizobia* co-inoculated with *Trichoderma* compared to the control. This increase could be attributed to both direct and indirect interactions between the host plant with *Rhizobia* and *Trichoderma*. Indirectly, *Rhizobia* and *Trichoderma* play some roles in modifying the fertility of the soil mainly through their participation in the decomposition of organic matter, mineral nutrient transformations such as phosphorus solubilisation (Mweetwa et al., 2014b) and modification of soil reaction. Directly, *Trichoderma* increases uptake of nutrients and water due to its ability to increase contact of the plant with the rhizospheric soil through enhanced plant lateral root production and mycelial extensions from the roots (Samolski, Rincón, Pinzón, Viterbo, & Monte, 2012; Pereira et al., 2014). This increases the surface area from which the plant can absorb nutrients required for the formation of nodules and for their functioning. In this paper, we report a 4 fold increase in the number of nodules per plant due to inoculation with *Trichoderma* alone. *Rhizobia* on the other hand, infect the host root system and therefore, directly have an influence on the nodulation process. Inoculating legume seeds at planting with *Rhizobia* has been shown to increase number of nodules per plant (Mweetwa et al., 2014a). Our results show a 4 fold increase in the number of nodules per plant and corroborate with earlier findings that inoculating common bean with *Rhizobia* can cause a significant increase in the number of nodules per plant (Otieno et al., 1997). It must be noted, however, that inoculation with *Rhizobia* may not always result in enhanced nodulation due to many other factors; these include the presence of many but ineffective *Rhizobia* strains; the ensuing competition between the introduced and the indigenous strains; poor quality of the inoculum and unfavourable soil conditions (Mcloughlin & Dunican, 1985; Montealegre & Graham, 1996; Mohammadi et al., 2012). This study shows that there is added value with respect to number of nodules per plant as well as their effectiveness when *Rhizobia* and *Trichoderma* are co-inoculated compared to when they are applied singly. The co-inoculation, brings together both the direct and indirect influences of the individual microbial symbionts.

Total nitrogen and phosphorus accumulated in common bean plants differed significantly among amendments. Inorganic N and P fertilization provided high amounts of available forms of these nutrients explaining why significantly higher amounts were accumulated during plant growth. On the other hand, inoculating with *Rhizobia* or *Trichoderma* alone or together did not result in significantly higher amounts of total nitrogen at harvest. Our results agree with earlier reports that have shown that while inoculating groundnuts, cowpea and soybean with *Rhizobia* resulted in increased nodule number and fresh weigh per plant, there was no corresponding increase in total nitrogen accumulated at harvest (Mungai & Karubiu, 2010; Mweetwa et al., 2014a). The observations in this study suggest that apart from nodule number, other nodule factors such as nodule efficiency play a crucial role in influencing the amount of total accumulated N at harvest (Mungai & Karubiu, 2010). In addition, inoculation with *Rhizobia* potentially introduces moderately effective strains that occupy the nodules, resulting in unimproved nitrogen accumulation or biomass yield (Singleton & Tavares, 1986).

In this study, total nitrogen accumulated was strongly and positively correlated with total phosphorous accumulated (r^2 = 0.97). This relationship was deemed important because of the role of phosphorus in nodule function and subsequent nitrogen fixation and accumulation. Unlike nitrogen accumulation, total phosphorus accumulation was significantly higher in microbial amendments than in the unamended. However, results did not show any added value in phosphorus accumulation by inoculating with *Trichoderma* alone or in the presence of *Rhizobia* when compared to *Rhizobia* inoculation alone. This result was rather unexpected; being relatively immobile in the soil, phosphorus acquisition can be enhanced by increasing the contact of root surface area with the soil. While the extent to which *Trichoderma* colonises and responds to the plant varies among species, the hyphae of the fungi when in symbiosis with the host plant have the potential to greatly increase the absorbing surface area of the roots and to increase the amount of phosphorus taken up by the plant (Rubio et al., 2014). In addition, *Trichoderma* has been shown to provide an alternative phosphorus uptake pathway which contributes to the total phosphorus plant uptake. This unexpected observation in this study requires further investigation by looking at the extent of *Trichoderma* colonization and the accompanying changes in root volume and span.

Biomass yield varied across amendments with the highest being observed in plants treated with inorganic P and N fertilizers. This observation has been previously reported by Otieno et al. (1997) and Mungai and Karubiu (2010). Biomass yield was positively and strongly correlated with both total N and P accumulation in the plants (r^2 = 0.90 and r^2 = 0.98, respectively; data not shown). Phosphorus and nitrogen as plant nutrients are important in plant metabolism as components of chlorophyll and amino acids; they are therefore key in photosynthesis, respiration, energy storage and transfer, cell division, cell enlargement and several other processes in plants. This explains why plants that had accumulated the most P and N also had the most biomass yield. In this study, inoculation with *Rhizobium* either alone or in combination with *Trichoderma* resulted in significant increases in biomass yield; this observation is incongruent with earlier reports of non-responsiveness of biomass yield to *Rhizobia* inoculation in other legumes (Mweetwa et al., 2014a).

In the current study, inoculating plants with *Trichoderma* alone, *Rhizobia* alone and *Rhizobia* with *Trichoderma* resulted in non-significantly higher grain yields than the unamended plants. Application of *Rhizobia* alone has previously been shown not to significantly influence common bean grain yields (Musandu & Joshua, 2001; Otieno et al., 1997); our results corroborate these earlier reports. In this study, grain yield was positively and strongly correlated with nitrogen accumulation (r^2 = 0.97); therefore, the low grain yield was attributed to the poor responsiveness of total nitrogen accumulation to the various microbial amendments.

Growth of legumes is typically known to depress soil pH due to their high uptake of bases such as calcium, magnesium and potassium in order to meet plant development demands (Liu, Lund, & Page, 1989). The results of this study show a decline in soil pH at harvest across all the amendments. Potassium levels in the soils at harvest had declined from the initial 0.35 cmol+/kg soil in all cases including the unamended soils, while calcium levels declined from their original levels in soils amended with *Rhizobia* alone and the unamended soils. Magnesium levels were generally unchanged by the time of harvest except for soils amended with *Rhizobium* alone. The decline in pH observed across all treatments can therefore be attributed to uptake of bases as indicated by lower soil solution concentrations. While *Trichoderma* has been suggested to buffer pH fluctuations in the soil, this was not true for soils amended with *Trichoderma* alone or in combination with *Rhizobia*.

In conclusion, amending soils with inorganic N at high rates can depress nodulation even in the presence of high levels of inorganic phosphorus which is typically beneficial for nodule initiation and development. Inoculating common bean with *Rhizobia* and *Trichoderma* either singly or in combination can enhance nodule number per plant but may not result in higher nitrogen and phosphorus accumulation, or in an increase in subsequent biomass or grain yields. The poor phosphorus accumulation in *Trichoderma* inoculated plants observed in this study needs to be investigated further by studying the extent of colonization and changes in root volume.

References

Attar, H. A., Blavet, D., Selim, E. M., Abdelhamid, M. T., & Drevon, J.-J. (2012). Relationship between phosphorus status and nitrogen fixation by common beans (*Phaseolus vulgaris* L.) under drip irrigation. *International Journal of Environmental Sciences and Technology, 9*, 1-13. http://dx.doi.org/10.1007/s13762-011-0001-y

Beebe, S. (2009). Improvement of common bean for mineral nutritive content at CIAT. In R. M. Welch, & I. Cakmaka (Eds.), *Impacts of agriculture on human health and nutrition* (Vol. 1). Encyclopaedia of Life Support Systems (EOLSS). Retreived May 9, 2016, from http://www.eolss.net/Eolss-sampleAllCpater.aspx

Beebe, S., Gonzalez, A. V., & Rengifo, J. (2000). Research on trace minerals in the common bean. *Food Nutr. Bulletin, 21*, 387-91.

CSO. (2012). *Agriculture Analytical Report for the 2010 Census of population and housing.* MoFNP, Central Statistical Office. Retrieved from http://zambia.opendataforafrica.org

da Silva, D. A., Esteves, J. A. D. F., Messias, U. A., Teixeir, A., Goncalves, J. G. R., Chiorato, A. F., & Carbonell, S. A. M. (2014). Efficiency in the use of phosphorus by common bean genotypes. *Scientia Agricola, 71*(3). http://dx.doi.org/10.1590/S0103-9016201400030008

Doll, E. C., & Lucas, R. E. (1975). Testing soil for K, Ca and Mg. In L. M. Walsh, & J. D. Beaton (Eds.), *Soil testing and plant analysis* (pp. 133-152). Soil Analysis Society of America, Inc., Madison, Wisconsin, USA.

El-Katany, M. H. (2010). Enzyme production and nitrogen fixation by free, immobilized and co-immobilized inoculants of *Trichoderma harzianum* and *Azospirillum brasilense* and their possible role in growth promotion in tomato. *Food Technol. Biotechnol., 48*(2), 161-174.

Gupta, P. K. (2011). *A handbook of soil, fertilizer, and manure* (2nd ed.). Jodhpur, India: AGROBIOS.

Hamazakaza, P., Katungi, E., Reyes, B., Maredia, M., Muimui, K., & Ojara, M. (2014). *Assessing access and adoption of common bena improved varieties in Zambia.* Research Technical Report-CIAT & MSU.

Izquierdo, M. (1990). *Effects of N and P fertilizers on common bean (Phaseolus vulgaris L.) grown in a p-fixing Nicaraguan mollic andosol* (Master's thesis, Swedish University of Agricultural Sciences, Uppsala). Retrieved from http://www.vaxteko.nu/html/sll/slu/reports_diss_markvetenskap/RDV08/RDV08.HTM

Jayaraj, J., & Ramabadran, R. (1999). *Rhizobium Trichoderma* interaction *in vitro* and *in vivo*. *Indian Phytopathol, 52*, 190-192.

Kabahuma, M. K. (2013). *Enhancing biological nitrogen fixation in common bean (Phaseolus vulgaris L.).* (Graduate Theses and Dissertations, Paper 13162, Iowa State University). Retrieved from http://lib.dr.iastate.edu/cgi/viewcontent.cgi?article=4169&context=etd

Kalra, Y. P., & Maynard, D. G. (1991). *Methods manual for forest soils and plant analysis* (p. 120).

Kouas, S., Labidi, N., Debez, A., & Abdelly, C. (2005). Effect of P on nodule formation and N fixation in bean. *Agron. Sustain. Dev, 25*, 389-393. http://dx.doi.org/10.1051/agro:2005034

Liu, W. C., Lund, L. J., & Page, A. L. (1989). Acidity produced by leguminous plants through symbiotic dinitrogen fixation. *J. Environ. Qual, 18*, 529-534. http://dx.doi.org/10.2134/jeq1989.00472425001800040025x

Mclean, E. O. (1982). Soil pH and lime requirement. In A. L. Page, R. H. Miller & D. R. Keeney (Eds.), *Methods of soil analysis, Part 2: Chemical and microbiological Properties* (2nd ed.). ASA, SSA, Madison, Wisconsin, USA.

Mcloughlin, J. J., & Dunican, L. K. (1985). Competition studies with *Rhizobium trijohi* in laboratory experiments. *Plant and Soil, 88*, 139-143. http://dx.doi.org/10.1007/BF02140673

Mohammadi, K., Shohrabi, Y., Gholamreza, H., Khalesro, S., & Majidi, M. (2012). Effective factors on biological nitrogen fixation. *African J. Agricult. Res., 7*(12), 1782-1788. http://dx.doi.org/10.5897/AJARX11.034

Montealegre, C., & Graham, P. H. (1996). Preference in the nodulation of *Phaseolus vulgaris* cv RAB 39.11. Effect of delayed nodulation or cell representation in the inoculant nodule occupancy by *Rhizobium tropici* UMR 1899. *Canadian Journal of Microbiology, 41*(42), 844-850. http://dx.doi.org/10.1139/m96-106

Muller, S. H., & Pereira, P. A. A. (1995). Nitrogen fixation of common bean (*Phaseolus vulgaris* L.) as affected by mineral nitrogen supply at different growth stages. *Plant and Soil, 177*(1), 55-61. http://dx.doi.org/10.1007/BF00010337

Mungai, N. W., & Karubiu, N. M. (2010). Effectiveness of rhizobia isolates from Njoro soils (Kenya) and commercial inoculants in nodulation of common beans (*Phaseolus vulgaris*). *Journal of Agriculture, Science and Technology, 12*(1), 47-59.

Musandu, A. A. O., & Joshua, O. O. (2001). Response of common bean to *Rhizobium* inoculation and fertilizers. *The Journal of Food and Technology in Africa, 6*(4). http://dx.doi.org/10.4314/jftaa.v6i4.19303

Muthamia, J., Kimani, P. M., Chemining'wa, G., & Esilaba, A. O. (2015). Effects of *Rhizobium, Azospirillum,* and *Rhizobium Azospirillum* coinoculation inoculants on climbing beans growth and development in low and high phosphorus sand media. *Proceedings of the 5th Post graduates PhD. Seminar on communication of research findings at the University of Nairobi, Faculty of Agriculture.* Retrieved from http://agriculture.uonbi.ac.ke/sites/default/files/cavs/agriculture/larmat/Proceedings%20of%20the%205th%20PhD%20seminar%20HELD%20on%2018th%20SEPTEMBER%202015.pdf

Mweetwa, A. (2011). Potential for increased yield and market opportunities for beans. *SanBio, 2*(2).

Mweetwa, A. M., Eckhardt, E. A., Stott, D. E., Chilombo, G., Armstrong, A. P., Schulze, D., & Nakatsu, C. H. (2014). Isolation and characterization of Chilembwe and Sinda rock phosphate solubilizing soil microorganisms. *African Journal of Microbiology Research, 8*(34), 3191-3203. http://dx.doi.org/10.5897/AJMR2014.6923

Mweetwa, A. M., Mulenga, M., Mulilo, X., Ngulube, M., Banda, J. S. K., Kapulu, N., & Ngandu, S. H. (2014). Response of cowpea soya beans and groundnuts to non-indigenous legume inoculants. *Sustainable Agriculture Research, 3*(4), 84-95. http://dx.doi.org/10.5539/sar.v3n4p84

Nakitto, A. M., Muyona, J. H., & Nakimbungwe, D. (2015). Effects of combined processing methods on the nutritional quality of beans. *Food Science & Nutrition, 3*(3), 233-241. http://dx.doi.org/10.1002/fsn3.209

Nyau, V. (2014). Nutraceuical perspectives and utilization of common beans (*Phaseolus vulgaris* L.): A review. *African Journal of Food, Agriculture Nutrition Development, 14*(7).

Otieno, P. E., Muthomi, J. W., Cheminig'wa, G. N., & Nderitu, J. H. (2007). Effect of rhizobia inoculation, farmyard manure and nitrogen fertilizer on growth, nodulation and yield of selected grain legumes. *African Crop Science Conference Proceedings, 8*, 305-312.

Panda, S. C. (2011). *Soil management and organic farming.* Jodhpur, India: AGROBIOS.

Pereira, J. L., Queiroz, R. M. L., Charneau, S., Felix, C. R., Ricart, C. A., Lopes da Silva, F., ... Noronha, E. F. (2014). Analysis of Phaseolus vulgaris response to its association with *Trichoderma harzianum* (ALL-42) in the presence or absence of the phytopathogenic fungi *Rhizoctonia solani* and *Fusarium solani.* http://dx.doi.org/10.1371/journal.pone.0098234

Rodino, A. P., Santalla, M., De Ron, A. A. M., & Drevon, J. J. (2005). Variability in symbiotic nitrogen fixation among white landraces of common bean from the Iberian peninsula. *Symbiosis, 40*, 69-78.

Rubio, M. B., Quijada, N. M., Pérez, E., Domínguez, S., Monte, E., & Hermosa, R. (2014). Identifying beneficial qualities of *Trichoderma parareesei* for plants. *Appl. Environ. Microbiol., 80*(6), 1864-1873. http://dx.doi.org/10.1128/AEM.03375-13

Samavat, S., Samavat, S., Mafakheri, S., & Shakour, J. (2012). Promoting common bean growth and nitrogen fixation by the co-inoculation of *Rhizobium* and *Pseudomonas fluorescens* isolates. *Bulgarian Journal of Agricultural Science, 18*(3), 387-395.

Samolski, I., Rincón, A. M., Pinzón, L. M., Viterbo, A., & Monte, E. (2012). The qid74 gene from *Trichoderma harzianum* has a role in root architecture and plant biofertilization. *Microbiology, 158*, 129-138. http://dx.doi.org/10.1099/mic.0.053140-0

Sharma, A., Jadeja, K. B., Kataria, G. K., Ananika, M., & Kumar, D. J. (2014). Plant growth promoting effect of *Trichoderma* on groundnut. *AGRES-An International e-Journal, 3*(4), 324-352.

Singleton, P. W., & Tavares, J. W. (1986) Inoculation response of legumes in relation to the number and effectiveness of indigenous Rhizobium populations. *Appl Environ Microbiol., 51*(5), 1013-1018.

Stajkovic, O., Delic, D., Josic, D., Kuzmanovic, D., Rasulic, N., Knezevic-Vukcevic, J. (2011). Improvement of common bean growth by co-inoculation with *Rhizobium* and plant growth-promoting bacteria. *Romanian Biotechnological Letters, 16*(1), 5919-5926.

Tatarani, G., Dichio, B., & Xiloyannis, C. (2012). Soil fungi-plant interaction. *Advances in selected plant physiology aspects.* Retrieved from http://cdn.intechopen.com/pdfs-wm/35827.pdf

Thomas, G. W. (1982). Exchangeable Cations. In A. L. Page, R. H. Miller & D. R. Keeney (Eds.), *Methods of Soil Analysis, Part 2: Chemical and Microbiological Properties* (2nd ed., pp. 159-165). ASA, SSA, Madison, Wisconsin.

Yadegari, M., & Rahmani, H. A. (2010). Evaluation of bean (*Phaselous vulgaris*) seeds inoculation with *Rhizobium phaseoli* and plant growth promoting Rhizobacteria (PGPR) on yield and yield components. *African Journal of Agricultural Research, 5*(9). http://dx.doi.org/10.3923/pjbs.2008.1935.1939

9

Urea Fertilizer Placement Impacts on Corn Growth and Nitrogen Utilization in a Poorly-Drained Claypan Soil

Frank E. Johnson II[1], Kelly A. Nelson[2] & Peter P. Motavalli[1]

[1] Department of Soil, Environmental and Atmospheric Sciences, University of Missouri, Columbia, Missouri, USA

[2] Division of Plant Sciences, University of Missouri, Novelty, Missouri, USA

Correspondence: Frank E. Johnson II, Department of Soil, Environmental and Atmospheric Sciences, University of Missouri, Columbia, Missouri 65211, USA. E-mail: fej526@mail.missouri.edu

The research is financed by the Missouri Fertilizer and Ag Lime Board.

Abstract

Practices to increase nitrogen (N) use efficiency (NUE) include selecting appropriate N fertilizer sources and application methods, but minimal research has focused on these practices in poorly-drained claypan soils which are prone to N loss. This research assessed the impact of different urea fertilizer placement practices on corn (*Zea mays* L.) production and N utilization in a poorly-drained claypan soil. Field trials were conducted in 2014 and 2015 in Missouri. Treatments consisted of pre-plant deep banding (20 cm) urea at 202 kg N ha^{-1} or urea plus a nitrification inhibitor (NI) (nitrapyrin) compared to pre-plant urea broadcast surface-applied or incorporated to a depth of 8 cm. In 2014, incorporating urea, deep banding urea, and deep banding urea plus NI had higher yields (> 10%) of corn compared to the control with grain yields ranging from 13.73 to 14.05 Mg ha^{-1}. In 2015, grain yields were lower than in 2014, ranging from 4.1 to 7.9 Mg ha^{-1}. Deep placing banded urea with a NI yielded an increase in grain yield up to 48% compared to the other treatments. Rainfall amounts were higher in 2015, which could have resulted in poorer root growth and greater N loss in deep banded treatments. In 2014, deep banding urea with a NI produced the highest NUE. Similar to NUE, silage tissue N concentrations in 2014 were greater with deep banded urea plus NI, while in 2015 silage tissue N concentrations were higher with surface applied urea. The results suggest that urea fertilizer incorporation including deep banding may improve corn grain production, N uptake, and NUE, but response was affected by climatic conditions. The addition of an NI may be an important safeguard when deep banding urea in years with excessive precipitation.

Keywords: nitrogen use efficiency, deep banding, nitrification inhibitor, urea, nitrogen uptake

1. Introduction

1.1 Nitrogen Management of Poorly-Drained Soils

Nitrogen fertilizer management strategies to increase NUE and corn production have focused on enhancing plant N availability at critical growth stages and minimizing environmental N losses. These management strategies have included changes in crop genetic traits (Hirel et al., 2011), N fertilizer sources (Nelson et al., 2008), timings and methods of fertilizer application (Nash et al., 2013), and the spatial placement of N fertilizer in soil (Drury et al., 2006) or across agricultural fields (Motavalli et al., 2012; Roberts et al., 2012).

In poorly-drained soils, such as claypan soils, there is a high potential for N loss through denitrification which can reduce the available N that can be retrieved by the plant and increase loss to the environment (Nash et al., 2012). Claypans usually are situated 20 to 40 cm below the soil surface (Jung et al., 2006) and encompass an area of approximately 4 million ha across Missouri, Illinois, and Kansas in the Midwestern U.S. (Anderson et al., 1990).

Research on a poorly-drained claypan soil in Missouri has indicated that strip tillage and deep banding of N fertilizer to a depth of 15 cm increased corn yields 1.57 to 5.39 Mg ha^{-1} compared to no-till, broadcast surface application of N fertilizer (Nash et al., 2013). Lehrsch et al. (2000) observed a 6% increase in tissue N concentration in corn silage with banded N fertilizer compared to a broadcast N application. This can be attributed to the higher concentration of nutrients within the root zone with banded fertilizer compared to the lower but more uniformly distributed nutrient concentrations resulting from broadcasting the fertilizer (CAST, 2004). Reeves and Touchton (1986) observed that deep banding urea fertilizer in the tillage row produced the highest corn yield over a three-year study.

Nitrification inhibitors can reduce N loss by slowing down the nitrification process, which converts ammonium to nitrite and then nitrate (Pfab et al., 2012). Reductions in nitrification rates can lower environmental nitrate loss by processes such as nitrate leaching and denitrification thereby increasing N availability to plants (Di & Cameron, 2002; Ferguson et al., 2003).

1.2 Research Needs and Objective

Little research has examined the effectiveness of deep banding N fertilizers and combining that deep placement with a NI in poorly-drained soils. In a well-drained soil, Reeves and Touchton (1986) observed no increase in grain yield when a NI was added to deep placed urea at the time of subsoiling. Therefore, it is important to determine in poorly-drained claypan soils whether or not coupling NIs with deep placement of urea can increase crop production as compared to other urea placement practices. The objective of this research was to assess the impact of different urea fertilizer placement practices, including deep banding with and without a NI, on corn production and crop N utilization in a poorly-drained claypan soil.

2. Method

2.1 Site Location and Experimental Design

This study was conducted in 2014 and 2015 in Northeast Missouri at the University of Missouri Greenley Memorial Research Center (40°1'17" N, 92°11'24.9" W) near Novelty on a poorly-drained Putnam silt loam (fine, smectitic, mesic Vertic Albaqualfs). The claypan subsurface layer at this research location has been observed to be as shallow as 31 cm (Nash et al., 2012). Daily precipitation data were obtained from a nearby automated weather station located at the Greenley Center.

Two different field locations were used in 2014 and 2015. Experimental plots were planted with a corn (*Zea mays* L.) hybrid, DeKalb 61-88, in fields with soybean (*Glycine max* L. Merr.) the previous year. Plots were approximately 3 by 61 m and contained four rows of corn. Planting was done with a John Deere 7000 planter (Deere and Co, Moline, IL) on 76 cm row spacing at 79,000 seeds ha^{-1}. Planting and harvest dates as well as other management practices are listed in Table 1. The experimental design was a randomized complete block with five replications. Treatments consisted of a non-treated control, urea deep banded (UDB) at a depth of 20 cm, urea deep banded (20 cm) plus a nitrification inhibitor (UDB+NI), urea incorporated after surface broadcast application (UAA) at a depth of approximately 8 cm, and urea broadcast surface applied after incorporation (USA).

Table 1. Field information and management operations in 2014 and 2015

Year	Field information and management	Date and/or rate
2014	N application	9 May
	Planting date	9 May
Herbicide		13 May
	S-metolachlor[†]	1.48 kg ha^{-1}
	Atrazine[‡]	1.48 kg ha^{-1}
	Mesotrione[§]	0.19 ha^{-1}
	Glyphosate [N-(phosphonomethyl)glycine	0.87 kg ha^{-1}
	Topramezone[¶]	11 Jun/0.05 kg ha^{-1}
Fungicide		10 Jul
	Propiconazole[#]	0.11 kg ha^{-1}
	Azoxystrobin[††]	0.13 kg ha^{-1}
	Harvest Date	8 Oct.
2015	N application	30 Apr.
	Planting date	30 Apr.
Herbicide		2 Jun.
	S-metolachlor	1.19 kg ha^{-1}
	Glyphosate [N-(phosphonomethyl)glycine	1.19 kg ha^{-1}
	Mesotrione	0.12 kg ha^{-1}
	Harvest Date	22 Sept.

Note. [†] (2-Chloro-N-(2-ethyl-6-methylphenyl)-N-[(1S)-2-methoxy-1-methylethyl]acetamide).

[‡] (2-Chloro-4-ethylamino-6-isopropylamino-s-triazine).

[§] (2-[4-(Methylsulfonyl)-2-nitrobenzoyl]cyclohexane-1,3-dione).

[¶] (2-methyl-4-(methylsulfonyl)phenyl](5-hydroxy-1-methyl-1H-pyrazol-4-yl)methanone).

[#] (1-[[2-(2,4-dichlorophenyl)-4-propyl-1,3-dioxolan-2-yl]methyl]-1,2,4-triazole).

[††] (Methyl (2E)-2-(2-{[6-(2-cyanophenoxy)pyrimidin-4-yl]oxy}phenyl)-3-methoxyacrylate).

All urea treatments were applied at 202 kg N ha^{-1}. The NI (Instinct II®, Dow AgroSciences, Indianapolis, IN) was applied at 2.7 L ha^{-1}. The active ingredient in the NI was nitrapyrin (2-chloro-6-(trichloromethyl) pyridine).

The deep banded N fertilizer treatments were applied using a custom-designed strip-till conservation C-jet unit (Advance, MO). The UDB and UDB+ NI treatments were banded to a depth of 20 cm below the planted row using a Montag (Montag Manufacturing, Inc., Emmetsburg, IA) dry fertilizer air delivery system. After the banded fertilizer application, the entire soil surface was surface-tilled with a field cultivator (John Deere 1000, Moline, IL) to remove the possible physical effects of strip tillage on crop response among treatments.

2.2 Soil Sampling and Analytical Procedures

The initial soil properties for each year (Table 2) were determined from analysis of soil samples taken at 0-15, 15-30, and 30-45 cm depths from the non-treated control plots in all the five replicates. Samples were taken using a stainless steel push probe with four subsamples per plot composited to make one sample. All soil samples were air-dried and ground to pass through a sieve with 2 mm openings. The initial soil samples were analyzed by the University of Missouri Soil and Plant Testing Laboratory using standard soil testing procedures (Nathan et al., 2006). Soil bulk density was determined using the core method (Blake et al., 1986).

Table 2. Means (± 1 standard deviation) of selected initial soil properties in 2014 and 2015

Soil properties	Depth (cm)		
	0-15	15-30	30-45
2014			
pH (0.01 M CaCl$_2$)	6.7 ± 0.1	6.0 ± 0.4	5.0 ± 0.2
Neut. acidity (cmol$_c$ kg^{-1})	0.4 ± 0.2	2.4 ± 0.9	7.0 ± 1.5
Organic matter (g kg^{-1})	28.2 ± 2.8	21.4 ± 1.8	22.4 ± 1.9
Bray I P (kg ha^{-1})	86 ± 14	17 ± 3	17 ± 12
Exc. Ca (kg ha^{-1})	5509± 500	4985 ± 502	6182 ± 375
Exch. Mg (kg ha^{-1})	404 ± 102	518 ± 189	1205 ± 206
Exch. K (kg ha^{-1})	325± 41	167 ± 13	225 ± 23
CEC (cmol$_c$ kg^{-1})	14.6 ± 1.5	15.6 ± 2.3	25.6 ± 2.9
Bulk density (g cm^{-3})	1.02 ± 0.09	1.24 ± 0.10	1.34 ± 0.05
Total N (g kg^{-1})	1.80 ± 0.10	1.33 ± 0.21	1.18 ± 0.07
Total organic C (g kg^{-1})	19.16 ± 2.37	11.54 ± 0.94	10.70 ± 1.08
2015			
pH (0.01 M CaCl$_2$)	6.1± 0.1	5.9 ± 0.2	4.8 ± 0.3
Neut. acidity (cmol$_c$ kg^{-1})	1.2 ± 0.3	1.8 ± 0.6	6.4 ± 2.0
Organic matter (g kg^{-1})	25.8 ± 1.3	17.8 ± 2.6	20.6 ± 3.7
P Bray I (kg ha^{-1})	79 ± 41	9 ± 5	8 ± 1
Exch. Ca (kg ha^{-1})	4111 ± 2571	4224 ± 658	3851 ± 700
Exch. Mg (kg ha^{-1})	372 ± 40	358 ± 114	521 ± 140
Exch. K (kg ha^{-1})	306 ± 28	105 ± 16	104 ± 15
CEC (cmol$_c$ kg^{-1})	12.1 ± 0.9	12.7 ± 2.4	17.1 ± 4.1
Bulk density (g cm^{-3})	1.11 ± 0.02	1.30 ± 0.07	1.52 ± 0.03
Total N (g kg^{-1})	1.18 ± 0.65	0.92 ± 0.06	0.87 ± 0.13
Total organic C (g kg^{-1})	16.03 ± 1.33	8.38 ± 0.87	7.53 ± 1.64

2.3 Plant Tissue Characterization and Yield Determinations

Stand counts for the two inner rows of each plot were recorded each year over a row length of 15.2 m. Corn silage was harvested at physiological maturity by cutting plants from 3.1 m of one corn row and obtaining the total wet weight of the plants using a hanging scale. These plants were then chopped using a mechanized brush chopper (Vermeer Corp., BC700XL, Pella, IA) and subsamples were collected to determine silage moisture for adjusting calculated silage yield on a dry weight basis and for tissue analysis. Silage samples were dried at 70 °C and ground in a Wiley-Mill (Swedesboro, NJ) to pass through a 1 mm sieve.

Total N in the tissue was analyzed using the combustion method with a total carbon-nitrogen analyzer (LECO Corp., Township, Michigan). Tissue N concentration and silage yield were used to calculate N uptake and subsequently the apparent N recovery efficiency (Baligar et al., 2001). Apparent N recovery efficiency is one measure of N use efficiency (NUE) and was calculated using the equation:

$$\text{NUE (\%)} = ((\text{N uptake}_{(treatment)} - \text{N uptake}_{(control)})/\text{N applied}) \times 100 \qquad (1)$$

Corn grain yields were determined using a plot combine (Wintersteiger Delta, Salt Lake City, UT) that harvested the two center rows in each plot. All grain yields were adjusted to 150 g kg^{-1} moisture. Grain samples were collected from each plot for determination of test weight, grain moisture (Harvest Master, Logan, UT), and analysis of starch, oil, and protein concentration (Foss 1241 Infratec, Eden Prairie, MN).

2.4 Statistical Analysis

Analysis of variance (ANOVA) was performed using PROC GLM on grain yields, grain moisture, N uptake, and NUE for each year using the Statistical Analysis Software (SAS) v9.4 (SAS Institute, 2014). Fischer's Protected LSD at $P < 0.10$ was used to separate means.

3. Results

3.1 Precipitation

Cumulative rainfall amounts for the 2014 and 2015 growing seasons were 705 and 839 mm, respectively (Figures 1A and 1B). The average ten-year precipitation amount from 2000 to 2010 during the growing season (from April through September) was 660 mm (Nash et al., 2012). The cumulative rainfall during the 2014 growing season exceeded this ten-year average by 45 mm (7%). In 2015, the cumulative rainfall exceeded the ten year average by 179 mm (27%) and was especially high at the beginning of the growing season resulting in visually observed soil waterlogging.

3.2 Corn Plant Populations and Silage and Grain Yields

There were no significant treatment effects on observed corn plant population (data not shown). The mean plant population was 69,240 plants ha^{-1} in 2014 and 75,230 plants ha^{-1} in 2015. Grain yields due to the treatments varied between 2014 and 2015, possibly due to differences in rainfall distribution since 2015 had a very wet spring (Figure 2). In 2014, corn grain yields ranged from 12.1 to 14.1 Mg ha^{-1} (Figure 2). Deep banded urea with and without a NI provided a 9 and 8% increase in yield, respectively, compared to the USA treatment. The UAA treatment had a 7% increase compared to the USA treatment.

In 2015, grain yields were much lower for all treatments compared to 2014 (Figure 2). Grain yields in 2015 ranged from 4.1 to 7.9 Mg ha^{-1} (Figure 2). As expected, the NTC treatment resulted in the lowest yield at 4.1 Mg ha^{-1}. The UDB+NI treatment increased yields compared to UDB by approximately 30%. The USA treatment produced grain yields of 7.4 Mg ha^{-1} which was similar to the UAA treatment at 6.5 Mg ha^{-1}.

In 2014, silage yields ranged from 43 Mg ha^{-1} in the NTC to approximately 55 Mg ha^{-1} for UDB+NI (Figure 3). All treatment means were significantly greater than the NTC treatment, but none of the fertilizer placement treatments were significantly different from each other in silage yield (Figure 3).

Silage yields were lower in 2015 compared to 2014 (Figure 3). In 2015, silage yields ranged from 18 Mg ha^{-1} in the NTC to 27 Mg ha^{-1} in the USA (Figure 3). Surface applying urea during wetter soil conditions increased silage yields 4 to 22% compared to the other N placement treatments. Mean silage yield for USA was significantly higher than for NTC and UDB, but was not significantly different from the UAA and UDB+NI plots.

3.3 Plant N

In 2014, plant N concentrations in the corn silage tissue ranged from 0.73% in the NTC to 1.21% in the UDB+NI (Table 3). The mean silage tissue N concentration for UDB+NI was significantly higher than the USA and UAA treatments. Every treatment except for UAA had tissue N concentrations that were significantly greater than that of the NTC treatment.

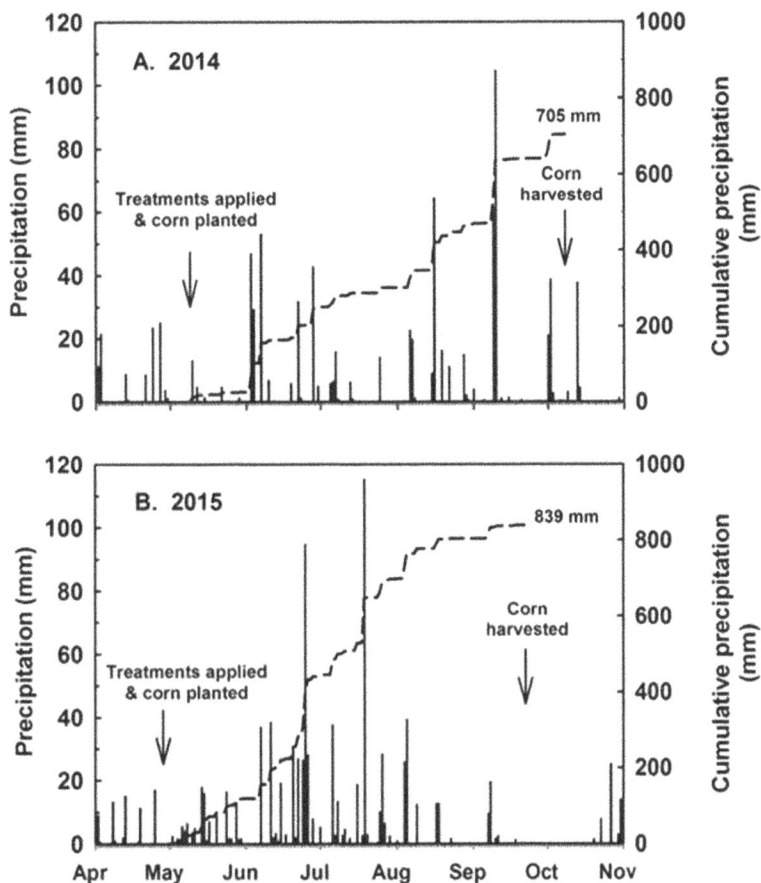

Figure 1. A&B. Daily (bars) and cumulative precipitation (line) in (A) 2014 and (B) 2015

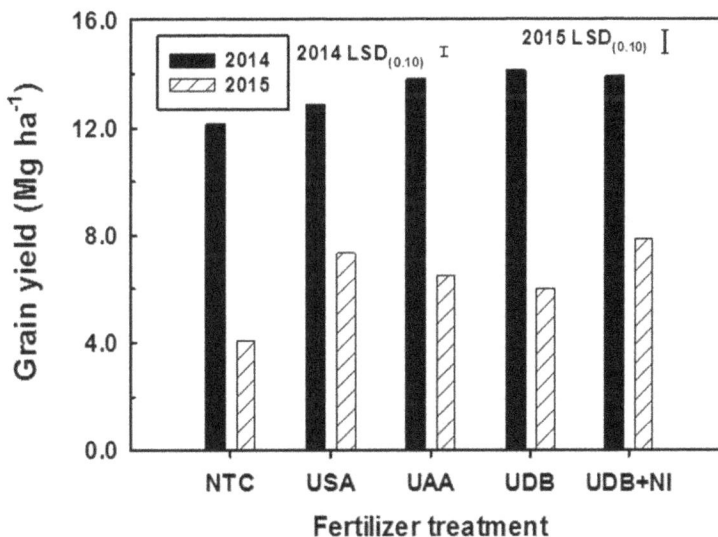

Figure 2. Grain yield response to urea placement and inclusion of a nitrification inhibitor in 2014 and 2015

Note. NTC = non-treated control; USA = urea, surface-applied; UAA = urea incorporated after surface broadcast application; UDB = urea deep banding; UDB + NI = urea deep banding plus nitrification inhibitor. Vertical bar indicate $LSD_{(0.10)}$ values for individual years.

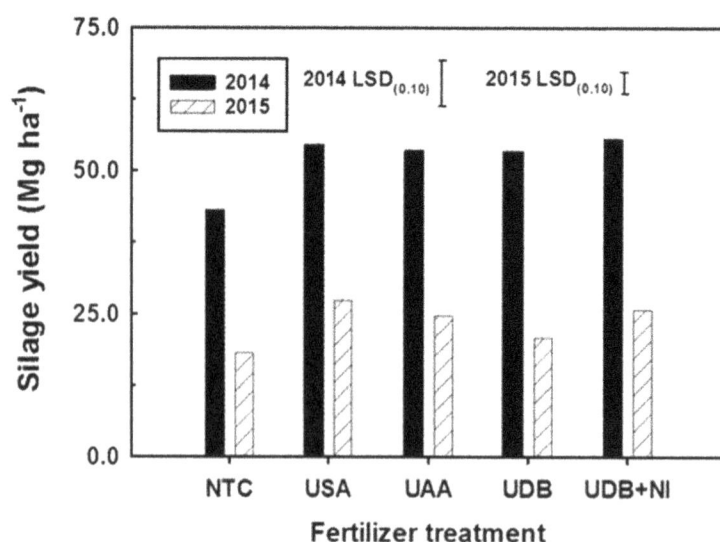

Figure 3. Silage yield response to urea placement and inclusion of a nitrification inhibitor in 2014 and 2015

Note. NTC = non-treated control; USA = urea, surface-applied; UAA = urea incorporated after surface broadcast application; UDB = urea deep banding; UDB + NI = urea deep banding plus nitrification inhibitor. Vertical bar indicate $LSD_{(0.10)}$ values for individual years.

Table 3. Silage tissue N concentration, N uptake and apparent N recovery efficiencies in response to urea placement and inclusion of a nitrification inhibitor in 2014 and 2015

Treatment	Tissue N concentration	Nitrogen Uptake	Apparent N Recovery Efficiency
	--------------- % ---------------	------- kg N ha^{-1} -------	------------------ % --------------------
2014			
NTC[†]	0.73	177	-
USA	0.96	278	47
UAA	0.92	267	44
UDB	1.03	278	50
UDB+NI	1.21	357	79
$LSD_{(0.10)}$[‡]	0.20	80	NS
2015			
NTC	0.70	62	-
USA	1.33	167	52
UAA	0.75	102	20
UDB	0.77	75	7
UDB+NI	0.77	80	9
$LSD_{(0.10)}$	NS	83	NS

Note. [†]NTC = non-treated control; USA = urea, surface-applied; UAA = urea incorporated after surface broadcast application; UDB = urea deep banding; UDB + NI = urea deep banding plus nitrification inhibitor; NS = not significant.

[‡]Fishers protected least significant difference at $P \leq 0.10$; NS = Not significant.

Mean tissue N concentrations in silage for the 2015 season ranged from 0.70% in the NTC to 1.34% in the USA (Table 3). Mean N concentration for the USA treatment was at least 43% greater than all other treatments; however, all of the means were not significantly different.

The UDB+NI treatment had 31 to 33% higher N uptake than USA and UAA, respectively (Table 3). However, significantly higher N uptake was observed in the USA treatment compared to the UDB and UDB+NI treatments in 2015. In 2014, N uptake ranged from 177 kg N ha^{-1} in the NTC treatment to 357 kg N ha^{-1} in the UDB+NI treatment. Nitrogen uptake values for the USA, UAA, and UDB treatments were 278, 267, and 278 kg N ha^{-1}, respectively. There were no significant differences between either of the deep banded urea treatments. However, the addition of a NI resulted in a 22% increase in N uptake compared to UDB alone.

Nitrogen uptake for each treatment was lower in 2015 compared to 2014 (Table 3). In 2015, N uptake ranged from 62 kg N ha^{-1} in the NTC to 167 kg N ha^{-1} for USA (Table 3). Surface-applying urea (USA) resulted in a 52% increase in N uptake compared to UDB+NI, and a 55% increase in N uptake compared to UDB. No differences in N uptake were observed between the UAA, UDB+NI, UDB, and NTC treatments.

Calculations of apparent N recovery efficiency, which is a measure of N use efficiency, showed no significant differences among the treatments in 2014 and 2015 (Table 3). In 2014, apparent recovery efficiency ranged from 44 to 79% and in 2015 from 7 to 52%. The relatively lower N use efficiency in 2015, especially with the UAA, UDB and UDB+NI treatments is consistent with the lower N uptake observed with those treatments that included deeper fertilizer placement during a year with excessively high precipitation (Figure 1B).

3.4 Grain Moisture and Quality

Corn grain moisture was also significantly different between 2014 and 2015, which could have been due to differences in rainfall distribution and possible N fertilizer effects on plant maturity (Table 4). In 2014, the overall average grain moisture content for the experimental field was 196.6 g kg^{-1}, while in 2015, the average moisture content for the experimental field was 104.6 g kg^{-1}. In 2014, grain from the UAA plots contained the lowest moisture content (192.4 g kg^{-1}), while grain from UDB+NI contained the highest moisture content (200.4 g kg^{-1}) (Table 4). Grain moistures were not significantly different between UDB and UDB+NI plots, but were significantly different from other treatments. The means from the NTC, USA, and UAA were not significantly different from each other.

Table 4. Selected corn grain quality characteristics in response to urea placement and inclusion of a nitrification inhibitor in 2014 and 2015

Treatment	Corn Grain Content			Grain Moisture
	Starch	Protein	Oil	
	-- g kg^{-1} --			
2014				
NTC[†]	738.4	71.4	40.0	195.8
USA	730.4	86.6	39.4	194.0
UAA	727.6	86.6	40.8	192.4
UDB	729.2	88.0	39.2	200.2
UDB+NI	725.8	88.4	40.9	200.4
LSD$_{(0.10)}$	5.4	2.8	NS	3.4
2015				
NTC	745.0	61.2	38.6	90.0
USA	742.0	62.0	39.4	97.2
UAA	743.8	60.2	38.7	102.6
UDB	736.0	64.8	41.4	113.0
UDB+NI	738.8	61.2	41.2	120.4
LSD$_{(0.10)}$[‡]	6.8	NS	2.4	11.4

Note. [†]NTC = non-treated control; USA = urea, surface-applied; UAA = urea incorporated after surface broadcast application; UDB = urea deep banding; UDB + NI = urea deep banding plus nitrification inhibitor.

[‡]Fisher's protected least significant difference at $P \leq 0.10$; NS = not significant.

Grain moisture content in 2015 ranged from 90.0 g kg^{-1}, in the NTC treatment to 120.4 g kg^{-1} in the UDB+NI treatment (Table 4), which was much lower than what was observed in the 2014 season. The grain moisture content resulting from the UDB, USA, and UAA treatments were 113.0, 97.2, and 102.6 g kg^{-1}, respectively. The UDB+NI treatment was not statistically significantly different from the UDB treatment; however, it was greater than the UAA, USA, and NTC.

Treatment and year were significant factors influencing grain starch and protein concentrations over the two growing seasons (Table 4). There also were significant interactions between the two seasons concerning protein concentrations (P = 0.0005). A higher cumulative rainfall amount in 2015 (Figure 1B) can explain the influence year has on grain quality and suggests that N fertilizer placement can react differently depending on the growing season conditions and precipitation. Starch concentrations in corn grain from the 2014 growing season ranged from 725.8 g kg^{-1} in the UDB+NI plots to 738.4 g kg^{-1} in the NTC treatment (Table 4). All treatments were significantly lower than the NTC treatment by at least 10 g kg^{-1}, but were not significantly different from each other. In 2015, mean starch concentrations ranged from 736.0 g kg^{-1} in the UDB treatment to 745.0 g kg^{-1} in the NTC treatment. The UDB, UDB+NI and USA treatments had similar grain starch concentrations. However, there were significant differences between the two years. Mean starch concentrations were greater in 2015 than in 2014.

Protein concentrations in 2014 ranged from 71.4 g kg^{-1} in the NTC to 88.4 g kg^{-1} in the UDB+NI (Table 4). All treatments had protein concentrations that were significantly higher than the NTC treatment, but the fertilizer placement treatments were not significantly different from each other. In 2015, no significant differences among protein concentrations were observed. Oil concentration was similar among treatments in 2014. However, oil concentration was greatest with UDB, UDB+NI, and USA.

4. Discussion

Deep banding N fertilizers with or without a NI has been observed to reduce N loss and increase NUE possibly due to fertilizer placement in closer proximity to roots and reduction in N loss due to runoff and other possible N loss processes. For example, Reeves and Touchton (1986) observed that deep banding N fertilizer with strip tillage produced the highest grain yields compared to surface-applied urea. A similar result of higher corn yields with the combined use of strip tillage and deep banded urea compared to a no-till surface urea application was observed in poorly-drained claypan soils (Nash et al., 2013). These corn grain yield results were similar to the 2014 results found in this study in which the UAA, UDB and UDB+NI fertilizer placement treatments resulted in higher grain yields than that of the USA treatment. In contrast to other studies, the effect of tillage on yield response was not a factor. However, this pattern was not observed in 2015, possibly because of the increased rainfall in the 2015 growing season (124 mm of rainfall more than in 2014) especially in the early part of the growing season. During the 2015 growing season, the USA treatment had higher grain production. These results indicate that the saturated soil conditions in the beginning of the growing season in 2015 may have limited the growth of corn roots with depth so that roots were unable to absorb the deep banded fertilizer N, but were able to obtain the surface and shallow-placed N fertilizer. Under these conditions, the use of an NI was also effective for increasing yields with the deep banded urea fertilizer placement treatment possibly due to reduced nitrate leaching.

Lehrsch et al. (2000) observed at least a 5% increase in corn grain yield from banded N fertilizer compared to that of broadcast fertilizer which was similar to the results in this study. In 2014, the UDB and UDB+NI treatments produced grain yields that were 10% and 8% more than the USA treatment, respectively. In 2015, UDB+NI yielded 7% more than the USA treatment. Furthermore, Randall et al. (2003) observed a 5% increase in grain yield when nitrapyrin was utilized in corn production. These results were also further supported by Mengel et al. (1982) who observed an increase in grain yield when N fertilizer was injected below the soil surface compared to surface applied N sources. Grant et al. (2001) also observed that banding fertilizer increased grain yield. In wheat, Blackshaw et al. (2004) observed a 12% increase when banding ammonium nitrate compared to broadcast ammonium nitrate during spring application. It is possible that in their study, banding N fertilizer reduced N loss by providing more opportunity for N uptake due to the proximity of the fertilizer to the plant roots.

Even though grain yields were higher in 2014, corn population was lower compared to 2015. This result was unexpected since the 2015 growing season was much wetter than in the 2014 growing season. With the wetter spring in 2015, there was a higher probability for poor seed germination and stand establishment, but this result based on final corn plant population was not observed. However, corn plant populations for both 2014 and 2015 were higher than the average population of 56,000 plants ha^{-1}, which North American farmers have utilized to optimize profit (Williams II, 2012).

Multiple factors may affect corn grain moisture content including the time of harvest, climate, the tillage system used, corn hybrid grown, and application of banded starter fertilizer (Vetsch & Randall, 2000; Wolkowski, 2000; Vetsch & Randall, 2002). Effects of starter fertilizer banding placement on lowering corn grain moisture when they occur are often attributed to a more rapid early season corn growth and rate of maturation with banding at planting (Beegle et al., 2007). Greater plant N availability also tends to promote more rapid rates of corn maturation thereby resulting in relative lower grain moisture content. Reduced grain water content is favored by corn growers due to lower costs associated with grain drying (Wolkowski, 2000).

In this research, grain moisture content did not directly relate to N uptake resulting from the N fertilizer treatments and was generally lower with deep banding compared to surface or shallow incorporation. From an economic standpoint, the lower grain moisture in 2015 would have been cheaper to dry for farmers. A possible reason the grain moisture was much lower is because of the interaction of rainfall distribution and the effects of N on corn maturation. The dry conditions experienced in the late summer in 2015 could also have increased the rate of grain drying.

Nitrogen fertilizer placement methods did not consistently affect silage yields and this response appeared to be affected by differences in precipitation during the two years of this research. Lehrsch et al. (2000) observed a 5 to 26% increase in silage yields for banded fertilizer compared to when the N fertilizer was surface broadcast-applied. Their research results may differ from this research because their research was primarily conducted under irrigated and not rain-fed conditions and the soils in this research were poorly-drained. Greater yield response to banded starter fertilizer under irrigated versus rain-fed conditions has also been observed by Wortmann et al. (2006). The difference in silage yields among the two years of this research could be explained by the increased early season and cumulative rainfall amount in the 2015 growing season compared to that of 2014 which favored surface N fertilizer application.

Nitrogen concentration in the corn silage was not affected by treatment or year. In accordance with these results, Nash et al. (2013) compared corn treatments of no-till/surface broadcast N fertilizer versus strip-till/deep banded N fertilizer and observed minimal differences in earleaf N concentrations among these two treatments that differed in both tillage and N fertilizer placement.

In 2014, UDB and UDB+NI increased N uptake and N use efficiency. This was similar to the results of Lehrsch et al. (2000) in which they observed a 6% increase in N uptake in banded treatments compared to surface broadcasted treatments. Takahashi et al. (1991) reported that deep placed N fertilizer increased NUE by 22% compared to top-dress applied N in fields grown under soybean. These studies suggested that combining multiple management practices can increase NUE. The 2015 results did not follow the same pattern as 2014. In 2015, the USA treatment produced the highest NUE compared to UDB and UDB+NI. The 2015 results were opposite of the findings of Moraghan et al. (1984), who observed greater plant N amounts (4%) for banded urea compared to surface-applied or incorporated urea. In their study, sorghum treated with banded urea was efficient in recovering N almost 18% more than sorghum treated with surface-applied or incorporated urea. These observations could explain why they observed an increase in grain yield production, which was similar to the 2014 results in this study. The difference in response among the two years for the study could have occurred because this study was performed on poorly-drained claypan soils, which are susceptible to gaseous N loss (Nash et al., 2012). This explanation can further be supported by the fact that cumulative rainfall during 2015 was greater and more intense early in the growing season which may have inhibited deep root growth as compared to the conditions experienced in 2014.

The addition of a NI to deep banded urea resulted in a 23% increase in N uptake in 2014 and a 6% increase in 2015. These findings are similar to that of Randall et al. (2003) in which they observed a 23% increase in N uptake in corn with the addition of nitrapyrin to anhydrous ammonia. Even though their study was conducted in the autumn months, these results were similar to this study in which fertilizer was applied prior to planting. In 2015, there was also a significant increase (52%) in N uptake in the USA treatment compared to UDB and UDB+NI. Higher cumulative rainfall during the 2015 growing season could have made deep banding N treatments less effective due to poor early root growth and development. Practicing multiple management practices including applying deep banded urea with a NI has the ability to provide many benefits to crop production including an increase in N uptake and NUE.

Some variability in NUE values were observed in 2014 and a high NUE was observed in 2014 for UDB+NI. In general, NUE was highest for the UDB+NI treatment in 2014 and for the USA treatment in 2015 illustrating the interactive effect of climatic conditions on response to N fertilizer placement.

The higher NUE in 2014 provides an explanation for the higher grain protein concentrations in 2014 because protein content is increased by N application when N is not the most limiting factor of growth (Gauer et al., 1992). This is further supported by decreased NUE in 2015. Treatments did not influence corn grain oil content in 2014, but were a significant factor in 2015. The 2014 results are similar to Miao et al. (2006), in which they observed that N fertilization did not heavily influence grain oil content. Corn grain oil results in 2015 could have been influenced by increased precipitation, environmental N loss, and decreased NUE. However, these results suggest there are other factors not observed in this study influencing oil concentration.

5. Conclusions

These results indicate that the effectiveness of different N fertilizer placement strategies for corn production in a poorly-drained claypan soil was influenced by climatic conditions. Deep-banding urea fertilizer was more effective in a relatively-well distributed high rainfall year, but a surface urea application increased corn grain yields and N uptake in a high rainfall year which had excessive precipitation early in the growing season. We speculate that this difference in response was due to the effects of excessive early season rainfall on inhibiting root growth deeper into the soil. Deep-banded urea with a NI may further benefit crop production due to the delay in nitrification. Results of this study suggest that combining multiple management practices in poorly-drained claypan soils, such as deep banding and use of a NI, could increase crop production under certain climatic conditions. In areas where high cumulative rainfall limits root growth, deep banding N fertilizer may not be as effective and, if deep banding is utilized, addition of a NI with the deep banded fertilizer may be an important safeguard to limit corn yield losses.

References

Anderson, S. H., Gantzer, C. J., & Brown, J. R. (1990). Soil physical properties after 100 years of continuous cultivation. *J. Soil Water Conserv., 45*, 117-121.

Baligar, V. C., Fageria, N. K., & He, Z. L. (2001). Nutrient use efficiency in plants. *Comm. Soil Sci. Plant Analysis, 32*, 921-950. http://dx.doi.org/10.1081/CSS-100104098

Beegle, D. B., Roth, G. W., & Lingenfelter, D. D. (2007). Starter fertilizer. *Agronomy Facts 51*. Penn State Extension, State College, PA.

Blackshaw, R. E., Molnar, L. J., & Janzen, H. H. (2004). Nitrogen fertilizer timing and application method affect weed growth and competition with spring wheat. *Weed Sci., 52*(4), 614-622. http://dx.doi.org/10.1614/WS-03-104R

Blake, G. R., & Hartge, K. H. (1986). Bulk density. In A. Klute (Ed.), *Methods of soil analysis: Part 1. Physical and mineralogical methods* (pp. 363-375). Madison, WI: Soil Sci. Soc. of Am.

CAST. (2004). Council for Agricultural Science and Technology. *Climate change and greenhouse gas mitigation, challenges and opportunities for agriculture*. K. Paustian and B. Babcock (Cochairs). Report 141.

Di, H. J., & Cameron, K. C. (2002). The use of a nitrification inhibitor, dicyandiamide (DCD), to decrease nitrate leaching and nitrous oxide emissions in a simulated grazed and irrigated grassland. *Soil Use and Manag., 18*, 395-403. http://dx.doi.org/10.1111/j.1475-2743.2002.tb00258.x

Drury, C. F., Reynolds, W. D., Tan, C. S., Welacky, T. W., Calder, W., & McLaughlin, N. B. (2006). Emissions of nitrous oxide and carbon dioxide: influence of tillage type and nitrogen placement depth. *Soil Sci. Soc. Am. J., 70*, 570-581. http://dx.doi.org/10.2136/sssaj2005.0042

Ferguson, R. B., Lark, R. M., & Slater, G. P. (2003). Approaches to management zone definition for use of nitrification inhibitors. *Soil Sci. Soc. Am. J., 67*, 937-947. http://dx.doi.org/10.2136/sssaj2003.0937

Gauer, L. E., Grant, C. A., Gehl, D. T., & Bailey, L. D. (1992). Effects of nitrogen fertilization on grain protein content, nitrogen uptake, and nitrogen use efficiency of six spring wheat (*Triticum aestivum* L.) cultivars, in relation to estimated moisture supply. *Can. J. Plant Sci., 72*, 235-241. http://dx.doi.org/10.4141/cjps92-026

Grant, C. A., Brown, K. R., Racz, G. J., & Bailey, L. D. (2001). Influence of source, timing, and placement of nitrogen on grain yield and nitrogen removal of durum wheat under reduced and conventional-tillage management. *Can. J. Plant Sci., 81*, 17-27. http://dx.doi.org/10.4141/P00-091

Hirel, B., Tétu, T., Lea, P. J., & Dubois, F. (2011). Improved nitrogen use efficiency in crops for sustainable agriculture. *Sustainability, 3*, 1452-1485. http://dx.doi.org/10.3390/su3091452

Jung, W. K., Kitchen, N. R., Sudduth, K. A., & Anderson, S. H. (2006). Spatial characteristics of claypan soil properties in an agricultural field. *Soil Sci. Soc. Am. J., 70*, 1387-1397. http://dx.doi.org/10.2136/sssaj2005.0273

Lehrsch, G. A., Sojka, R. E., & Westermann, D. T. (2000). Nitrogen placement, row spacing, and furrow irrigation water positioning effects on corn yield. *Agron. J., 92*, 1266-1275. http://dx.doi.org/10.2134/agronj2000.9261266x

Mengel, D. B., Nelson, D. W., & Huber, D. M. (1982). Placement of nitrogen fertilizer for no-till and conventional till corn. *Agron. J., 74*, 515-518. http://dx.doi.org/10.2134/agronj1982.00021962007400030026x

Miao, Y., Mulla, D. J., Robert, P. C., & Hernandez, J. A. (2006). Within-field variation in corn yield and grain quality responses to nitrogen fertilization and hybrid selection. *Agron. J., 98*, 129-140. http://dx.doi.org/10.2134/agronj2005.0120

Moraghan, J. T., Rego, T. J., & Buresh, R. J. (1984). Labeled nitrogen fertilizer research with urea in the semi-arid tropics. *Plant and Soil., 82*, 193-203. http://dx.doi.org/10.1007/BF02220246

Motavalli, P. P., Nelson, K. A., & Bardhan, S. (2012). Development of a variable source nitrogen fertilizer management strategy using enhanced efficiency nitrogen fertilizers. *Soil Sci., 177*, 708-718.

Nash, P. R., Motavalli, P. P., & Nelson, K. A. (2012). Nitrous oxide emissions from claypan soils due to nitrogen fertilizer source and tillage/fertilizer placement practices. *Soil Sci. Soc. Am. J., 76*, 983-993. http://dx.doi.org/10.2136/ssaj2011.0296

Nash, P. R., Nelson, K. A., & Motavalli, P. P. (2013). Corn yield response to polymer- and non-coated urea placement and timings following red clover. *Int. J. Plant Prod., 7*(3), 373-392.

Nathan, M., Stecker, J., & Sun, Y. (2006). *Soil testing in Missouri: A guide for conducting soil tests in Missouri.* Univ. of Missouri, Columbia, MO. Retrieved September 26, 2013, from http://soilplantlab.missouri.edu/soil/ec923.pdf

Nelson, K. A., Scharf, P. C., Bundy, L. G., & Tracy, P. (2008). Agricultural management of enhanced efficiency fertilizers in the north-central United States. *Crop Management.* http://dx.doi.org/10.1094/CM-2008-0730-03-RV

Pfab, H., Palmer, I., Buegger, F., Fiedler, S., Muller, T., & Ruser, R. (2012). Influence of a nitrification inhibitor and of placed N-fertilization on N_2O fluxes from a vegetable cropped loamy soil. *Ag. Ecosys. Env., 150*, 91-101. http://dx.doi.org/10.1016/j.agee.2012.01.001

Randall, G. W., Vetsch, J. A., & Huffman, J. R. (2003). Corn production on a subsurface-drained mollisol as affected by time of nitrogen application and nitrapyrin. *Agron. J., 95*, 1213-1219. http://dx.doi.org/10.2134/agronj2003.1213

Reeves, D. W., & Touchton, J. T. (1986). Subsoiling for nitrogen applications to corn grown in a conservation tillage system. *Agron. J., 78*, 921-926. http://dx.doi.org/10.2134/agronj1986.00021962007800050035x

Roberts, D. F., Ferguson, R. B., Kitchen, N. R., Adamchuk, V. I., & Shanahan, J. F. (2012). Relationships between soil-based management zones and canopy sensing for corn nitrogen management. *Agron. J., 104*, 119-129. http://dx.doi.org/10.2134/agronj2011.0044

SAS Institute. (2014). *SAS 9.4.* SAS Inst., Cary, NC.

Takahashi, Y., Chinushi, T., Nagumo, Y., Nakano, T., & Ohyama, T. (1991). Effect of deep placement of controlled release nitrogen fertilizer (coated urea) on growth, yield, and nitrogen fixation of soybean plants. *Soil Sci. Plant Nutr., 37*(2), 223-231. http://dx.doi.org/10.1080/00380768.1991.10415032

Vetsch, J. A., & Randall, G. W. (2000). Enhancing no-tillage systems for corn with starter fertilizers, row cleaners, and nitrogen placement methods. *Agron. J., 92*, 309-315. http://dx.doi.org/10.2134/agronj2000.922309x

Vetsch, J. A., & Randall, G. W. (2002). Corn production as affected by tillage system and starter fertilizer. *Agron. J., 94*, 532-540. http://dx.doi.org/10.2134/agronj2002.5320

Williams, II, W. M. (2012). Agronomics and economics of plant population density on processing sweet corn. *Field Crops Research, 128*, 55-61. http://dx.doi.org/10.1016/j.fcr.2011.12.007

Wolkowski, R. P. (2000). Row-placed fertilizer for maize grown witn an in-ro crop residue management system in souther Wisconsin. *Soil and Till. Res., 54*, 55-62. http://dx.doi.org/10.1016/S0167-1987(99)00114-2

Wortmann, C. S., Xerinda, S. A., Mamo, M., & Shapiro, C. A. (2006). No-till row crop response to starter fertilizer in Eastern Nebraska: I. Irrigated and rainfed corn. *Agron. J., 98*, 156-162. http://dx.doi.org/10.2134/agronj2005.0015

10

Ion Exchange Resin Membrane Sensitivity Analysis of Selected Parameters for Soil Nutrient Extraction

Plinio L. Kroth[1], Clesio Gianello[1], Leandro Bortolon[2], Jairo A. Schlindwein[3] & Elisandra S. O. Bortolon[2]

[1] Soil Science Department, Federal University of Rio Grande do Sul, Porto Alegre, Brazil

[2] Embrapa Pesca e Aquicultura, Palmas, Brazil

[3] Soil Science Department, Federal University of Rondônia, Porto Velho, Brazil

Correspondence: Leandro Bortolon, Embrapa Pesca e Aquicultura, Palmas, TO, Brazil.
E-mail: Leandro.bortolon@embrapa.br

The research is financed by CAPES, UFRGS Soil Testing Laboratory.

Abstract

The ion exchange resin method has received considerable attention as an alternative soil test method to evaluate plant available nutrients. This study sought to investigate the effect of changes to the resin standard method in the capacity of the resin to extract soil P, K, Ca, and Mg in soils with different texture. We tested the following modifications: soil grinding levels (< 2 mm; < 0.3 mm), shaking time (8 h; 16 h; 24 h), reciprocation level (12.5 rpm; 25 rpm; 50 rpm), solution and elution saturation concentration (0.25/0.25 mol L^{-1}; 0.5/0.5 mol L^{-1}; 1.0/1.0 mol L^{-1}), soil:solution ratio (1:5 v:v; 1:10 v:v; 1:16 v:v; 1:25 v:v), room temperature (10 ^{o}C, 15 ^{o}C, 25 ^{o}C, 40 ^{o}C), and resin amount (1 n; 2 n). When one factor was changed all the others were kept the same as the standard procedure. We selected the five most representative soil orders used for crop production in Southern Brazil which have a wide range of clay, organic matter, Mehlich-1 extractable P and K, and KCl exchangeable Ca, and Mg contents. Results showed that modifications on the standard extraction procedure affected the amounts of soil P, K, Ca, and Mg extracted. Temperature was the main factor affecting the amount of P extracted from the soil with ion exchange membrane resin. Our results can be useful to other regions that might be interested in adopting the resin soil test method, allowing others to identify the impacts of similar method modifications on soil nutrient availability according to soil type, soil management, and temperature conditions.

Keywords: phosphorus, soil testing, potassium, nutrient management, resin

1. Introduction

Ion exchange resin membranes (resin) have received considerable attention as an alternative method to extract soil nutrients (Amer et al., 1955; Takahashi, 2013). The majority of resin-based research evaluated soil phosphorus (P) availability to plants (Kovar et al., 2009). The high correlation between extracted resin soil P and plant P uptake (or crop yield), and due to its multi-element simultaneous extraction capacity, led some researchers to conclude that the resin methodology is adequate for soil multi-element extraction in a single extraction procedure (Raij et al., 1986; Schoenau & Huang, 1991; Qian et al., 1992; Bortolon & Gianello, 2008).

There are two broad methodologies used to evaluate soil nutrient availability with exchange resins. The first method involves two steps: (i) nutrient extraction with resin by shaking a soil-resin-solution suspension for a pre-determined period of time and; (ii) the recovery of the adsorbed ions eluted with a solution. The second method incubates resin membranes in situ with soils for a pre-determined amount of time, and then the nutrients are recovered from the resins (Skogley et al., 1990; Yang et al., 1991b; Cooperband & Logan, 1994; Dobermann et al., 1994; Kovar, 2003; Kovar et al., 2009; Schlindwein & Gianello, 2009). Independent of the procedure used; there is a wide range of variables that affect the efficacy of exchange resins in extracting nutrients from the soil. These variables, include but are not limited to, the type and amount of resin (Sibbesen, 1978; Vaidyanathan & Talibudeen, 1970), shaking time (Bache & Ireland, 1980; Qian et al., 1992), soil-solution ratio (Cooke & Hislop, 1963), concentration and counter-ions (Sibbesen, 1978), temperature (Yang et al., 1991a), and resin extraction capacity (Somarisi & Edwards, 1992).

The time consumed to perform a soil test is an important factor that should be taken in consideration when evaluating the suitability of a new method for use in a commercial laboratory. The resin method (Tedesco et al., 1995) used in many Brazilian laboratories requires a 16 h resin-soil-solution suspension shaking time. Schoenau and Huang (1991) compared different shaking times (1 h, 6 h, and 16 h) and found that P concentration increased with shaking time, but shaking time did not influence the interpretation of the soil P levels. Qian et al. (1992) showed that P extracted with 15 min of shaking was nearly 40% of the P amount extracted with a shaking time of 60 min. The authors concluded that the difference did not impact the interpretation of the P level relative to fertilizer recommendation. However, statistical differences do exist in the quantities of nutrients that are extracted from soils and are directly affected by shaking time, resin type, and soil type (Sibbesen, 1978; Bache & Ireland, 1980).

The amount of soil-P that can be extracted is dependent on the type and amount of resin used for extraction (Sibbesen, 1978). However, the major factor that affects the resin suitability in evaluating soil-P availability to plants is the resin's ionic saturation during the extraction period (Sibbesen, 1978). Bicarbonate is the most common counter-ion used due to its buffering effect during the extraction process, and it does increase soil P extractability.

Temperature affects soil nutrient availability to plants through ion activity, soil buffer capacity, and ion exchange and diffusion (Sparks, 2003). However, relatively few studies have looked at the effects of temperature on soil nutrient content extracted by resin (Yang et al., 1991b; Haagsma & Miller, 1963). Laboratory temperature effects during extraction are specific to each nutrient, and temperature must be kept constant during the extraction period and analysis (Yang et al., 1991b; Simonis, 1996).

Factors such as resin ion saturation, eluent volume, and shaking time are important and their effects on the content of nutrients extracted vary among nutrients (Somasiri & Edwards, 1992). Variations on soil grinding level also affects extracted nutrient amounts, and therefore, must be quantified for both resin and diluted-acid extractants (Takahashi, 1996). The influence of these factors on the resin extraction procedure must be investigated to improve soil analysis quality and to avoid potential mistakes on crop fertilizer recommendations, which could impact both farmer's profitability and the environment.

Since 1955 (Amer et al., 1955), the use of the resin method to evaluate soil-P has increased due to advances in technology, such as impregnating a resin onto a plastic membrane (Mallarino & Atia, 2005; Vandecar et al., 2010). There is a wide range of synthetic resins with different physical-chemical properties and reticulation degrees to keep the resin structure stable during the extraction process. Stability is important for resins made in membrane form, because some of their properties (physical-chemical) are modified during the polymerization process to keep the resin stable, which can affect the resin's capacity to extract nutrients from the soil.

The objective of this study was to determine the sensitivity of resin membrane extraction capacity during simultaneous extraction of P, K, Ca, and Mg, to the following modifications to the standard procedure used by soil testing laboratories in Southern Brazil: (i) soil particle size; (ii) shaking time; (iii) oscillations of soil:resin suspension; (iv) concentrations of both saturation and elution solution; (v) soil:solution ratio; (vi) room temperature during the extraction process and; (vii) quantity of resin. We also investigated the effect of laboratory temperature on phosphorus extracted with ion exchange resin membrane and the impact on nutrient recommendation to corn in Southern Brazil.

2. Method

The standard resin procedure adopted by soil testing laboratories in Southern Brazil at the time of this study, uses a 7.5 cm^2 ion resin exchange membrane strips (CRG1CZR42—strong cationic and AR103DQP 434—strong anionic; Ionics, Inc., Watertown, MA), separated by a nylon screen. Before using the strips, the resin strips are saturated with 0.5 mol L^{-1} NaHCO$_3$ for a 24 h period and then washed with distilled water 2-fold with the same volume of sodium bicarbonate. For P, K, Ca, and Mg extraction, 2.5 cm^3 of soil, ground to pass a 2-mm sieve, 40 mL distilled water, and two exchange membrane strips are placed into a 50 mL glass snap-cap flask and shaken at 27 rpm for 16 h on an end-over-end shaker (Tecnal, Piracicaba). The resin strips are then removed, washed with a minimum amount of distilled water (about 50 mL) to remove excess of sodium bicarbonate from resin, and eluted with 40 mL of 0.5 mol L^{-1} HCl. The flasks are allowed to stand for 30 min, and then shaken for 90 min on a reciprocating shaker at 110 oscillations per minute (rpm). Phosphorus in the resin strip extracts is determined colorimetrically (1-amino-2-naftol-4-sulfonic and ammonium molybdate method, Tedesco et al., 1995) using a FEMTO 600 spectrophotometer (FEMTO S.A., São Paulo, Brazil), K determined by flame photometry, and Ca and Mg determined by atomic spectrometry using a AAnalyst 200 Atomic Absorption Spectrometer (Perkin Elmer spectrometer, Waltham, MA).

We investigated the following modifications of the standard procedure: (i) soil grinding-particle size; (ii) shaking time; (iii) oscillations of soil:resin suspension; (iv) concentration of both saturation and elution solutions; (v) soil:solution (water) ratio; (vi) environmental (room) temperature during the extraction period and; (vii) quantity of resin used (Table 1). When one factor was changed all the others were kept the same as the standard procedure. Five most representative soil orders used for crop production in southern Brazil were selected, with a wide range of chemical, physical, and mineralogical properties (Table 2). Within each soil type, we had three soil-P levels (Very Low, Adequate, and Very High) and four replicates. The change in interpretation of available soil P, K, Ca, and Mg was determined for each factor tested. In addition, the extraction of P, K, Ca, and Mg using resin beads with temperature as the single variable were conducted according to the procedure described in Raij et al. (1987). The use of resin spheres by several laboratories across Brazil prompted this testing.

Table 1. Alterations applied to the standard ion exchange resin membrane procedure used for soil analysis in Southern Brazil

Factor	Standard[1]	Modifications
Soil particle size - grinding (mm)	< 2	< 0.3
Shaking time (h)	16	8; 24
Oscillations (rpm)	25	12.5; 50
Concentrations of saturation/elution solutions (mol L^{-1})	0.5/0.5	0.25/0.25; 1.0/1.0
Soil:solution ratio (v:v)	1:16	1:5; 1:10; 1:25
Laboratory temperature ($^{\circ}$C)	25	10; 15; 40
Resin amounts (n)	1	2

Note. [1] Tedesco et al. (1995).

Table 2. Orders and characterization of soils used to grow crops in Southern Brazil

Soil Order	Particle size			Organic Matter	pH in water	SMP Index	Soil available[1]			
	Sand	Silt	Clay				P	K	Ca	Mg
	------------------ g kg^{-1} ------------------						--- mg dm^{-3} ---		-- cmol$_c$ dm^{-3} --	
Oxisol	50	330	620	61	6.1	6.0	2.6	123	10.7	3.6
				60	6.3	6.1	14.7	98	12.1	3.7
				52	6.1	6.1	11.5	94	11.7	3.8
Inceptisol	110	410	480	81	6.5	6.2	1.6	101	17.1	4.5
				83	5.9	5.6	8.3	114	18.2	4.8
				78	6.6	6.2	18.3	134	17.9	4.5
Spodosol	470	210	320	21	6.8	6.6	9.1	46	3.5	1.8
				18	6.4	6.5	33.2	45	4.3	1.9
				22	6.4	6.6	58.2	46	4.4	1.7
Alfisol	650	90	260	21	6.1	6.4	5.5	42	4.0	1.2
				22	6.0	6.3	14.2	46	4.1	1.2
				18	6.0	6.3	63.0	44	3.6	1.2
Acrisol	930	30	40	6	6.3	6.9	3.1	38	1.3	0.4
				6	5.9	6.9	41.8	28	1.6	0.4
				5	6.1	6.8	72.8	29	1.4	0.4

Note. [1] Soil P and K extracted with Mehlich-1 solution and exchangeable Ca and Mg extracted with 1.0 mol L^{-1} KCl.

The influence of laboratory temperature on P recommendation to crop was also investigated. We considered 25 $^{\circ}$C as the standard laboratory temperature and compared its influence on soil-P, changes in soil-P status, P applied, and P balance. Soil P status was classified based on soil-P extracted with membrane resin according to Southern Brazil Manual for Crop Nutrient and Lime Recommendations (SBCS/NRS, 2004). Soil P classes according to soil-P (mg dm^{-3}) extracted with membrane resin are: Very Low (VL = soil-P ≤ 5.0); Low (L = soil-P

Ion Exchange Resin Membrane Sensitivity Analysis of Selected Parameters for Soil Nutrient...

121

range of 5.1-10.0); Medium (M = soil-P range of 10.1-20.0); High (H = soil-P range of 20.1-40.0); Very High (VH = soil-P $\geq$ 40.0). Soil-P critical level which is the soil-P level in soil that is adequate to crop production in Southern Brazil is 20.0 mg dm^{-3}. We tested the P recommendation for corn (*Zea mays* L.) considering an expected yield of 10 Mg ha^{-1}, and we verified the P amounts recommended according to soil-P extracted at each different laboratory temperature. We calculated P applied balance (%), that is the P amount applied based on soil-P extracted at 25 $^{\circ}$C laboratory temperature (100%), compared to P amount applied base on soil-P extracted at 10 $^{\circ}$C, 15 $^{\circ}$C and 40 $^{\circ}$C laboratory temperature.

The statistical analyses were performed comparing standard versus procedure modifications by ANOVA, and the mean differences in each factor were evaluated with Tukey test (0.05) using SAS (SAS, 2002).

3. Results

3.1 Effect of Soil Grinding Size

Soil clay content affected the efficiency of resin P extraction and higher P amounts were found in high sand content soils compared to soils with high clay content (Table 3). The difference in higher soil P amounts extracted in sandy soils is probably due to the soil grinding processes and also because the sandy soils had more available P (Table 2). The finer screen (< 0.3 mm) passed the soil fractions with both high soil clay and organic matter to which P can be adsorbed, increasing the soil P amounts extracted (Bortolon et al., 2011). The finer constituents also increased surface sorption area. High clay content soils probably had greater amounts of Ca bonded with clays that are highly resistant to detachment, which increased the tortuosity of the path of P from the soil solution through resin exchange surface, reducing P in the solution (Bortolon et al., 2011). In clayey soils, the finer sieve decreased the sized particle content passing through the sieve, which may also have preferentially selected for Fe and Al oxides, increasing P adsorption, and consequently decreased the amount of P extracted. On the other hand, the high concentrations of bicarbonate ion, released from the resin during the extraction period, could increase the pH of the suspension and promote the precipitation of calcium phosphate (Raij et al., 1986), leading to a decrease in the P available to react with the resin. Contrasting results were found by Rolim et al. (2008), where the authors concluded that soil particle size did not affect amounts of soil-P extracted by resin in both fertilized and non-fertilized soils. However, the differences observed in our study can be explained by a 2 h shaking time at 150 rpm (Rolim et al., 2008) as compared to 16 h at 27 rpm on our procedure.

Soil particle size did affect the amount of Ca extracted with resin (Table 3). The greatest amounts were observed in soils with high clay content. Higher Ca amount extracted in clayey soils can be explained by the higher soil exchange capacity, which could compete with the resin on Ca adsorption. Calcium could also be precipitated as calcium phosphate. This process can also explain the reduction of extracted K due to particle size (Table 3). On average, extracted K amounts were different between grinding sizes and they did not affect the interpretation of K levels in the soil or the fertilizer recommendations to crops. The results showed that extraction of P with resin was affected by the soil grinding level, clay content, and amounts of P and Ca in soils. The obstruction of the resin surface by attached fine soil particles making some resin sites unavailable to adsorption may explain the effects on P and K; however, detailed studies must be carried out to validate this hypothesis.

Table 3. Phosphorus, potassium, calcium and magnesium extracted with ion exchange resin membrane affected by soil grinding size

Soil Order	Soil grinding size (mm)							
	< 2.0	< 0.3	< 2.0	< 0.3	< 2.0	< 0.3	< 2.0	< 0.3
	P		K		Ca		Mg	
	---------------------- mg dm^{-3} --------------------				----------------- cmol$_c$ dm^{-3} ---------------------			
Oxisol	4.9 a	3.9 a	58.7 a	59.5 a	5.22 a	3.74 a	2.11 a	1.83 a
	31.5 a	27.9 a	41.5 a	37.7 a	6.47 a	5.16 a	2.37 a	2.13 a
	60.3 a	55.9 a	50.8 a	48.0 a	4.67 a	4.11 a	2.27 a	2.26 a
Inceptisol	6.3 a	6.5 a	40.9 a	38.9 a	10.09 a	8.88 a	3.03 a	3.14 a
	18.1 a	20.6 a	39.8 a	37.1 a	4.49 a	4.91 a	2.68 a	3.10 a
	58.8 a	57.5 a	39.0 a	37.1 a	9.99 a	10.08 a	2.82 a	3.17 a
Spodosol	19.0 a	19.2 a	38.6 a	37.6 a	2.38 a	2.43 a	1.26 a	1.36 a
	50.8 a	49.7 a	34.3 a	31.0 a	2.66 a	2.67 a	1.36 a	1.43 a
	81.9 a	82.4 a	35.2 a	33.4 a	3.03 a	3.04 a	1.25 a	1.38 a
Alfisol	8.2 a	9.8 a	32.8 a	32.9 a	2.09 a	2.56 a	0.84 a	1.11 a
	20.0 b	24.5 a	29.1 a	31.0 a	1.86 a	2.42 a	0.75 a	0.99 a
	73.0 b	87.0 a	29.1 a	29.2 a	2.27 a	2.61 a	0.80 a	1.05 a
Acrisol	2.7 a	3.0 a	12.7 a	12.2 a	0.37 a	0.55 a	0.14 a	0.21 a
	28.5 b	36.3 a	15.0 a	14.0 a	0.53 a	0.54 a	0.12 a	0.19 a
	62.3 b	81.1 a	16.2 a	16.4 a	0.51 a	0.73 a	0.14 a	0.23 a
Average	35.1 b	37.7 a	34.2 a	33.1 b	3.63 a	3.77 a	1.48 a	1.57 a

Note. Means followed with the same lower case letter within a row did not differ statistically by Tukey test (5%); C.V. (%) P = 4.1; C.V. (%) K = 8.4; C.V. (%) Ca = 15.3; C.V. (%) Mg = 4.9.

3.2 Shaking Time

Shaking time is an important issue during the development of a soil test method. The goal during the shaking process is to increase the contact of soil particles and the soil extractant. As shaking time increased, there was an increase in extracted soil P (Table 4), which was greater in clayey than in sandy soils, probably due to the high P adsorption capacity associated with most clay soils. Ascertaining soil-P using the resin method is a lengthier process compared to Mehlich-1 or Mehlich-3, due to the long shaking time involved, which is independent of resin type (Sibbesen, 1978). Qian et al. (1992) showed that although longer shaking times extracted more soil P, shaking time did not affect the interpretation of soil P levels and P fertilizer recommendations. Longer shaking time can extract insoluble soil P forms that have no bearing on plant P availability (Schoenau & Huang, 1991) and may lead to a lower recommendation for crops which could reflect in lower yields. The increase of shaking time affected Ca extraction, and more Ca was extracted from soil with longer shaking time (Table 4), especially on those soils with higher Ca content (Table 2). For both P and K, the greatest extracted amounts were found with the 24 h shaking time. As the shaking time increased, the amount of Ca and Mg extracted increased and K decreased (Table 4). This finding can be explained by different charges among the two cations in the solution even though in similar concentration; resin affinity favors high valence cations (Sparks, 2003). The results showed that 16 h of shaking time is not sufficient to extract soil exchangeable Ca and Mg, and 24 h shaking extracted more Ca and Mg than 16 h shaking. In addition, soil-P was greater with 24 h shaking in some cases 2-fold compared to 8 h shaking.

Table 4. Phosphorus, potassium, calcium, and magnesium extracted with ion exchange resin membrane affected by shaking time

Soil Order	Shaking time (h)											
	8	16	24	8	16	24	8	16	24	8	16	24
		P			K			Ca			Mg	
	------------------------- mg dm^{-3} -------------------------						--------------------- cmol$_c$ dm^{-3} ----------------------					
Oxisol	2.9 b	4.9 ab	6.5 a	67.0 a	58.7 b	51.4 c	2.00 b	5.22 a	5.38 a	1.23 b	2.11 a	2.42 a
	23.0 c	31.5 b	50.4 a	48.1 a	41.5 b	35.1 c	2.97 b	6.47 a	7.04 a	1.56 b	2.88 a	3.36 a
	43.1 c	60.3 b	88.0 a	56.0 a	50.8 b	44.8 c	2.39 c	4.67 b	6.37 a	1.63 c	2.27 b	3.05 a
Inceptisol	4.4 b	6.3 b	9.7 a	45.1 a	40.9 ab	36.9 b	5.33 c	10.09 b	12.31 a	2.28 c	3.03 b	3.78 a
	11.9 c	18.1 b	27.9 a	39.0 a	39.8 a	32.7 b	2.22 c	4.49 b	6.46 a	1.94 c	2.68 b	3.64 a
	38.9 c	58.8 b	85.0 a	42.0 a	39.0 ab	35.7 b	5.78 c	9.99 b	13.01 a	2.24 b	2.82 b	3.61 a
Spodosol	15.9 b	19.0 b	25.2 a	33.5 a	38.6 a	35.7 a	1.24 b	2.38 a	2.86 a	0.99 b	1.26 ab	1.69 a
	41.2 c	50.8 b	64.4 a	29.2 a	34.3 a	32.1 a	1.28 b	2.66 a	3.34 a	1.04 b	1.36 ab	1.85 a
	68.1 c	81.9 b	103.6 a	28.0 b	35.2 a	33.9 a	1.54 b	3.03 a	3.45 a	0.97 b	1.25 ab	1.63 a
Alfisol	7.7 a	8.2 a	10.0 a	23.8 b	32.8 a	30.8 a	0.98 b	2.09 a	2.66 a	0.63 a	0.84 a	1.16 a
	18.3 b	20.0 b	26.3 a	22.5 b	29.1 a	29.0 a	1.13 b	1.86 b	2.69 a	0.62 a	0.75 a	1.07 a
	61.4 c	73.0 b	97.2 a	22.5 b	29.1 a	29.6 a	1.41 c	2.27 b	3.38 a	0.66 a	0.80 a	1.13 a
Acrisol	2.9 a	2.7 a	3.2 a	7.9 a	12.7 a	12.7 a	0.15 a	0.37 a	0.47 a	0.13 a	0.54 a	0.21 a
	24.9 c	28.5 b	35.8 a	7.3 b	15.0 a	13.9 a	0.10 a	0.53 a	0.56 a	0.08 a	0.12 a	0.13 a
	56.3 c	62.3 b	90.3 a	9.1 b	16.2 a	17.5 a	0.13 a	0.51 a	0.55 a	0.13 a	0.14 a	0.20 a
Average	28.1 c	35.1 b	48.2 a	32.1 b	34.2 a	31.4 b	1.91 c	3.77 b	4.68 a	1.08 c	1.55 b	1.90 a

Means followed with the same lower case letter within a row did not differ statistically by Tukey test (5%); C.V. (%) P = 4.6; C.V. (%) K = 8.2; C.V. (%) Ca = 12.0; C.V. (%) Mg = 21.5.

3.3 Oscillations of Soil:Resin Suspension

The shaking process accelerates nutrient extraction that promotes a fast and constant contact between soil solid phase and resin, in addition to the initial nutrient in the soil solution. The application of different oscillation rates based on the end-over-end shaker were not different to soil-P compared to 25 and 50 rpm, and in almost all soils orders soil-P concentrations extracted with 12.5 rpm were similar to those extracted at 50 rpm (Table 5). However, the differences due to this factor showed little effect among soil orders and were related to the increased contact time between soil P and resin. We verified from the data shown in Table 5 that it was not necessary to have vigorous oscillations for extracting soil Ca and Mg with resin. However, the lowest oscillation rate (12.5 rpm) did reduce the extraction of Ca and Mg and it would affect the calculation of soil CEC that is used to classify the K status in soils from Southern Brazil. Increasing the number of oscillations per minute reduced extractable K concentrations (Table 5) in four soils following polyvalence cations selectivity, which increased Ca and Mg amounts extracted and reduces K amounts extracted (Sparks, 2003). The results show that oscillation rate affects extractable nutrient concentrations. Also, the differences found for P and K had little or no effect on the nutrient recommendations to crops (SBCS/NRS, 2004).

Table 5. Phosphorus, potassium, calcium, and magnesium extracted with ion exchange resin membrane affected by oscillation of soil:solution:resin

Soil Order	Oscillation (rpm)											
	12.5	25	50	12.5	25	50	12.5	25	50	12.5	25	50
	P			K			Ca			Mg		
	------------------------ mg dm^{-3} -----------------------						----------------------- cmol$_c$ dm^{-3} ----------------------					
Oxisol	3.7 a	4.9 a	3.5 a	60.5 a	59.3 a	60.0 a	3.64 b	5.22 a	4.51 ab	1.42 b	2.11 a	2.12 a
	31.3 a	31.5 a	32.7 a	42.3 a	44.2 a	37.9 b	4.91 b	6.47 a	5.93 a	1.68 b	2.37 a	2.52 a
	56.9 b	60.3 ab	63.0 a	53.3 a	53.4 a	48.6 b	4.52 a	4.67 a	4.96 a	1.81 b	2.27 a	2.58 a
Inceptisol	6.1 a	6.3 a	5.6 a	41.6 b	47.2 a	42.3 ab	8.77 b	10.09 a	9.82 a	2.33 b	3.03 a	3.26 a
	18.1 a	18.1 a	16.3 a	40.7 ab	43.1 a	37.3 b	4.46 a	4.49 a	4.47 a	2.19 b	2.68 a	2.86 a
	54.7 b	58.8 a	56.7 ab	42.3 ab	49.2 a	36.6 b	9.55 b	9.99 ab	10.49 a	2.24 b	2.82 a	3.13 a
Spodosol	18.1 a	19.0 a	19.1 a	39.7 a	31.5 b	34.7 b	1.81 a	2.38 a	2.62 a	0.91 b	1.26 ab	1.45 a
	45.5 b	50.8 a	47.5 ab	34.8 a	30.6 b	31.6 b	2.14 a	2.66 a	2.95 a	0.98 b	1.36 a	1.54 a
	80.2 b	81.9 ab	84.6 a	34.5 a	34.8 a	33.5 a	2.85 a	3.03 a	3.13 a	1.00 b	1.25 ab	1.40 a
Alfisol	8.6 a	8.2 a	8.7 a	33.8 a	33.8 a	29.7 a	1.72 a	2.09 a	2.34 a	0.67 a	0.84 a	1.01 a
	20.3 a	20.0 a	22.4 a	31.9 a	30.6 a	29.0 a	1.98 a	1.86 a	2.28 a	0.60 a	0.75 a	0.90 a
	74.0 b	73.0 b	79.3 a	29.9 a	28.7 a	27.8 a	2.26 a	2.27 a	2.73 a	0.65 a	0.80 a	0.99 a
Acrisol	3.1 a	2.9 a	2.7 a	15.7 a	13.5 ab	10.8 b	0.21 a	0.37 a	0.57 a	0.13 b	0.14 a	0.23 ab
	26.6 a	28.5 a	29.4 a	17.6 a	12.5 b	15.1 ab	0.26 a	0.53 a	0.50 a	0.10 a	0.12 a	0.17 a
	66.2 b	62.3 c	69.9 a	17.6 a	12.9 b	15.8 ab	0.29 a	0.51 a	0.66 a	0.12 a	0.14 a	0.19 a
Average	34.2 b	35.1 ab	36.2 a	35.8 a	34.3 ab	32.7 b	3.29 b	3.77 a	3.86 a	1.12 c	1.46 b	1.62 a

Note. Means followed with the same lower case letter within a row did not differ statistically by Tukey test (5%); C.V. (%) P = 5.0; C.V. (%) Ca = 13.2; C.V. (%) Mg = 13.2; C.V. (%) K = 9.3.

3.4 Concentration of Saturation and Elution Solutions

Several solutions have been studied for the saturation of the resin and eluting the adsorbed ions after soil extraction. For saturation of anion exchange resin, the bicarbonate anion has been preferred due the buffering effect, which increases soil extraction of P (Sibbesen, 1978; Raij et al., 1986), while Na is preferable for saturation of cation exchange resins. There was significant interaction between concentration of extraction solution and soil-P concentration on sandy soils with high soil-P, where the most concentrated solution extracted greater amounts of soil-P (Table 6). There were differences in only for Alfisol and Acrisol soils (both with very high soil-P content) where the standard method (0.5 mol L^{-1} solution) extracted the lowest amount of P. Extraction of these soils does not follow the same behavior as the solution concentration. This behavior is probably due to similar mineralogical properties on both Alfisol and Acrisol. Since this finding was only observed in soils with high P concentrations, the use of a more concentrated solution would not impact fertilizer recommendations to crops, since soil-P level is considered adequate to crops and the fertilizer recommendation on that case is based only on the crop P removal (SBCS/NRS, 2004).

The results on Table 6, for soil Ca, Mg, and K extracted by ion-exchange resin membrane saturated with different concentrations of sodium bicarbonate, show that in soils with greater Ca content, the amount of Ca extracted decreased as extraction solution concentration increased. On average, the extracted concentrations of soil K and Mg were higher with increasing concentration of both saturation and elution solutions, regardless of the element content in soil, and were statistically different for both elements in the three concentrations. Although the results for the four elements (P, K, Ca, and Mg) in soil by using different solution concentrations showed statistical differences, the use of the most concentrated solution (1.0 mol L^{-1}) would double the amount of chemicals required for the analyzes compared to the standard method. In addition, the 1.0 mol L^{-1} concentration would also have a negative effect on laboratory equipment and waste management.

3.5 Soil:Solution Ratio

There was significant interaction between soil:solution ratio and soil-P (Table 7). On average, the extracted soil-P concentrations were greatest for 1:25 soil:solution ratio. Similar results were obtained by McLaughlin et al. (1993). At a soil:solution ratio of 1:16, there was a decrease in soil-P concentrations. Whereas in other relationships studied such as soil grinding and shaking time, there was little difference in the results. For the 1:10 and 1:25 soil solutions ratio the effect, soil-P extraction although statistically significant different for some soils, is not pronounced enough in terms of soil-P value to affect the results if the soil:solution remains close to that established for the method.

Table 6. Phosphorus, potassium, calcium, and magnesium extracted with ion exchange resin membrane affected by solution concentration

Soil	Solution Concentration (mol L^{-1})											
	0.25	0.5	1.0	0.25	0.5	1.0	0.25	0.5	1.0	0.25	0.5	1.0
		P			K			Ca			Mg	
	---------------------- mg dm^{-3} ----------------------						----------------------- cmol$_c$ dm^{-3} ----------------------					
Oxisol	4.1 a	4.9 a	3.7 a	55.8 a	59.3 a	49.4 b	5.20 a	5.22 a	4.24 a	1.68 c	1.82 b	1.92 a
	33.6 ab	31.5 b	35.5 a	39.4 a	44.2 a	40.2 a	6.08 a	6.47 a	5.60 a	2.05 c	2.26 b	2.41 a
	61.3 a	60.3 a	63.5 a	48.7 b	53.4 a	53.0 a	5.85 a	4.67 b	5.09 ab	2.19 b	2.05 b	2.60 a
Inceptisol	5.6 a	6.3 a	6.0 a	40.3 b	47.2 a	47.4 a	10.64 a	10.09 a	8.82 b	2.83 b	2.96 b	3.12 a
	17.8 a	18.1 a	18.1 a	38.7 b	43.1 a	42.8 a	4.73 a	4.49 a	4.88 a	2.52 b	2.69 b	3.07 a
	56.8 a	58.8 a	57.2 a	37.6 c	49.2 a	40.8 b	12.02 a	9.99 b	9.46 b	2.87 b	2.89 b	3.02 a
Spodosol	18.5 a	19.0 a	19.4 a	35.2 b	31.5 b	42.8 a	2.81 a	2.38 a	2.49 a	1.11 b	1.40 a	1.45 a
	50.4 a	50.8 a	49.8 a	31.1 b	30.6 b	36.4 a	2.57 a	2.66 a	2.68 a	1.19 b	1.51 a	1.53 a
	83.2 a	81.9 a	84.3 a	32.3 b	34.8 a	36.4 a	3.58 a	3.03 a	3.16 a	1.14 b	1.48 a	1.44 a
Alfisol	7.5 a	8.2 a	8.1 a	28.8 b	33.8 a	31.9 a	2.03 a	2.09 a	2.23 a	0.74 b	0.97 a	0.98 a
	20.7 a	20.0 a	21.9 a	27.0 b	30.6 a	32.5 a	2.26 a	1.86 a	2.23 a	0.67 b	0.91 a	0.93 a
	76.1 ab	73.0 b	78.4 a	27.6 a	28.7 a	31.3 a	1.89 a	2.27 a	2.49 a	0.69 b	0.98 a	0.96 a
Acrisol	2.9 a	2.7 a	3.1 a	14.1 b	13.5 b	17.2 a	0.41 a	0.37 a	0.49 a	0.14 a	0.18 a	0.18 a
	29.1 a	28.5 a	30.6 a	14.1 b	12.5 b	20.1 a	0.44 a	0.53 a	0.41 a	0.12 a	0.14 a	0.14 a
	70.0 a	62.3 b	71.0 a	17.0 b	12.9 c	19.2 a	0.47 a	0.51 a	0.57 a	0.15 a	0.16 a	0.14 a
Average	35.8 ab	35.1 b	36.7 a	32.5 c	34.3 b	36.4 a	4.06 a	3.77 ab	3.65 b	1.34 c	1.49 b	1.59 a

Note. Means followed with the same lower case letter within a row did not differ statistically by Tukey test (5%); C.V. (%) P= 4.8; C.V. (%) Ca = 15.5; C.V. (%) Mg = 12.5; C.V. (%) K = 8.2.

Table 7. Phosphorus, potassium, calcium, and magnesium extracted with ion exchange resin membrane affected by soil:solution ratio

Soil	Soil:solution ratio															
	1:5	1:10	1:16	1:25	1:5	1:10	1:16	1:25	1:5	1:10	1:16	1:25	1:5	1:10	1:16	1:25
	P				K				Ca				Mg			
	$------------------ mg\ dm^{-3} ------------------$								$------------------ cmol_c\ dm^{-3} ------------------$							
Oxisol	5.9a	5.4a	4.9a	7.5a	35.8d	48.6c	58.8b	89.4a	9.13a	9.76a	3.81b	3.58b	3.71a	2.65b	1.82b	2.13b
	42.7a	42.8a	31.5b	44.5a	29.0c	32.8be	41.5b	65.0a	10.7a	8.22a	4.85b	5.18b	3.82a	3.18b	2.26b	2.62b
	72.8b	73.8b	60.3c	77.9a	29.6d	41.3c	50.8b	77.2a	9.76a	9.13a	4.47b	4.26b	4.18a	3.35b	2.05b	2.74b
Inceptisol	8.2a	7.7a	6.3a	8.5a	29.6c	34.6be	40.9b	66.la	13.8a	12.3a	8.97b	9.92b	3.89a	3.77a	2.96b	3.74a
	22.9a	23.2a	18.1b	26.7a	29.6c	32.2be	39.8b	61.0a	9.19a	9.97a	4.47b	4.16b	4.76a	4.04ab	2.69b	3.32b
	67.0b	70.5b	58.8c	76.4a	24.1c	32.8be	39.0b	64.la	14.0a	13.5a	9.75b	10.4b	3.56a	3.73a	2.89b	3.61a
Spodosol	21.5a	21.la	19.0a	22.3a	34.0b	35.2b	38.6b	57.9a	3.86a	3.41a	2.16b	1.79b	2.25a	1.86a	1.40b	1.50b
	55.0a	56.3a	50.8b	55.6a	25.9b	30.9b	34.3b	51.8a	4.36a	3.80a	2.57b	2.41b	2.37a	2.06a	1.51b	1.61b
	91.3a	92.5a	81.9b	90.6a	27.2b	29.8b	35.2b	50.8a	4.82a	4.28a	2.99b	2.51b	2.13a	1.87ab	1.48b	1.45b
Alfisol	9.la	8.9a	8.2a	10.0a	25.3b	28.6b	32.8b	44.7a	3.06a	3.20a	1.63b	1.95b	1.47a	1.28ab	0.97b	0.99b
	23.3a	22.5a	20.0a	23.9a	25.9b	26.7b	29.1b	42.7a	3.58a	3.09a	2.07b	1.94b	1.42a	1.20ab	0.91b	0.94b
	86.5a	84.5a	73.0b	82.9a	24.8b	27.3b	29.1b	44.7a	3.69a	3.56a	2.27b	2.22b	1.41a	1.24a	0.98a	0.96b
Acrisol	3.8a	3.6a	2.7a	4.7a	12.4b	9.8b	12.7b	22.4a	0.24a	0.39a	0.38a	0.33b	0.23a	0.21a	0.18a	0.22a
	32.4ab	32.8a	28.5b	34.la	12.4b	12.8b	15.0ab	22.4a	0.04b	0.54a	0.36b	0.44a	0.13a	0.16a	0.14a	0.18a
	74.8a	75.9a	62.3b	77.la	15.4a	15.8a	16.3a	19.8a	0.42b	0.75a	0.38b	0.55a	0.18a	0.20a	0.16a	0.21a
Average	41.1b	41.4b	35.1c	42.8a	25.4d	29.3c	34.3b	52.0a	6.04a	6.04a	3.77b	3.46b	2.78a	2.08ab	1.49b	1.75b

Note. Means followed with the same lower case letter within a row did not differ statistically by Tukey test (5%); C.V. (%) P= 4.8; C.V. (%) K = 12.2; C.V. (%) Ca = 5.6; C.V. (%) Mg = 6.

We observed an interaction between soil dilution and the K content in the soil (Table 7), where increasing soil:solution ratio increased the amount of soil K extracted compared to the standard method. The soil dilution effect was less pronounced in soils with low K amounts; however in soils with higher K content, there was a significant increase on the amount of soil-K extracted due to soil dilution increase. For Ca and Mg, the soil:solution ratio affected the amounts extracted in all soils, regardless the element content in the soil (Table 7). The amounts of soil Ca and Mg extracted decreased as soil dilution increased. The results (Table 7) showed that K had an opposite behavior to Ca and Mg; thereby, when the extraction of Ca and Mg decreased, the extraction of K increased, as previously observed in other studies. These results differ from Sparks (2003), who reported that higher solution dilution favors the exchange of cations of higher valence, namely Ca and Mg in this study compared to K.

3.6 Laboratory Temperature

As the temperature and soil-P increased, the amounts of P extracted increased significantly (Table 8). Extractable soil-K was not significantly affected by laboratory or room temperature, likely due to the low levels of K in most of the soils used in this study. Kinetic energy of the substances increases with temperature, and therefore the ion activity in the solution, as well as the buffering capacity and CEC. These factors are more important for the extraction of K, Ca, and Mg than for P. The greater extraction of P with increasing temperature is probably due to increased solubility of low solubility compounds. More importantly, the temperature increase also expands the resin matrix structure. The resin expansion exposes additional sites for sorption of ions present in the solution. In addition, increasing temperature increases ion activity with adsorbed ions of the solution, thereby, increases the concentration of elements in soil solution. Yang et al. (1991b) observed that the resin extraction of elements from the soil can affect nutrient extraction positively or negatively due to temperature range and is specific for each nutrient and soil type.

Table 8. Phosphorus, potassium, calcium, and magnesium extracted with ion exchange resin membrane affected by laboratory temperature

Soil	Environmental temperature (°C)															
	10	15	25	40	10	15	25	40	10	15	25	40	10	15	25	40
	P				K				Ca				Mg			
	------------------------- mg dm^{-3} -------------------------								------------------------- cmol$_c$ dm^{-3} -------------------------							
Oxisol	3.0a	4.5a	4.9a	7.2a	59.1a	63.0a	59.3a	58.5a	3.93b	4.42b	5.22a	5.70a	1.85b	2.19ab	2.11b	2.47a
	15.5d	24.3c	31.5b	59.4a	42.6a	43.5a	44.2a	42.2a	4.68c	6.02b	6.47b	7.44a	2.06c	2.45b	2.37be	2.95a
	32.2d	46.6c	60.3b	103.8a	51.5b	52.2b	53.4b	57.9a	3.29c	3.45c	4.67b	6.36a	1.90c	2.21be	2.27b	3.18a
Inceptisol	3.0b	6.2ab	6.3ab	9.3a	44.5a	43.5a	47.2a	39.8b	5.77c	9.80b	10.09b	12.30a	2.35c	3.57a	3.03ab	3.88a
	8.7c	14.1b	18.1b	32.1a	38.1b	43.5a	43.1a	38.0b	2.89c	3.43c	4.49b	6.94a	2.12c	2.63b	2.68b	3.93a
	29.7d	45.0c	58.8b	95.5a	40.7b	40.9b	49.2a	38.6b	6.43d	9.99c	11.01b	13.28a	2.39c	3.52a	2.82b	3.65a
Spodosol	12.1c	17.7b	19.0b	27.8a	38.8b	40.2a	31.5c	42.8a	1.68b	2.11b	2.38ab	2.92a	1.10b	1.36ab	1.26b	1.68a
	33.5d	42.9c	50.8b	70.7a	31.8b	35.4a	30.6b	37.4a	1.99b	2.20b	2.66ab	3.43a	1.18b	1.31b	1.36b	1.82a
	52.8d	71.0c	81.9b	118.9a	31.8b	39.5a	34.8b	38.0a	2.07c	2.45bc	3.03ab	3.75a	1.06b	1.26b	1.25b	1.69a
Alfisol	6.1b	7.5ab	8.2ab	11.8a	29.9b	36.8a	33.8a	35.6a	1.64b	1.60b	2.09ab	2.73a	0.72b	0.84ab	0.84ab	1.13a
	11.3c	20.0b	20.9b	28.6a	26.7b	33.4a	30.6a	32.6a	1.52b	1.60b	1.86ab	2.57a	0.58b	0.76ab	0.75ab	1.08a
	43.2d	61.4c	73.0b	100.1a	27.3b	32.8b	28.7b	43.4a	1.92b	1.69b	2.27ab	2.88a	0.68b	0.74ab	0.80ab	1.04a
Acrisol	1.9a	2.6a	2.7a	4.1a	13.3b	15.6ab	13.5b	18.1a	0.59a	0.35a	0.37a	0.49a	0.19b	0.16b	0.54a	0.20ab
	17.5c	26.6b	28.5b	40.9a	13.3b	16.1a	12.5b	18.1a	0.65a	0.21a	0.53a	0.41a	0.14a	0.09a	0.12a	0.16a
	39.0d	54.5c	62.3b	92.1a	13.3b	18.8a	12.9b	20.6a	0.48a	0.21a	0.51a	0.66a	0.17a	0.13a	0.14a	0.17a
Average	20.6d	29.7c	35.1b	53.5a	34.4b	37.0a	34.3b	37.3a	2.64c	3.37c	3.77b	4.79a	1.23c	1.55b	1.49b	1.93a

Note. Means followed with the same lower case letter within a row did not differ statistically by Tukey test (5%); C.V. (%) P = 7.1; C.V. (%) K = 10.6; C.V. (%) Ca = 10.3; C.V. (%) Mg = 10.5.

Due to the numerous changes occurring in soil as a result of temperature, the process of extraction with resin is always affected, regardless of the type of resin used in the extraction procedure. Data (Table 9) were obtained by the spherical resin method proposed by Raij et al. (1986). Just as in the extraction with resin membrane, the amounts extracted increased with increasing temperature. The same was observed for the extraction of K, particularly with higher levels in the soil. The extracted Ca and Mg concentrations were unaffected by higher temperatures (Table 9). These results are similar to those obtained with the resin membrane (Table 8). The results show that the extraction technique with resin is influenced by temperature and must be controlled throughout the extraction process. This finding is therefore a negative aspect of the use of resin in the determination of nutrient availability as a standard laboratory practice, especially in regions with a wide range temperature throughout the year, requiring a more rigorous control of this effect to achieve and maintain analytical reproducibility.

Table 9. Phosphorus, potassium, calcium, and magnesium extracted with ion exchange resin in sphere affected by environmental temperature

Soil	Environmental temperature (°C)							
	15	25	15	25	15	25	15	25
	P		K		Ca		Mg	
	------------------- mg dm^{-3} ---------------------				------------------ cmol$_c$ dm^{-3} ------------------			
Oxisol	12.8 b	17.0 a	42.9 b	74.1 a	2.43 a	2.38 a	0.85 a	0.84 a
	53.0 b	68.7 a	35.1 b	54.6 a	2.32 a	2.31 a	0.91 a	0.98 a
	66.1 b	91.1 a	109.2 b	187.2 a	2.42 a	2.39 a	0.98 a	0.93 a
Inceptisol	16.0 a	18.9 a	35.1 b	58.5 a	2.33 a	2.29 a	0.99 a	0.99 a
	33.7 b	43.1 a	136.5 b	198.9 a	2.45 a	2.48 a	0.75 a	0.77 a
	94.2 b	105.6 a	42.9 b	50.7 a	2.32 a	2.39 a	0.98 a	0.99 a
Spodosol	22.7 b	28.7 a	31.2 a	35.1 a	2.31 a	2.18 a	0.91 a	0.85 a
	54.0 b	69.2 a	31.2 b	39.0 a	2.45 a	2.38 a	0.99 a	0.96 a
	84.7 b	133.4 a	31.2 a	31.2 a	2.35 a	2.29 a	0.85 a	0.88 a
Alfisol	7.4 b	12.0 a	35.1 a	27.3 b	2.45 a	2.38 a	0.99 a	0.97 a
	21.0 b	28.5 a	15.6 b	23.4 a	2.32 a	2.33 a	0.89 a	0.85 a
	68.1 b	97.7 a	15.6 b	23.4 a	2.15 a	2.25 a	0.93 a	0.93 a
Acrisol	2.4 a	4.6 a	3.9 a	3.9 a	2.43 a	2.33 a	0.94 a	0.91 a
	27.8 b	37.9 a	3.9 a	3.9 a	2.41 a	2.29 a	0.98 a	0.95 a
	58.3 b	77.1 a	3.9 b	11.7 a	2.18 a	2.23 a	0.85 a	0.88 a
Average	41.6 b	55.6 a	38.3 b	54.6 a	2.36 a	2.33 a	0.92 a	0.91 a

Note. Means followed with the same lower case letter within a row did not differ statistically by Tukey test (5%); C.V. (%) P = 4.1; C.V. (%) K = 8.4; C.V. (%) Ca = 15.3; C.V. (%) Mg = 4.9.

3.7 Resin Amounts

There was interaction between the resin amounts and soil P concentration (Table 10). The use of two resin membranes extracted greater amounts of P from all soils. However, the effect was more pronounced in soils with high contents of both clay and P. This effect was also found by Sagger et al. (1990) who obtained 25-40% more P extracted by using two resin membranes. Increasing the amount of resin, besides increasing the exchange capacity, increases the contact surface between the resin and the ions in solution and therefore the activity of bicarbonate ions in the solution. According to Sibbesen (1978), the increase of bicarbonate ions creates more competition with phosphate ions for exchange sites, releasing the P.

The content of extracted Ca and Mg were affected by the amount of resin (Table 10). There was a significant difference in the amounts extracted in the soils with higher contents of these elements. The amount of soil-K extracted increased in all soil when two resin membranes were used, regardless the soil-K content. The increase of resin amount can be more important for extraction of all cations, whereas in the study of some factors we observed competition between K, Ca, and Mg for the resin.

3.8 Impact on Crop Nutrient Recommendations

Soil testing quality plays an important role on results that directly impact crop nutrient recommendations. Temperature in the laboratory is one important factor to be controlled during the soil testing routine. We observed that at 40 °C laboratory temperature the P recommendations to corn were underestimated in soils with very low and very high P (Table 11). In some cases, the P applied was reduced by 33% and the impact on farmer's profit could be great especially due the high costs involved in corn production, mainly in Brazil where P sources are costly. We observed a shift in soil P classes according to increased laboratory temperature (Table 11). Lower temperature, on average reduced the P recommended for corn by 33%. This large rate change surely could adversely affect corn yield. Temperatures from 15 to 25 °C did not affect the P recommendation, except for the Alfisol where the P level was originally classified as high (25 °C) and shifted to medium (15 °C). Based on the results, laboratory temperature must be taken into account and controlled in the soil testing laboratory to produce reliable P fertilizer recommendations for crops based upon a predictive soil-test result.

Table 10. Phosphorus, potassium, calcium, and magnesium extracted with ion exchange resin membrane affected by amount of resin

Soil	Amount of resin (n)							
	1	2	1	2	1	2	1	2
	P		K		Ca		Mg	
	-------------------- mg dm^{-3} -------------------				-------------------- cmol$_c$ dm^{-3} ---------------------			
Oxisol	4.9 a	6.4 a	59.3 b	69.3 a	5.22 a	5.83 a	2.11 b	2.66 a
	31.5 b	44.2 a	44.2 a	48.6 a	6.47 a	7.38 a	2.37 b	3.08 a
	60.3 b	76.8 a	53.4 a	57.0 a	4.67 a	5.44 a	2.27 b	3.45 a
Inceptisol	6.3 b	9.8 a	47.2 a	48.5 a	10.09 b	13.38 a	3.03 b	3.94 a
	18.1 b	26.8 a	43.1 a	41.4 a	4.49 b	6.66 a	2.68 b	4.07 a
	58.8 b	76.1 a	49.2 a	44.7 a	9.99 b	14.39 a	2.82 b	4.22 a
Spodosol	19.0 a	19.2 a	31.5 b	39.5 a	2.38 a	3.23 a	1.26 b	1.72 a
	50.8 b	54.0 a	30.6 b	36.3 a	2.66 a	3.49 a	1.36 b	1.91 a
	81.9 b	85.6 a	34.8 a	36.2 a	3.03 a	3.90 a	1.25 b	1.76 a
Alfisol	8.2 a	8.8 a	33.8 a	30.4 a	2.09 a	2.94 a	0.84a	1.19 a
	20.0 a	21.4 a	30.6 a	31.1 a	1.86 a	2.72 a	0.75 a	1.10 a
	73.0 b	82.3 a	28.7 b	34.3 a	2.27 a	3.03 a	0.80 a	1.09 a
Acrisol	2.7 a	3.4 a	13.5 a	14.2 a	0.37 a	0.68 a	0.54 a	0.19 a
	28.5 a	29.3 a	12.5 b	16.2 a	0.53 a	0.63 a	0.12 a	0.12 a
	62.3 b	72.1a	12.9 b	16.8 a	0.51 a	0.86 a	0.14 a	0.19 a
Average	35.1 b	41.1 a	34.3 b	37.6 a	3.77 b	4.97 a	1.49 b	2.05 a

Means followed with the same lower case letter within a row did not differ statistically by Tukey test (5%); C.V. (%) P = 4.6; C.V. (%) K = 8.0; C.V. (%) Ca = 13.0; C.V. (%) Mg = 14.

4. Conclusion

Modifications in the extraction procedure with ion-exchange resin membrane resulted in significant differences in the extracted amounts of soil P, K, Ca, and Mg. Among the variables studied, temperature had the greatest effect on the amounts of P extracted by resin ion exchange membrane and similarly with ion-exchange resin spheres. Temperature effects on soil P extraction can result in inadequate P recommendations to crops, which would have a larger impact on those soils with already low P availability to crops. Temperature must be controlled to avoid inaccurate results that can affect fertilizer recommendations to crops. Studies with soils from other regions and also with large variability in soil nutrient contents are necessary to better understand the factors that affect resin membrane efficiency in the soil testing analyzes.

Table 11. Phosphorus extracted with ion exchange resin membrane affected by laboratory temperature and the impact on nutrient recommendation to corn in Southern Brazil[§]

Soils	Laboratory temperature (°C)															
	10				15				25				40			
	P															
	Soil-P mg·dm⁻³	Status[†]	Applied[‡] kg·ha⁻¹	Balance[¥] %	Soil-P mg·dm⁻³	Status	Applied kg·ha⁻¹	Balance %	Soil-P mg·dm⁻³	Status	Applied kg·ha⁻¹	Balance %	Soil-P mg·dm⁻³	Status	Applied kg·ha⁻¹	Balance %
Oxisol	3.0a	VL	215	100	4.5a	VL	215	100	4.9a	VL	215	100	7.2a	L	175	81
	15.5d	M	165	122	24.3c	H	135	100	31.5b	H	135	100	59.4a	VH	90	67
	32.2d	H	135	150	46.6c	VH	90	100	60.3b	VH	90	100	103.8a	VH	90	100
Inceptisol	3.0b	VL	215	123	6.2ab	L	175	100	6.3ab	L	175	100	9.3a	L	175	100
	8.7c	L	175	106	14.1b	M	165	100	18.1b	M	165	100	32.1a	H	135	82
	29.7d	H	135	150	45.0c	VH	90	100	58.8b	VH	90	100	95.5a	VH	90	100
Spodosol	12.1c	M	165	100	17.7b	M	165	100	19.0b	M	165	100	27.8a	H	135	82
	33.5d	H	135	150	42.9c	VH	90	100	50.8b	VH	90	100	70.7a	VH	90	100
	52.8d	VH	90	100	71.0c	VH	90	100	81.9b	VH	90	100	1l89a	VH	90	100
Alfisol	6.1b	L	175	100	7.5ab	L	175	100	8.2ab	L	175	100	11.8a	M	165	94
	11.3c	M	165	122	20.0b	M	165	122	20.9b	H	135	100	28.6a	H	135	100
	43.2d	VH	90	100	61.4c	VH	90	100	73.0b	VH	90	100	100.1a	VH	90	100
Acrisol	1.9a	VL	215	100	2.6a	VL	215	100	2.7a	VL	215	100	4.1a	VL	215	100
	17.5c	M	165	122	26.6b	H	135	100	28.5b	H	135	100	40.9a	VH	90	67
	39.0d	H	135	100	54.5c	VH	90	100	62.3b	VH	90	100	92.1a	VH	90	100
Average	20.6d	-	90	67	29.7c	-	135	100	35.1b	-	135	100	53.5a	-	90	67

Note. §: Means in soil-P followed with the same lower case letter within a row in soil-P did not differ statistically by Tukey test (5%); C.V. (%) = 7.1;

†: P status in soil based on soil-P extracted (mg dm⁻³) with membrane resin according to Southern Brazil Manual for Crop Nutrient and Lime Recommendations. Very Low (VL = soil-P ≤ 5.0); Low (L = soil-P range of 5.1-10.0); Medium (M = soil-P range of 10.1-20.0); High (H = soil-P range of 20.1-40.0); Very High (VH = soil-P ≥ 40.0);

‡: P applied to corn to an expected yield of 10 Mg ha⁻¹ according to Southern Brazil Manual for Crop Nutrient and Lime Recommendations.

¥: P balance is amount of P (in percent) above or below recommendation P applied considering 25 °C as laboratory temperature standard (100%).

References

Amer, F., Bouldin, D. R., Black, C. A., & Duke, F. R. (1955). Characterization of soil phosphorus by anion exchange resin adsorption and P32 equilibration. *Plant Soil, 6*, 391-408. http://dx.doi.org/10.1007/BF01343648

Bache, B. W., & Ireland, C. (1980). Desorption of phosphate from soils using anion exchange resins. *Journal of Soil Science, 31*, 297-306. http://dx.doi.org/10.1111/j.1365-2389.1980.tb02082.x

Bortolon, L., & Gianello, C. (2008). Interpretação de resultados analíticos de fósforo pelos extratores Mehlich-1 e Mehlich-3 em solos do Rio Grande do Sul. *Revista Brasileira de Ciência do Solo, 32*, 2751-2756. http://dx.doi.org/10.1590/S0100-06832008000700019

Bortolon, L., Gianello, C., & Kovar, J. L. (2011). Phosphorus availability to corn and soybean evaluated by three soil test methods for southern Brazilian soils. *Communications in Soil Science and Plant Analysis, 42*, 39-49. http://dx.doi.org/10.1080/00103624.2011.528488

Cooke, I. J., & Hislop, J. (1963). Use of anion-exchange resin for the assessment of available soil phosphate. *Soil Science, 96*, 308-312. http://dx.doi.org/10.1097/00010694-196311000-00004

Cooperband, L. R., & Logan, T. J. (1994). Measuring in situ changes in labile soil phosphorus with anion-exchange membranes. *Soil Science Society of America Journal, 58*, 105-114. http://dx.doi.org/10.2136/sssaj1994.03615995005800010015x

Dobermann, A., Langer, H., & Mutcher, H. (1994). Nutrient adsorption kinetics of ion exchange resin capsules: A study with soils of international origin. *Communications in Soil Science and Plant Analysis, 25*, 1329-1353. http://dx.doi.org/10.1080/00103629409369119

Haagsma, T., & Miller, M. H. (1963). The release of non-exchangeable soil potassium to cation exchange resins as influenced by temperature, moisture and exchanging ion. *Soil Science Society of America Proceedings, 27*, 153-156. http://dx.doi.org/10.2136/sssaj1963.03615995002700020020x

Kovar, J. L. (2003). Searching for more effective corn starters in conservation-till. *Fluid Journal, 11*, 8-11.

Kovar, J. L., Bortolon, L., & Karlen, D. L. (2009). Distribution of phosphorus and potassium following surface banding of fertilizer in conservation tillage systems. *Plant Nutrition Colloquium Proceedings*. Retrieved from http://escholarship.org/uc/item/8j0224dq

Mallarino, A. P., & Atia, A. M. (2005). Correlation of a resin membrane soil phosphorus test with corn yield and routine soil tests. *Soil Science Society of America Journal, 69*, 266-272. http://dx.doi.org/10.2136/sssaj2005.0266

McLaughlin, M. J., Lancaster, P. A., & Sale, P. W. G. (1993). Use of cation/anion exchange membranes for multi-element testing of acid soils. *Plant Soil, 22*, 249-252.

Qian, P., Schoenau, J. J., & Huang, W. Z. (1992). Use of ion exchange membrane in routine soil testing. *Communications in Soil Science and Plant Analysis, 23*, 1791-1804. http://dx.doi.org/10.1080/00103629209368704

Raij, B., Quaggio, J. A., & Silva, M. N. (1986). Extraction of phosphorus, potassium, calcium, and magnesium from soils by an ion-exchange resin procedure. *Communications in Soil Science and Plant Analysis, 17*, 547-566. http://dx.doi.org/10.1080/00103628609367733

Rolim, M. V., Novais, R. F., Nunes, F. N., & Alvarez, V. H. V. (2008). Efeito da moagem do solo no teor de fósforo disponível por Mehlich-1, resina em esferas e em lâmina. *Revista Brasileira de Ciência do Solo, 32*, 1181-1190. http://dx.doi.org/10.1590/S0100-06832008000300026

Saggar, S., Hedley, M. J., & White, R. E. (1990). A simplified resin membrane technique for extracting phosphorus from soils. *Nutrient Cycling in Agroecossystem, 24*, 173-180. http://dx.doi.org/10.1007/bf01073586

SBCS/NRS (Sociedade Brasileira de Ciência do Solo/Núcleo Regional Sul). (2004). *Manual de adubação de de calagem para os estados do Rio Grande do Sul e Santa Catarina*. Porto Alegre, UFRGS. Retrieved from http://www.sbcs-nrs.org.br/docs/manual_de_adubacao_2004_versao_internet.pdf

Schlindwein, J. A., & Gianello, C. (2009). Fósforo disponível determinado por lâmina de resina enterrada. *Revista Brasileira de Ciência do Solo, 33*, 77-84. http://dx.doi.org/10.1590/S0100-06832009000100008

Schoenau, J. J., & Huang, W. Z. (1991). Anion-exchange membrane, water, and sodium bicarbonate extractions as soil tests for phosphorus. *Communications in Soil Science and Plant Analysis, 22*, 465-492. http://dx.doi.org/10.1080/00103629109368432

Sibbesen, E. A. (1978). Simple ion-exchange resin procedure for extracting plant-available elements from soil. *Plant Soil, 46*, 665-669. http://dx.doi.org/10.1007/BF00015928

Simonis, A. D. (1996). Effect of temperature on extraction of phosphorus and potassium from soils by various extracting solutions. *Communications in Soil Science and Plant Analysis, 27*, 665-484. http://dx.doi.org/10.1080/00103629609369586

Skogley, E. O., Geortis, S. J., & Yang, J. E. (1990). The phytoavailability soil test - PST. *Communications in Soil Science and Plant Analysis, 21*, 1229-1243. http://dx.doi.org/10.1080/00103629009368301

Somarisi, L. L. W., & Edwards, A. C. (1992). An ion exchange resin method for nutrient extraction of agricultural advisory soil samples. *Communications in Soil Science and Plant Analysis, 23*, 645-657. http://dx.doi.org/10.1080/00103629209368616

Sparks, D. L. (2003). *Environmental soil chemistry*. San Diego: Academic Press.

Takahashi, S. (1996). Influence of soil particle size and grinding on Bray-2 extractable phosphorus. *Communications in Soil Science and Plant Analysis, 27*, 2829-2835. http://dx.doi.org/10.1080/00103629609369744

Takahashi, S. (2013). Soil phosphorus extracted by three anion exchange membranes compared to water-soluble and truog-extractable phosphorus. *Communications in Soil Science and Plant Analysis, 44*, 3231-3234. http://dx.doi.org/10.1080/00103624.2013.840835

Tedesco, M. J., Gianello, C., Bissani, C. A., Bohnen, H., & Volkweiss, S. J. (1995). *Análise de solo, plantas e outros materiais*. Porto Alegre, UFRGS.

Vaidyanathan, L. V., & Talibuden, O. (1970). Rate process in the desorption of phosphate from soils by ion-exchange resins. *Journal of Soil Science, 21*, 173-183. http://dx.doi.org/10.1111/j.1365-2389.1970.tb01165.x

Vandecar, K. L., Lawrence, D., & Clarck, D. (2010). Phosphorus Sorption Dynamics of Anion Exchange Resin Membranes in Tropical Rain Forest Soils. *Soil Science Society of America Journal, 75*, 1520-1529. http://dx.doi.org/ 10.2136/sssaj2010.0390

Yang, J. E., Skogley, E. O., & Geortis, S. J. (1991a). Phytoavailability soil test: development and verification of theory. *Soil Science Society of America Journal, 55*, 1358-1365. http://dx.doi.org/10.2136/sssaj1991.03615995005500050027x

Yang, J. E., Skogley, E. O., & Schaff, B. E. (1991b). Nutrient flux to mixed-bed ion exchange resin: temperature effects. *Soil Science Society of America Journal, 55*, 762-767. http://dx.doi.org/10.2136/sssaj1991.03615995005500030021x

11

Hydrocarbon-Contamination of Soils: The Potential of Sodium Clays to Decelerate Soil Toxicants

Mark Anglin Harris[1]

[1] College of Natural & Applied Sciences, Northern Caribbean University, Mandeville, WI, Jamaica

Correspondence: Mark Anglin Harris, College of Natural & Applied Sciences, Northern Caribbean University, Mandeville, WI, Jamaica. E-mail: mark.harris@ncu.edu.jm

Abstract

Hydrocarbon-contamination can change hydraulic conductivity (HC) in soils, and hence increase the spreading rate of aqueous toxicants in the ground. A constant head permeameter used in the laboratory to measure HC of soils taken from near the Pitch Lake in Trinidad determined that the HC exceeded that of a reference soil having "normal HC" for a loam. Although water moved rapidly through it, the Pitch Lake soil (PLS) remained dry due to water repellence. Treatment consisted of either of two red mud bauxite wastes mixed at 25 and 50% w/w with PLS at air dry. One of the bauxite wastes had undergone treatment with gypsum several years before and hence contained a greater proportion of calcium ions compared to the other red mud which contained more sodium ions. At 25% w/w the non-gypsum-treated red mud waste decreased HC of the PLS by 50%, and at 50% w/w caused a 10-fold decrease of HC on the PLS. The gypsum-treated red mud waste had no effect on the HC of the PLS. The drastic decrease in HC of the hydrocarbon-contaminated soil implies blocking of hydraulic channels by inorganic particles. The high levels of Na^+ released in the Bayer beneficiation process dispersed and released fine < 5 mμ clay particles from the non-gypsum-treated red muds. This suggests that the rapid movement of aqueous pollutants in such hydrocarbon-polluted soils could be similarly curtailed under field conditions.

Keywords: hydrophobic soils, saturated hydraulic conductivity, sodicity, water repellence

1. Introduction

Water repellent soils normally exhibit low infiltration rates (Blackwell, 2000). Hydrophobicity can reduce the affinity for soils to water such that infiltration or wetting may be delayed for periods ranging from as little as a few seconds to in excess of weeks (Hall, 2009). Soil hydrophobicity is thought to be caused primarily by a coating of long-chained hydrophobic organic molecules on individual soil particles, thereby influencing soil hydrological and ecological functions (Takawira et al., 2014). A study by Lourenco et al. (2015) confirmed the hypothesis that hydrocarbon contamination induces water repellence and reduces soil moisture retention at low suction (< 100 kPa) for laboratory contaminated soils. Water is not easily absorbed by such non-wetting soils. Yet, in a pilot study of the Trinidad Pitch Lake soils at La Brea:

(1) Water moved rapidly through the soil fabric, and;

(2) The soil remained completely dry, despite prodigious movement of water through it;

(3) Soil crumbs (> 2 mm), though denser than water, literally floated on water;

(4) Soil crumbs > 2 mm exhibited mutual repellence when floated in water.

Therefore, quite apart from the last three more esoteric above-mentioned reactions, it was concluded that aqueous contaminants released in such soils, could move just as quickly. There exists therefore, the potential (Figure 1) to rapidly contaminate large volumes of soil during toxic aqueous spills.

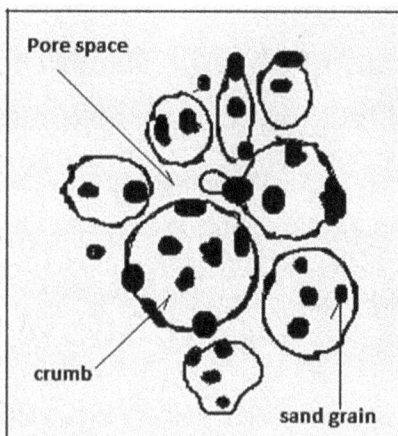

Figure 1. A typical soil entity with no barriers to internal wetting of individual crumbs

However when crumbs are coated and sealed with hydrophobic materials water fails to enter the crumb, not wetting them. Slow wetting does not occur. For the hydrocarbon contaminated pitch-lake soil of Trinidad, water through-flow rate abnormally increases as it bypasses the soil crumbs.

Surface seepages of petroleum hydrocarbons cause hydrophobicity in soils adjacent to the Pitch Lake in southern Trinidad. In other fractured fields where surface seepages occur, there may be potentially similar results. Non-wetting soils are often prone to dispersion and structural breakdown (Ward et al., 2015). Water repellence poses challenges to stability of building foundations (Pietruszczak & Oulapour, 1999), pesticide concentration, and leaching (Blackwell, 2000), and for crop production in terms of crop establishment, nutrition, and weed control (Ward et al., 2015). The exact chemical composition of substances responsible for the development of water repellence in soils is difficult to identify, but they are generally organic compounds that accumulate on and between soil particles (Dekker et al., 2009). Two approaches to the modification of soil hydrophobicity, and hence water movements, are (1) inorganic and (2) organic.

1.1 Inorganic Approach

Water retention in soil is critical for remediation of contaminated sites (Lourenco et al., 2015). According to Roy and McGill (2002), soil water repellence is a function of soil surface chemistry. For example, abrasion of sand particles during light sieving had only small effects on repellence, but more vigorous abrasion through rotational movement of the sand reduced repellence markedly (King, 1981). Alternatively, soil hydraulic conductivity (HC) is related to the composition and concentration of exchangeable cations and soluble electrolytes (Frenkel et al., 1992). Thus Quirk and Schofield (1955) showed that the HC of a soil decreased with increasing exchangeable sodium percentage (ESP) and decreasing electrolyte concentration in the soil solution. Essington (2004) described dispersive conditions as mutual repulsion of tactoids fully surrounded by associated Na^+ and waters of hydration (ESP < 15). McNeal et al. (1966) found a linear relationship between reduction of HC and the degree of macroscopic swelling of the extracted soil clay. Thus de-flocculation and movement of clay into the conducting pores was proposed as the second mechanism for explaining the reduction in HC. The decrease in soil HC due to swelling and dispersion of clay is explained by the presence of monovalent exchangeable cations. Such changes were responsible for the loss of nearly all drainable macro-pores in the soil, which was evidenced by the considerable reduction in cumulative infiltration amounts and rates (Lourenco et al., 2015).

Clay mineralogy has also been shown to have a large influence on reductions in HC (Churchman et al., 1993; McNeal & Coleman, 1966). Due to their 2:1 layer structure which accommodates a high amount of exchangeable Na+ within its interlayer space (Arienzo et al., 2012; Churchman et al., 1993), smectites show extensive swelling and dispersion. Na^+ is a large monovalent ion and more effectively forces clay tactoids (*i.e.*, particles) apart than Ca^{2+} or Mg^{2+} (Quirk, 1986). Swelling occurs with increasing Na^+ concentration, as hydration of Na^+ leads to the expansion of the interlayer (ESP > 15).

1.2 Organic Approach

Rainfall infiltration into water repellent soils is characteristically patchy, resulting in patchy and staggered crop germination of crops and of weeds (Ward et al., 2015). They speculated that improvements in soil water availability, and increases in organic matter, might encourage microbial activity, leading to increased degradation of the compounds causing water repellence.

Roper et al. (2013) showed that soil water contents in water repellent sand were greater in crop rows where crop residue was retained. Based on observations, Ward et al. (2015) determined that soils in crop rows showed lower levels of water repellence when crop rows were established on or close to previous crop rows, compared with crop rows established on the inter-row spaces. This was attributed to the likelihood of greater organic matter and nutrient accumulation, the former increasing water availability, which could have encouraged microbial activity (Blackwell et al., 2014). Such microbial activity could have led to increased degradation of the compounds causing water repellence (Walden, 2015). Nevertheless, compared to naturally-occurring and fire-induced hydrophobicity, limited information is available on the impacts of hydrocarbon contamination on water repellence and hydraulic properties (Lourenco et al., 2015). Increased saturated hydraulic conductivity associated with laboratory contaminated soils contradicted their original hypothesis (Lourenco et al., 2015) such that their findings imply that storms falling on initially dry recently contaminated soils may trigger contaminant transport and erosion via enhanced surface runoff, and rapid spreading of contaminants once they reach the groundwater systems.

1.3 Aim & Hypothesis

The aim of this study is to determine, for a hydrocarbon-contaminated non-wetting soil, (1) the hydraulic conductivity and hence (2) the likelihood of rapid spread of aqueous pollutants, and (3) a potential method to reduce the spreading rate of such contaminants in that hydrocarbon-contaminated soil.

Sources of Na^+ bentonite, which can reduce infiltration rates are rare in the Caribbean. Therefore it is hypothesized that abundant Na^+-rich dried red muds from bauxite waste (of relatively low toxicity) could release reduce very high infiltration rates in hydrophobic hydrocarbon-contaminated soils.

2. Materials and Methods

2.1 Site Details

The Pitch Lake at La Brea occupies 36 ha in southwestern Trinidad at 62° W and 10.5° N (Figure 2). Average annual rainfall at La Brea is 1250-1750 mm, most of which falls between May and December. Temperature is lowest in January when it averages 25 °C, rarely rising higher than 33 °C in June through September, which are the hottest months.

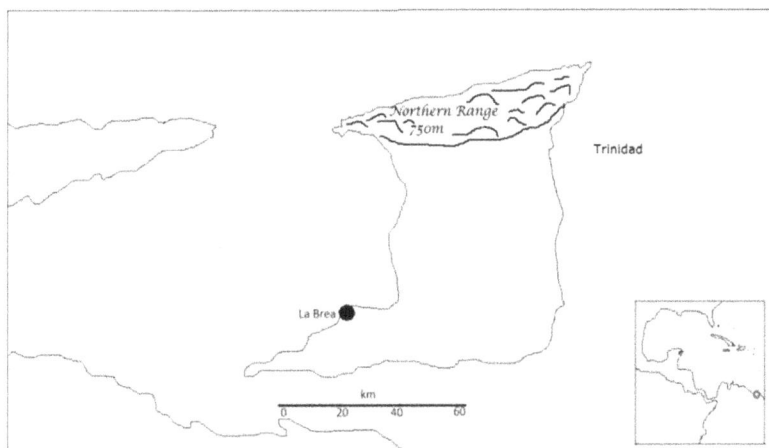

Figure 2. Location of the Pitch Lake at La Brea (circular dot) in Trinidad & Tobago

2.2 Soil Sampling

Soil cores were collected on the periphery of the Pitch Lake along a transect perpendicular to the centre in April 2012, to a depth of 20 cm and brought to a laboratory on the same day. The soil was brought to air dry after exposure for three days. Black petroleum tar contaminated some soil aggregates (Figure 3) at the site. Table 1 shows that organic carbon greatly dominated this soil, at > 10 times the normally expected levels.

Figure 3. Hydrocarbon-contaminated soil on periphery of Pitch Lake in LaBrea, south-western Trinidad
Note. The diameter of each circular indentation in the background at right is 1 mm.

Table 1a. Some properties of an undisturbed reference soil from Mount Nelson, Jamaica, and a Pitch Lake soil (Trinidad)

Property	Reference soil	Pitch Lake soil
Organic matter (%)	4.2	56.6
pH	5.3	6.9

Note. Soil samples were taken 100 m at depth 0-15 cm from the periphery of the pitch lake at LaBrea, Trinidad.

2.3 Water Repellence Determination

Water repellence was measured by counting the number of seconds required for water to be absorbed into the soil (WPDT) after the method of DeBano (1981). Fifty replicates were used, where water repellence is classified as follows: lower than 3 seconds = highly wetting (0); 3-6 seconds = wetting (1); 6-60 seconds = slightly water repellent (2); 60-600 seconds moderately water repellent (3); > 600 seconds = extremely water repellent (4). Samples were measured at air-dry were crushed and sieved to less than 2.0 mm to remove large organic matter and gravel.

2.4 Dispersion and Hydraulic Conductivity

Measurement of spontaneous dispersion and hydraulic conductivity were conducted respectively using the Modified Emerson Water Dispersion Test (Emerson, 1967). Based on the objective of potentially using the treatment as a barrier for aqueous contaminants in field soils, potentially large volumes of soil additive could be required to affect interflow under field conditions. Therefore, the constant head method used for hydraulic conductivity Klute (1983) was modified to include soil columns of thin cross-sections, *i.e.*, of 2.5 cm thickness.

2.5 Bauxite Wastes

Samples of gypsum-treated bauxite waste (O'Callaghan et al., 1998) were collected from Kirkvine Pond 6 during 2007 from two depths: 0-15 and 15-30 and stored at air-dry. Samples collected from an unmined, undisturbed bauxite soil within the same locality were used a reference soil. Bardossy (1982) characterized this reference soil as dominantly kaolinite, and freely draining. At air-dry, samples were grinded and passed through a 1.0 mm diameter sieve.

Table 1b. Some properties of bauxite wastes from Jamaica, and of a Pitch Lake soil from Trinidad

Bauxite wastes	pH	Organic carbon
Dried red mud	10.8	0.1
Gypsum-treated red mud	8.3	0.2

Note. Soil samples were taken at a depth of 0-15 cm for the gypsum-treated red mud, and 15-30cm for the dried red mud, subjacent to the gypsum-treated material.

2.6 Bauxite Waste Treatments

Two depths were selected as potential treatment additives because only the 0-15 cm layer was gypsum treated. It has been shown (Harris, 2008) that the gypsum applied to a depth of 15 cm had little or no physical effect on soil below 15 cm depth. Therefore the original chemistry of the RMW at 15-30 cm was not as altered as that of the 0-15 depth (GRMW). It was therefore postulated that the greater ESP and SAR of material from 15-30 cm may enhance displacement of inorganic ions in a hydraulic column. Removal of divalent ions from the exchange complex can contribute to reductions in aggregate stability and decreases in HC (Kopittke et al., 2006).

2.7 Red Mud Safety Concerns

Red mud can contain elevated levels of metals, and the pH of red mud characteristically ranges from 10 to 12 due to the use of caustic soda during the extraction process. Red mud is also known to contain technologically enhanced naturally occurring radioactive material (TENORM), including thorium and uranium. TENORM is naturally occurring radioactive material (NORM) that has been processed in such a manner that its concentration has increased (EPA, 2012).

However, concerns about the radioactivity in red mud waste were allayed because (Pinnock, 1991) showed that using 100% bauxite residue gave a dose equivalent to just over 2 m Sv y^{-1} and was judged to be acceptable. However, other work on Hungarian bauxites has recommended a maximum addition of 15% bauxite residue to avoid exceeding a level of 0.3 m Sv y^{-1} (Pinnock, 1991).

2.8 Statistics

Sample means compared using Tukey's Honestly Significant Difference ($P \leq 0.05$) to examine differences in HC between treatments.

3. Results

3.1 Water Repellence

Water repellence of the PLS was found to be at the highest value, *i.e.*, extremely water repellent (4), compared with the reference soil, which had a value of highly wetting (0), *i.e.*, showing very rapid water absorbance. Rapid water absorbance is unusual for hydrophobic soils. Further, the fact that soil aggregates after several days in water, and even aggregates as small as < 2 mm remained dry after several hours of water infiltration showed the extremely hydrophobic nature of the Pitch Lake soils (Figure 4).

Table 2. Water repellence rating* of soil near a pitch lake, and that of bauxite wastes

Reference soil	PLS	RM	GR
0	4	0	0

Note. *Water repellence, measured by counting the number of seconds required for water to be absorbed into the soil after the method of DeBano (1981), is classified as follows: lower than 3 seconds = highly wetting (0); 3-6 seconds = wetting (1); 6-60 seconds = slightly water repellent (2); 60-600 seconds moderately water repellent (3); > 600 seconds = extremely water repellent (4).

Figure 4. Pitch Lake soil agglomerates after being subjected to through-flow in a water column

Note. Soil particles remained dry even after water flowed for several (> 30) minutes through the mass. The agglomerates are floating in the water, as proven by the lack of shadows under the soil particles at left. Each circular depression in background cloth is 1 mm in diameter.

3.2 Changes in Hydraulic Conductivity

There was little, if any, restriction to water-flow through the hydrocarbon contaminated samples. The hydrocarbon-contaminated soils (HCS) all exhibited high values for hydraulic conductivity, exceeding 150 cm min^{-1} and exceeding that of the reference soil by > 35% (Table 3). This flow rate is similar to that through a fine-to-medium (< 2 mm) quartz sand.

When RMW was added at 25% of the volume of the PLS in the column, the RMW effect on the HC of the PLS was expected to be proportional to its volume (25%). Yet the 25% treatment of RMW changed PLS hydraulic conductivity, not, as was expected, by 25%, but by a far larger proportion, *i.e.*, by 400% (decrease) in HC (Table 3).

Table 3. Changes to hydraulic conductivity (cm min^{-1}) of Pitch Lake soils after treatment with dried red muds

	N & C	HN25 & HC25	HN50 & HC50
Na$^+$ & Ca$^+$-rich waste (N)	0	47	20
Ca$^+$ rich waste (C)	202	202	202

Note. Reference soil = 55 cm min^{-1}, H = hydrophobic soil, HN25 = H + 25% Na$^+$ & Ca$^+$-rich red mud, HC25 = hydrophobic soil + 25% Ca$^+$-rich red mud, HN50 = H + 50% Na$^+$ & Ca$^+$ -rich red mud, HC50 = hydrophobic soil + 50% Ca$^+$ dominated red mud.

Even more dramatically, RMW added at the rate of 50% to the PLS reduced HC in the column not by 50%, but by a 10-fold margin (1000%). Conversely, the GRMW in similarly added proportions of 25 and 50% had no effect on the HC of the PLS in the column. There was no observed or detected reaction. Though the reduction in porosity of the soil column was dependent on concentration of the additive, it must be noted that the particle diameters of all treatments, including the non-effective (gypsum-based) ones, were identical.

3.3 Hydrophobicity & Dispersion

Though all PLS soil samples along the transect taken across the periphery of the La Brea Pitch-Lake were water repellent, there was no observed pattern of soil hydrophobicity increasing inversely with distance from the centre of the Pitch Lake. This suggests a low threshold of hydrocarbon contamination required to produce hydrophobicity in a soil. The Modified Emerson Water Dispersion Test showed that the PLS exhibited slaking and dispersion at level 4 (the second highest level) after just 2 hours (Table 4). Contrastingly, no dispersive reaction was detected for the reference soil (Table 4), even after several weeks of contact with water. Though all the aggregates were water repellent, and, to varying degrees dispersive, those aggregates which visibly contained large amounts of high density black tar were the least dispersive, while the PLS with the lowest levels of hydrocarbon contamination were the most dispersive.

Figure 5. Picture shows air bubbles from explosive reaction when dried Na+ dominated red mud makes contact with water, culminating in total dispersion and destruction after just five minutes in water. The clay particles so released, blocked pores in a hydrophobic Pitch Lake soil

Table 4. Modified Emerson Dispersion Test levels of dried bauxite waste and Pitch Lake soils

	REF & REF	PLT & PHT	HN & HC
Na$^+$ rich waste (N)	0	4	4
Ca$^+$ rich waste (C)	0	4	0

Note. REF = reference soil, PLT = low-tar content pitch lake soil, PHT = high tar content pitch lake soil, HN = Na$^+$ & Ca$^+$-rich dominated red mud, HC = Ca$^+$-rich red mud.

Key: Emerson dispersion levels: *1* Slight milkiness *2* Obvious milkiness, less than 50% of the aggregate affected *3* Obvious milkiness, greater than 50% of the aggregate affected *4* Total dispersion leaving only sand grains.

Source: Emerson Aggregate Test (1967).

4. Discussion

4.1 Infiltration

On the assumption of high levels of hydrocarbon contamination in the PLS, the inherently high level of saturated hydraulic conductivity correlating with hydrocarbon-contamination of the soil in this study is corroborated by Lourenco et al. (2015), who, in a laboratory simulated hydrocarbon contamination, induced soil water repellence. Saturated hydraulic conductivity (K_s) increased linearly with level of (hydrocarbon) contamination ($p < 0.05$; r2 ≈ 0.8), indicating rapid flow of water attributed to a reduction of the dielectric constant, and hence water–soil matrix interactions (Lourenco et al., 2015). A humic soil is defined as soil in which organic carbon exceeds 50%. Yet, the PLS, with 56% organic carbon (Table 1) is not a humic soil, there being no visible humus (Figure 6). Therefore, it is reasonable to conclude that hydrocarbons represent most of the organic C in PLS. On close microscopic inspection of the mineral soil, there was no humus visible. In this study, the soil with lower observable levels of hydrocarbons (black tar) exhibited greater dispersion. Hence the less dispersive samples occurred closest to the centre of the Pitch Lake PLS. This was against expectations.

In the samples closer to the centre of the Pitch Lake, dark, tarry petroleum occupied a greater proportion of the soil. Asphalt generally contains some of the heaviest and least volatile fractions of petroleum distillates. In the aquatic environment, asphalt will sink to bottom as a dark tarry substance. Petroleum distillates in order of decreasing volatility include:

1. Petroleum ether or benzene

2. Gasoline

3. Naphtha

4. Mineral spirits

5. Kerosene

6. Fuel oil

7. Lubricating oils

8. Paraffin wax

9. Asphalt or tar

As gravitational forces act on all objects including those with horizontal movement, lighter, more mobile short-molecular chain hydrocarbon fractions would have been expected to have travelled furthest out from the centre of the Pitch Lake. It is plausible then, that the smaller hydrocarbon chains of the lighter fractions would have more easily penetrated the micro-pores and microstructure of those soil aggregates in direct variation with distance from the centre of the Pitch Lake. Thus, the greater the distance, the greater the sorting and concentration of contaminants in such soil samples. This leads thereby to (1) greater effective hydrophobicity and/or (2) greater dispersion amongst such (more distant) samples.

The results of this study show that a dispersive clay synergistically decreased hydraulic conductivity in a hydrophobic hydrocarbon-contaminated soil. This outcome supports the hypothesis of this study which was based on the activity of sodium-affected soils in an aqueous medium. Even as little as 1% of the total clay when dispersed, affects the hydraulic conductivity by blocking micro-channels in the soil mass (Goldberg & Glaubig, 1987). Tuffour et al. (2015) found that finer sediments were highly effective in altering soil properties even at low concentrations. They showed that after mixing with 10% dry Na^+ bentonite, wetting resulted in complete filling of skeletal pore space by a jell-like clay fabric. With a 3% level of sodium oxides in Jamaican red muds, the Na^+ levels of the RMW would have been sufficiently strong to power clay dislocation, and it is dislocated clay-sized particles that changed the HC in the Pitch Lake soils of this study.

The difference between the dried gypsum-treated red mud waste at 0-15 cm depth and that of the 15-30 cm depth was still large even after 10 years. For example, it can be seen that an explosive reaction (air bubbles observed) occurred when the Na^+-rich red muds were placed in water (hydrated, Figure 5), thus blocking pores in the PLS.

In contrast, the dried gypsum-treated red mud retained its integrity indefinitely (Figure 6) and thus had no effect on the hydraulic conductivity of the Pitch Lake soil. These samples had been in water not for 5 minutes but for five weeks.

These results suggest that the RMW released components which blocked hydraulic channels in the PLS. As described in previous studies, the negative effects of Na^+ are most pronounced in soils which are high in clay, especially high charge-density clays (Frenkel et al., 1978; Beulow et al., 2015). In the current study, the water-repellent soils were more responsive to soil with higher Na^+ quantities, than the less reactive gypsum-treated soil. Though the original bauxite soils are mainly kaolinite, red mud wastes contain an increased number of fine clay particles compared to that of the original soil.

Figure 6. Dried gypsum-treated red mud waste (GRMW) had no observable reaction with water, retaining its stability even after many weeks. Scale is supplied by 1-mm-diameter indented circles in the backgound

These provide extra negatively charged sites for adsorbing hydrated Na^+ cations. Similarly, the tendency of Na^+ ions to facilitate the formation of multiple layer hydrates in smectite clays leads to greater swelling and reduction of HC. On the contrary, Frenkel et al. (1992) showed that smectites did not release clay particles when added to a sandy soil unless anions were added, because the HC of smectite clay-sand mixtures decreased only following the addition of the various anions. Dispersed clay appeared in their effluent only upon addition of citrate or hexametaphosphate. Therefore, at 56% intrinsic organic matter (Table 1), such hydrocarbon contamination would have greatly increased the negative charges of the Pitch Lake soil. This could have increased repelling forces amongst the thoroughly mixed-in clay particles, thereby blocking hydraulic channels.

Further, this result conforms with the very high sensitivity of illite to even small amounts of exchangeable Na (Oster et al., 1980). They found that, for a given ESP, the critical flocculation concentrations (CFC) of illite was much larger than for smectite, and hypothesized that the explanation lies in the irregular nature of illite particles, which prevent good contact between edges and planar surfaces, thereby decreasing the potential for inter-particle attraction. Similarly, the red mud particles, having undergone beneficiation (extreme comminution), would have been altered to more irregular shapes. Compared with Sintering red mud, the Bayer Process (used in the Caribbean alumina processing) produces a relatively small particle diameter. The particle diameter of Bayer red mud is between 0.8 μm and 50 μm with an average value of 14.8 μm (Wang & Liu, 2012). In the case of the smectite cited by Frenkel et al. (1992), HC is decreased through partial blocking of pores by short distance migration of dispersed particles, provided the particle size of the sand (or host soil) is sufficiently fine to retain these particles. Being a clay-loam, the particle size of the PLS is smaller than that of fine sand and therefore would be susceptible to pore blockage by the fine clay particles released from the dried red mud.

5. Conclusion

By repelling water from pore channel walls, hydrophobicity can markedly reduce resistance to movements of aqueous liquids through soils. Finely crushing RMW from < 2 mm to < 1 mm produced an at least 4-fold increase in the effective surface area of the RMW. Soil water repellence is a function of soil surface chemistry. More specifically, it is a function of the free energy of the solid/gas interface in soil (γSG). In contrast to the low Na^+ RMW, the high Na^+ RMW produced dislocated clay particles which blocked hydraulic channels.

With a high energy surface exhibited by rapidly decreasing the initial advancing contact angle (θ), the high Na^+ RMW either expanded, or released clay particles, either of which could have blocked hydraulic channels in the PLS.

References

Arienzo, M., Christen E. W., Jayawardane, N. S., & Quayle, W. C. (2012). The relative effects of sodium and potassium on soil hydraulic conductivity and implications for winery wastewater management. *Geoderma, 173*, 303-310. http://dx.doi.org/10.1016/j.geoderma.2011.12.012

Bardossy, G. (1982). *Karst Bauxites*. Elsevier Scientific Publishing Co.

Blackwell, P. S. (2000). Management of water repellence in Australia, and risks associated with preferential flow, pesticide concentration and leaching. *Journal of Hydrology, 231*, 384-395. http://dx.doi.org/10.1016/S0022-1694(00)00210-9

Blackwell, P. S., Hagan, J., Davies, S. L., Bakker, D., Hall, D. J. M., Roper, M. M., ... Matthews, A. (2014). Smart no-till furrow sowing to optimise whole-farm profit on non-wetting soil. *GRDC Perth Crop Updates*. Perth: Department of Agriculture and Food, and Grains Research and Development Corporation.

Buelowa, M. C., Kerri, S., & Sanjai, J. P. (2015). The effect of mineral-ion interactions on soil hydraulic conductivity. *Agricultural Water Management, 152*, 277-285. http://dx.doi.org/10.1016/j.agwat.2015.01.015

Churchman, G. J., Skjemstad, J. O., & Oades, J. M. (1993). Influence of clay–Minerals and organic-matter on effects of sodicity on soils. *Australian Journal of Soil Research, 31*, 779-800. http://dx.doi.org/10.1071/SR9930779

DeBano, L. (1981). *Water repellent soils: A state of the art*. USDA Forest Service. General Technical Report PSW-46.

Dekker, L., Ritsema, C., Klaas, O., Moore, D., & Wesseling, J. (2009). Methods for determining soil water repellence on field-moist samples. *Water Resources Research, 45*, w00d33. http://dx.doi.org/10.1029/2008WR007070

Emerson, W. W. (1967). A classification of soil aggregates based on their coherence in water. *Australian Journal of Soil Research, 5*, 47-57. http://dx.doi.org/10.1071/SR9940173

EPA. (2012). *Radiation protection*. Retrieved June 13, 2016, from https://www.epa.gov/radiation

Essington, M. (2004). *Soil and Water Chemistry: An Integrative Approach*. CRC Press, Boca Raton, Florida.

Frenkel, H., Goertzen, J. O., & Rhoades, J. D. (1978). Effect of clay type and content, exchangeable sodium percentage and electrolyte concentration on clay dispersion and soil hydraulic conductivity. *Soil Science Society of America Journal, 42*, 32-39. http://dx.doi.org/10.2136/sssaj1978.03615995004200010008x

Frenkel, H., Levy, G. J., & Fey, M. V. (1992). Clay dispersion and hydraulic conductivity of clay-sand mixtures as affected by the addition of various anions. *Clays and Clay Minerals, 40*(5), 515-521. http://dx.doi.org/10.1346/CCMN.1992.0400504

Goldberg, S., & Glaubig, R. A. (1987). Effect of saturating cation, pH, and aluminium and iron oxides on the flocculation of kaolinite and montmorillonite clay. *Clay Mineralogy, 35*, 220-227. http://dx.doi.org/10.1346/CCMN.1987.0350308

Hall, D. (2009). Water repellence. *Managing South Coast Sandplain Soils to Yield Potential* (pp. 49-63, Bulletin 4773). Department of Agriculture and Food: W. Aust.

Harris, M. A. (2008). Structural improvement of age-hardened gypsum-treated bauxite red mud waste using readily decomposable phyto-organics. *Environmental Geology, 56*, 1517-1522. http://dx.doi.org/10.1007/s00254-008-1249-5

King, P. (1981). Comparison of Methods for Measuring Severity of Water Repellence of Sandy Soils and Assessment of Some Factors that Affect its Measurement. *Australian Journal of Soil Research, 19*(3). http://dx.doi.org/10.1071/SR9810275

Klute, A. (1965). Laboratory measurement of hydraulic conductivity of saturated soil. In C. A. Black (Ed.), *Methods of Soil Analysis* (Part 1, pp. 210-221).

Kopittke, P. M., So, H. B., & Menzies, N. W. (2006). Effect of ionic strength and clay mineralogy on Na–Ca exchange and the SAR–ESP relationship. *European Journal of Soil Science, 57*, 626-633. http://dx.doi.org/10.1111/j.1365-2389.2005.00753.x

Lourenco Sergio, D. N., Wakefield, C., & Morley, C. (2015). Wettability decay in an oil-contaminated waste-mineral mixture with dry-wet cycles. *Environmental Earth Sciences, 74*(3). http://dx.doi.org/10.1007/s12665-015-4276z

McNeal, B. L., & Coleman, N. T. (1966). Effect of solution composition on soil hydraulic conductivity. *Soil Science Society of America Proceedings, 30*, 308. http://dx.doi.org/10.2136/sssaj1966.03615995003000030007x

O'Callaghan, W. B., McDonald, S. C., Richards, D. M., & Reid, R. E. (1998). *Development of a topsoil-free vegetative cover on a former red mud disposal site*. Alcan Jamaica Rehabilitaion Project Paper.

Oster, J. D., Shainberg, I., & Wood, J. D. (1980). Flocculation value and gel structure of Na/Ca montmorillonite and illite suspension. *Soil Science Society of America Journal, 44*, 955-959. http://dx.doi.org/10.2136/sssaj1980.03615995004400050016x

Pietruszczak, S., & Oulapour, M. (1999). Assessment of Dynamic Stability of Foundations on Saturated Sandy Soils. *Journal of Geotechnology & Geoenvironmental Engineering, 125*(7), 576-582. http://dx.doi.org/10.1061/(ASCE)1090-0241(1999)125:7(576)

Pinnock, W. R. (1991). Measurement of Radioactivity in Jamaican Building Materials and Gamma Dose Equivalents in a Prototype Red Mud House. *Journal of Health Physics, 61*(5), 647-651. http://dx.doi.org/10.1097/00004032-199111000-00009

Quirk, J. P. (1986). Soil permeability in relation to sodicity and salinity. *Philosophical Transactions of the Royal Society London, 316*, 297-317. http://dx.doi.org/10.1098/rsta.1986.0010

Quirk, J. P., & Schofield, R. K. (1955). The effect of electrolyte concentration on soil permeability. *Journal of Soil Science, 6*, 163-178. http://dx.doi.org/10.1111/j.1365-2389.1955.tb00841.x

Roper, M. M., Ward, P. R., Keulen, A. F., & Hill, J. R. (2013). Under no-tillage and stubble retention, soil water content and crop growth are poorly related to soil water repellence. *Soil and Tillage Research, 126*, 143-150. http://dx.doi.org/10.1016/j.still.2012.09.006

Roy, J. M., & McGill, W. (2002). *Assessing Soil Water Repellence Using the Molarity of Ethanol Droplet (Med) Test*. Retrieved November 2, 2015, from http://www.researchgate.net/publication/232200700_Assessing_

Soil_Water_Repellence_Using_the_Molarity_of_Ethanol_Droplet_(Med)_Test

Takawira, A., Gwenzi, W., & Nyamugafata, P. (2014). Does hydrocarbon contamination induce water repellence and changes in hydraulic properties in inherently wetting tropical sandy soils? *Geoderma, 235-236*, 279-289. http://dx.doi.org/10.1016/j.geoderma.2014.07.023

Tuffour, H. O., Thomas, A.-G., Awudu, A., Caleb, M., David, A., & Abdul, A. K. (2015). Assessment of changes in soil hydro-physical properties resulting from infiltration of muddy water. *Applied Research Journal, 1*(3), 137-140.

Walden, L. L., Harper, R. J., Mendham, D. S., Henry, D. J., & Fontaine, J. B. (2015). Eucalyptus reforestation induces soil water repellence. *Soil Research, 53*, 168. http://dx.doi.org/10.1071/SR13339

Wang, P., & Dong-Yan, L. (2012). Physical and Chemical Properties of Sintering Red Mud and Bayer Red Mud and the Implications for Beneficial Utilization. *Materials, 5*, 1800-1810. http://dx.doi.org/10.3390/ma5101800

Ward, P., Roper, M., Jongepier, R., Micin, S., & Davies, S. (2015). On-row seeding as a tool for management of water repellent sands. Building Productive, Diverse and Sustainable Landscapes. *Proceedings of the 17th ASA Conference, September 20-24, 2015, Hobart, Australia.* Retrieved from http://www.agronomy2015.com.au

12

Allelopathic Effects of *Psychotria viridis* Ruiz & Pavon on the Germination and Initial Growth of *Lactuca sativa* L.

Amanda O. Andrade[1], Maria A. P. da Silva[1], Alison H. de Oliveira[1], Marcos Aurelio F. dos Santos[1], Lilian C. S. Vandesmet[1], Maria E. M. Generino[1], Helen K. R. C. Coelho[1], Hemerson S. Landim[1], Ana C. A. M. Mendonça[1] & Natália C. da Costa[1]

[1] Programa de Pós-graduação em Bioprospecção Molecular, Departamento de Ciências Biológicas, Universidade Regional do Cariri, Rua Cel. Antônio Luis, Crato, Ceará, Brazil

Correspondence: Amanda O. Andrade, Programa de Pós-graduação em Bioprospecção Molecular, Departamento de Ciências Biológicas, Universidade Regional do Cariri, Rua Cel. Antônio Luis, 1161, 63100-000, Crato, Ceará, Brazil. E-mail: amanda_crato@hotmail.com

Abstract

The effects of aqueous and ethanol extracts and leaf fractions of *Psychotria viridis* Ruiz & Pavon (chacrona) at different concentrations on the germination and initial growth of *Lactuca sativa* L. were tested, and the phenolic and flavonoid compounds of these extracts and fractions were assessed. The bioassays consisted of the following treatments: crude aqueous extract (CAE) at 25, 50, 75 and 100% concentration, crude ethanol extract (CEE) and ethyl acetate, dichloromethane and methanol fractions at 6.25, 12.5, 25, 50 and 100% concentration and a control group. All treatments consisted of five replicates. The CAE, CEE and the ethyl acetate fraction of *P. viridis* caused both positive and negative effects on the seeds and seedlings of *L. sativa*. By contrast, the dichloromethane and methanol fractions only caused negative effects on *L. sativa*. The following compounds were identified in the extracts and fractions: gallic acid, chlorogenic acid, caffeic acid, ellagic acid, catechin, orientin, vitexin, quercetin, apigenin, rutin and luteolin, and the presence of the alkaloid N,N-dimethyltryptamine (DMT) has also been reported in the literature. *P. viridis* had allelopathic effects in all types of plant extracts and fractions tested, and one of these compounds or their combined action may account for these effects.

Keywords: allelochemicals, chacrona, extracts, fractions, High-Performance Liquid Chromatography (HPLC)

1. Introduction

Psychotria viridis Ruiz & Pavon (Rubiaceae), commonly known chacruna or chacrona, occurs spontaneously in the Amazon forest (Taylor, 2014), Mexico, the Antilles, Bolivia, Argentina and southeastern Brazil, and it is also grown in several regions of the world for religious purposes (Taylor, 2007). Ayahuasca, an entheogenic beverage also known as daime, caapi, yajé, Hoasca or vegetal, commonly used in religious rituals, is produced from a mixture of *P. viridis* with *Banisteriopsis caapi* (Spruce ex Griseb.) C.V. Morton (Malpighiaceae) (Schultes & Hofmann, 1993; Pépin & Duffort, 2004).

Certain plant species produce chemical substances that positively or negatively affect the germination and/or growth of other plants when released into the environment. This phenomenon is known as allelopathy, and its main function is to decrease or eliminate competition (Rice, 1984; Ferreira & Borghetti, 2004; Fujii & Hiradate, 2007).

Bagchi, Jain, and Kumar (1997) considered allelochemicals as a resource for the development of natural herbicides to be used in organic farming to minimize the environmental impact caused by commercial herbicides or as plant growth stimulants given the variety of secondary metabolite activities, especially allelopathic activity. Accordingly, the present study aimed to evaluate the allelopathic effects of leaf extracts of *P. viridis* on the germination and growth of *Lactuca sativa* L. and to chemically identify the phenolic and flavonoid compounds present therein given the presence of the alkaloid N,N-dimethyltryptamine (DMT) in *P. viridis* leaves (Quinteiro, Teixeira, Moraes, & Silva, 2006) and the reports of alkaloids, flavonoids and phenolic compounds with allelopathic action, and because the occurrence of alkaloids has already been previously studied.

Allelopathic Effects of Psychotria viridis Ruiz & Pavon on the Germination and Initial Growth...

145

2. Materials and Methods

2.1 Collection and Identification of Botanical Material

P. viridis leaves were collected in the morning period in a planting area belonging to the União do Vegetal Beneficent Spiritualist Center (Centro Espirita Beneficente União do Vegetal–CEBUDV), Highway Vicente Teles s/n, District of Santa Fé, municipality of Crato, Ceará state (CE), Brazil, located at 7°11′S and 39°27′W.

The plant material was collected and treated according to the usual plant collection methods, identified and sent for confirmation. The voucher specimens were deposited in the Dárdano de Andrade-Lima Herbarium of the Regional University of Cariri (Herbário Caririense Dárdano de Andrade-Lima da Universidade Regional do Cariri, HCDAL-URCA) under the registration number 6159.

2.2 Preparation of Extracts

A total of 200 g of fresh leaves of P. viridis was blended in 427 ml of distilled water using an industrial blender to prepare the crude aqueous extract (CAE). The volume of distilled water used was set based on the ratio between the fresh matter weight (FMW) and dry matter weight (DMW) (Medeiros, 1989).

For the crude ethanol extract (CEE) preparation, 500 g of fresh leaves of P. viridis was ground and soaked in 3L of ethanol P.A. (99.3%) and stored in glass containers for seven days. That mixture was then filtered and the solvent was evaporated using a rotary vacuum evaporator and concentrated using a water bath. The ethanol extracte was vaporated fully.

The extract was fractionated by vacuum filtration using solvents of increasing polarity to prepare the fractions. Thus, 10.9 g of ethanol extract from P. viridis leaves was used for this purpose generating the following yields: hexane fraction: 0.42 g; dichloromethane fraction: 1.03 g; ethyl acetate fraction: 1.25 g and methanol fraction: 7.01. The hexane fraction was discarded because it failed to show sufficient yield to perform the bioassays. The dichloromethane (DCMF), ethyl acetate (EAF), and methanol (MF) fractions were used in the bioassays.

The (CEE) and fractions were dissolved in 66% ethanol in a ratio of 1:1, where 100 mg of the crude ethanol extract and 100 ml of ethanol were obtained, thus obtaining the stock solution of 100%, while the concentrations of 6.25, 12.5, 25, 50% were obtained by dilution (Mazzafera, 2003).

2.3 Bioassays

The aqueous extract bioassays consisted of four treatments at the concentrations of 100, 75, 50 and 25% and one control group at 0% (distilled water). The ethanol extract and the dichloromethane, ethyl acetate and methanol fractions were at concentrations of 100, 50, 25, 12.5 and 6.25%. Each treatment consisted of five replicates with 20 L. sativa seeds. The experimental design used was completely randomized (CRD).

According to Souza, Cattelan, Vargas, Piana, Bobrowski, and Rocha, (2005), the main advantage of using lettuce as the target of allelopathic studies lies in the sensitivity of the seeds of the species, therefore, even in low concentrations of allelochemicals its process of Germination can be compromised. In addition, germination is rapid in approximately 24 h, has linear growth, is insensitive to pH differencesIn wide range of variation and the osmotic potentials of the solutions (RICE, 1984).

The experiments were conducted in Petri dishes with two filter paper disks moistened with 3 ml of the extract and fractions in different extract concentrations. Conversely, the control was moistened in 3 ml of distilled water. The experiments were conducted in a biological oxygen demand (BOD) seed germination chamber at a temperature of 25 °C and photoperiod of 12 hours for seven days. The plates were left open for 48 hours to evaporate the alcohol completely for the bioassays with ethanol extract and fractions (Mazzafera, 2003).

The pH of the extracts and fractions at different concentrations was analyzed using a pH meter. The osmolarity of the crude aqueous extract at different concentrations was also analyzed using an osmometer. Solutions with pH higher than 6.0, which is the range considered optimal by Macias, Gallindo, and Molinillo (2000), were adjusted using 0.1 N KOH and 5% HCl solutions.

2.4 Parameters Assessed

The following parameters were assessed: germination percentage (GP), germination speed index (GSI; assessed every 24 hours) and hypocotyl and radicle length (evaluated after seven days of sowing). Five seedlings were used per replicate to measure the length of L. sativa hypocotyls and radicles, totaling 25 seedlings per treatment.

2.5 Statistical Analysis

The statistical analysis consisted of analysis of variance and regression analysis using the software ASSISTAT version 7.7 beta. The plots were disregarded for the regression equations with $R^2 \leq 0.7$. Data were transformed into ($X = 1/X$) or ($X = 1/\sqrt{X}$) where necessary.

2.6 Chemical Auantification by High-Performance Liquid Chromatography (HPLC)

The analyses were performed under gradient conditions, at were phenomenex C18 columns (4.6 mm × 250 mm) packed with particles 5 μm in diameter were used for the reverse phase chromatographic analyses. for the extract aqueous, the mobile phase was water containing 2% formic acid (A) and methanol (B), and the gradient composition was: 17% B to 10 min 40 with change; 20, 30, 50, 70 and 10% B at 20, 30, 40, 50 and 60 min, respectively, following the method described by Klimaczewski et al. (2014). For the extract etanolic, the mobile phase water containing 2% of acetic acid (A) and methanol (B) was used, and the composition of the gradient was: 5% of (B) to 2 min with change; 25% (B) to 10 min; 40, 50, 60, 70 and 80% (B) every 10 minutes; Following the method described by Barbosa Filho et al. (2014). And for de fractions elution binary eluent gradient A (0.05% trifluoroacetic acid in water) and eluent B (100% acetonitrile) was used for the detection of the compounds. The elution schedule was set as follows: 5% B between 0-5 minutes, 12% B between 5-50 minutes, 30% B between 50-51 minutes, 90% B between 51-56 minutes, and 5% B between 56-70 minutes (COLPO et al., 2014).

3. Results and Discussion

3.1 GP and GSI

The crude aqueous extract of *P. viridis* leaves stimulated the germination of *L. sativa* seeds at concentrations of 25% and 50% and inhibited it at 75% and 100% compared to the control (Figure 1). Conversely, the ethyl acetate fraction stimulated germination at 6.25% but caused inhibition at 50 and 100% (Figure 2). Such findings may result from the joint or isolated action of allelochemicals present in the extract and fraction. Reigosa, Sánchez-Moreiras, and González (1999) reported that the effects of allelochemicals on different plant physiological processes most likely depend on the concentration and usually act by promoting activation at low concentrations and inhibition at high concentrations.

Seeds subjected to the methanol fraction at 6.25, 12.5 or 50% concentration experienced inhibition of their germination. This result has been observed in other Rubiaceae species. Frescura (2012) tested the allelopathic effects of *Psychotria brachypoda* (Müll. Arg.) Briton and *Psychotria birotula* Smith (Rubiaceae) on *Eruca sativa* Mill. noting that the *P. brachypoda* extract significantly inhibited the germination rate and GSI of *E. sativa* at a concentration of 20 mg/L, corroborating the results found in the present study.

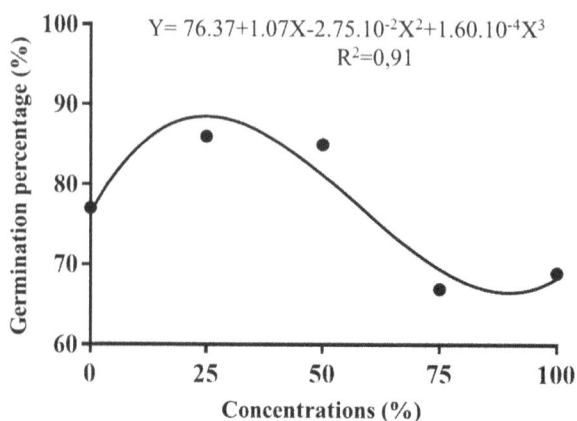

The equation in the figure reads: $Y = 76.37 + 1.07X - 2.75 \cdot 10^{-2}X^2 + 1.60 \cdot 10^{-4}X^3$, $R^2 = 0.91$

Figure 1. Germination Percentage (GP) of *L. sativa* (lettuce) subjected to different concentrations of crude aqueous extract (CAE) of *P. viridis* leaves

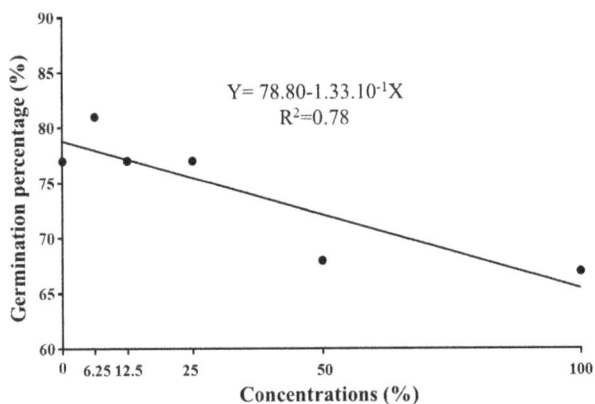

$$Y = 78.80 - 1.33.10^{-1}X$$
$$R^2 = 0.78$$

Figure 2. Germination percentage (GP) of *L. sativa* subjected to different concentrations of the ethyl acetate fraction from *P. viridis* leaves

The germination speed index (GSI) was reduced by the CAE and by the dichloromethane fraction (DCMF), decreasing with increasing extract concentrations (Figures 3 and 4).

Similar results were found by Pires et al. (2010), wherein the germination rate and the GSI of *Calopogonium mucunoides* Desv., *Stylosanthes capitata* Vogel and *L. sativa* were negatively affected by a high concentration of aqueous extract of dried *Coffea arabica* L. (Rubiaceae) leaves.

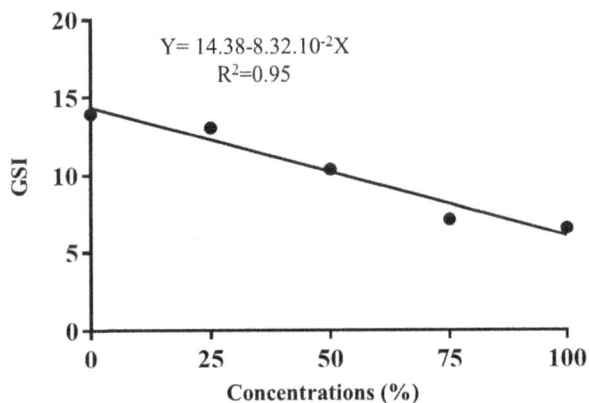

$$Y = 14.38 - 8.32.10^{-2}X$$
$$R^2 = 0.95$$

Figure 3. Germination speed index of *L. sativa* (lettuce) seeds subjected to different concentrations of crude aqueous extract of *P. viridis*

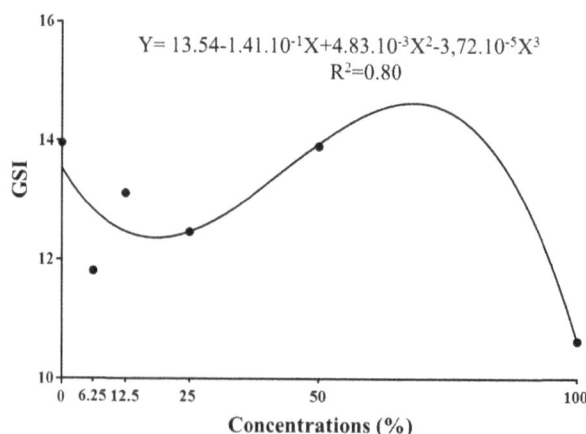

Figure 4. Germination speed index (GSI) of *L. sativa* seeds subjected to different concentrations of the dichloromethane fraction from *P. viridis*

The changes in the germination pattern, according to Ferreira (2004), can be resulting from the action of secondary metabolites on membrane permeability, transcription and translation of DNA, respiration, sequestration of oxygen (phenols), enzyme and receptor conformation, or the combination of these factors. Souza Filho (1997) points out that the same substance can perform several functions, depending on its concentration and form of translocation.

3.2 Hypocotyl and Radicle Biometrics

The hypocotyl growth of *L. sativa* seedlings was stimulated by the crude aqueous extract at 25% and 50% concentration and reduced at 75% and 100% (Figure 5). The crude ethanol extract also caused increased hypocotyl growth at 6.25, 50 and 100% concentration, but it caused decreased growth at 12.5 and 25% (Figure 6).

Seedling growth stimulation is often reported in studies related to allelopathy, and this process may be related to the effects of extracts on the phyto-hormone production of the target species or increased sensitivity of its tissues (Rice, 1984). The greater seedling growth at lower concentrations of extract may be a mechanism of protection according to Hong, Xuan, Eiji, and Khanh, (2004). Conversely, that may be related to the fact that a given chemical compound has an inhibitory or stimulating effect depending on its concentration in the environment according to Goldfarb, Pimentel, and Pimentel (2009).

Treatment with the dichloromethane fraction or methanol fraction inhibited *L. sativa* hypocotyl growth at all concentrations tested (Figures 7 and 8). Maraschin-Silva and Aqüila (2006) tested the allelopathic potential of five Brazilian native species. However, *Psychotria leiocarpa* Cham. & Schltdl. was the only species able to reduce the hypocotyl length parameter of lettuce seedlings, also inhibiting radicle length and causing a frail and brittle appearance.

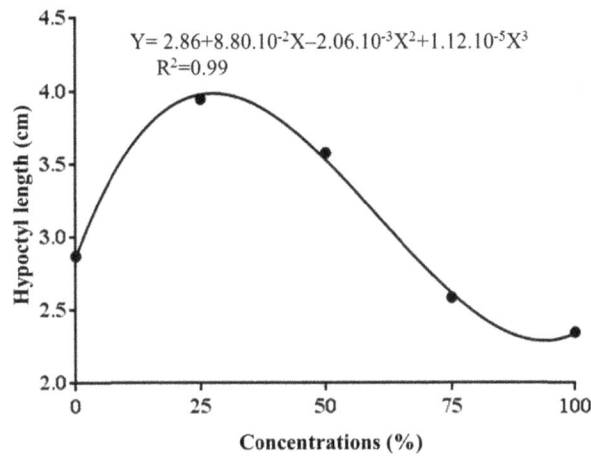

Figure 5. Mean length of *L. sativa* (lettuce) hypocotyls under the effect of different concentrations of crude aqueous extract of *P. viridis*

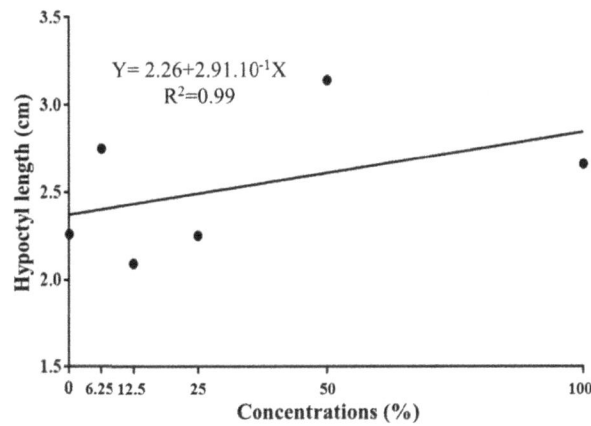

Figure 6. Hypocotyl length of *L. sativa* seedlings subjected to different concentrations of crude ethanol extract (CEE) of *P. viridis*

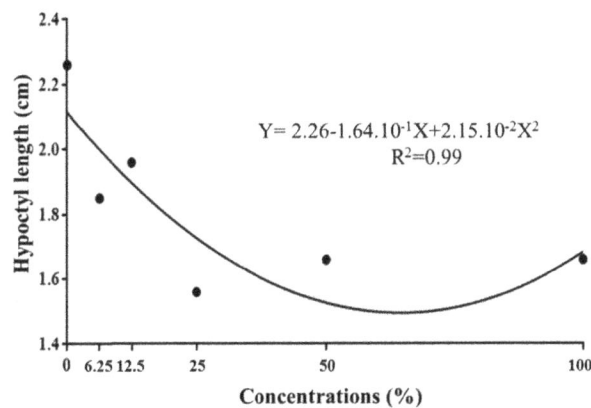

Figure 7. Hypocotyl length of *L. sativa* seedlings subjected to different concentrations of the dichloromethane fraction from *P. viridis*

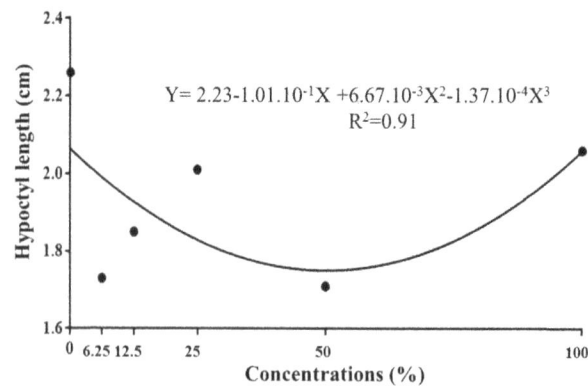

Figure 8. Hypocotyl length of *L. sativa* seedlings subjected to different concentrations of the methanol fraction from *P. viridis*

The radicle length of *L. sativa* seedlings was inhibited by all treatments from the lowest concentration tested (Figures, 9, 10, and 11). In general, roots are more sensitive to substances present in extracts and fractions than other seedling structures (Chon, Coutts, & Nelson, 2000). That results from the fact that roots are in direct and prolonged contact with allelochemicals compared to the other seedling structures (Chung, Ahn, & Yun, 2001) and/or from physiological differences between the structures (Aquila, Ungaretti, & Michelin, 1999).

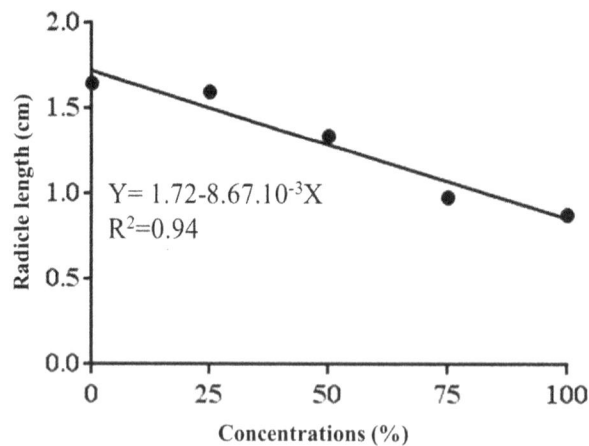

Figure 9. Mean length of *L. sativa* (lettuce) radicles under the effect of different concentrations of crude aqueous extract of *P. viridis*

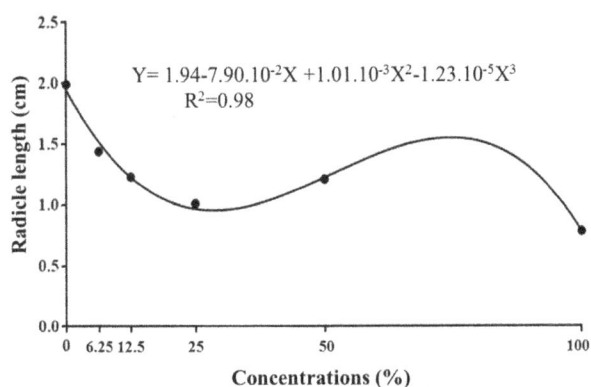

Figure 10. Radicle length of *L. sativa* seedlings subjected to different concentrations of crude ethanol extract (CEE) of *P. viridis*

$$Y= 1.94-7.90.10^{-2}X +1.01.10^{-3}X^2-1.23.10^{-5}X^3$$
$$R^2=0.98$$

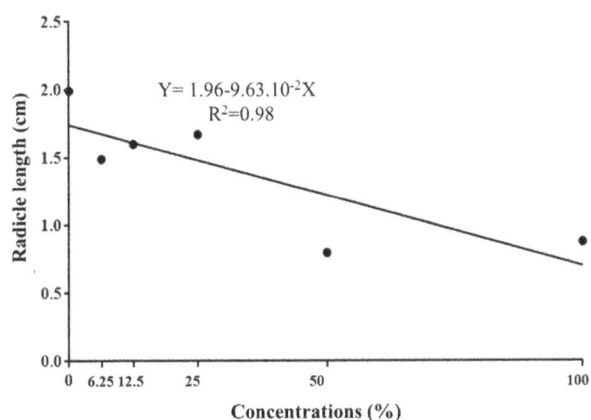

Figure 11. Radicle length of *L. sativa* seedlings subjected to different concentrations of ethyl acetate fraction of *P. viridis*

$$Y= 1.96-9.63.10^{-2}X$$
$$R^2=0.98$$

3.3 Osmolarity and pH

The pH values of Aqueous and Ethanol Extracts and Fractions were above the optimal range for seed germination and seedling growth as shown in Tables 1 and 2, which were adjusted to 6.0 or near that value.

Table 1. Physico-chemical characteristics of the crude aqueous extract of *P. viridis* leaves

Concentration	Normal pH	Adjusted pH	Osmolarity
25%	4.67	6.13	-0.063 MPa
50%	4.74	6.05	-0.093 MPa
75%	4.73	6.17	-0.143 MPa
100%	4.72	6.23	-0.181 MPa

Table 2. pH values according to the concentration of the ethanol extract and fresh leaf fractions of *P. viridis*

Treatment	Concentration (%)	Normal pH	Adjusted pH
Ethanol Fraction	6.25%	5.6	6.3
	12.5%	6.0	6.0
	25%	6.4	6.4
	50%	5.2	6.5
	100%	5.1	6.6
Dichloromethane Fraction	6.25%	6.1	6.1
	12.5%%	5.8	6.1
	25%	5.9	6.1
	50%	5.7	6.0
	100%	5.4	6.3
Ethyl Acetate Fraction	6.25%	5.5	6.0
	12.5%	5.8	6.0
	25%	7.0	6.7
	50%	6.2	6.2
	100%	5.4	6.7
Methanol Fraction	6.25%	4.9	6.4
	12.5%	4.9	6.3
	25%	5.0	6.6
	50%	5.3	6.3
	100%	4.5	6.3

Macias et al. (2000) recommend adjusting the pH of aqueous extracts to 6.0 because this is the optimal pH range for seed germination and observation of allelopathic effects. Both seedling germination and growth are affected when the pH is extremely alkaline or extremely acid (Roy, 1986), with deleterious effects observed under pH conditions below 4 and above 10 (Eberlein, 1987). An extract may contain solutes, including sugars, amino acids and organic acids, that may mask the allelopathic effect of the extract because they affect the pH according to Ferreira and Áquila (2000).

The osmotic potential of the aqueous extract was -0.06, -0.09, -0.14 and -0.18 MPa at concentrations of 25, 50, 75 and 100%, respectively. Such parameters constitute acceptable standards for seedling germination and growth in tests with potential allelopathics. Research studies, including those conducted by Mano (2006) and Gatti, Perez & Lima, (2004), have shown that those values are acceptable for allelopathic tests with seed germination.

3.4 Compound Quantification by HPLC

The presence of gallic acid (tR = 11.85 min; peak 1), catechin (tR = 16.27 min; peak 2), chlorogenic acid (tR = 20.63 min; peak 3), caffeic acid (tR = 23.81 min; peak 4), rutin (tR = 32.19 min; peak 5) and quercetin (tR = 42.05 min; peak 6; Figure 12) was detected in the aqueous extract of *P. viridis*.

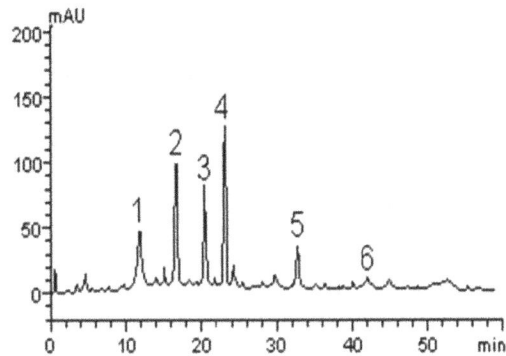

Figure 12. Phenolic and flavonoid compounds present in the aqueous extract of *P. viridis*

Note. Gallic acid (peak 1), catechin (peak 2), chlorogenic acid (peak 3), caffeic acid (peak 4), rutin (peak 5) quercetin (peak 6).

The presence of gallic acid (tR-10.67 min, peak 1), chlorogenic acid (tR = 20.07 min, peak 2), caffeic acid (tR = 24.91 min, peak 3), orientin (tR = 27.86 min, peak 4), vitexin (tR = 43.27 min, peak 5), quercetin (tR = 49.11 min, peak 6) and apigenin (tR = 62.73 min, peak 7; Figure 13) was detected in the crude ethanol extract.

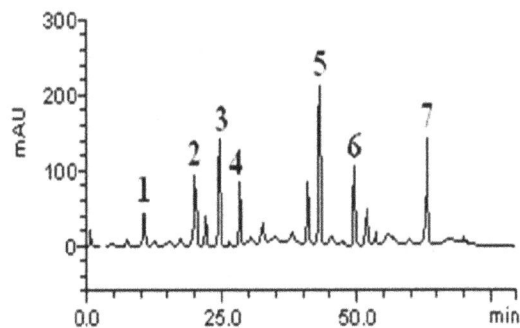

Figure 13. Phenolic and flavonoid compounds detected in the ethanol extract of *P. viridis*

Note. Gallic acid (peak 1), chlorogenic acid (peak 2), caffeic acid (peak 3), orientin (peak 4), vitexin (peak 5), quercetin (peak 6) and apigenin (peak 7).

Moreover, the presence of gallic acid (retention time (tR) = 10.73 min, peak 1), catechin (tR = 17.04 min, peak 2), chlorogenic acid (tR = 22.19 min, peak 3), caffeic acid (tR = 26.53 min, peak 4), ellagic acid (tR = 29.45 min, peak 5), rutin (tR = 38.91 min, peak 6), quercetin (tR = 49.11 min, peak 7), luteolin (TR = 54.30 min, peak 8) and apigenin (tR = 59/78 min, peak 9) was detected in the methanol, ethyl acetate and dichloromethane fractions (Figures 14).

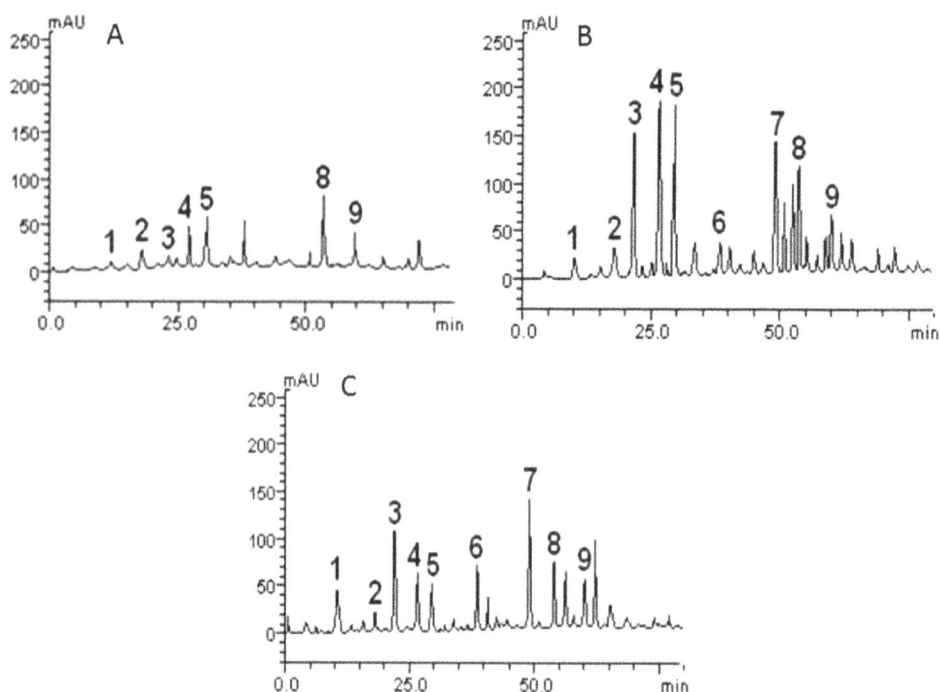

Figure 14. Phenolic and flavonoid compounds detected in the following fractions: A: dichloromethane, B: ethyl acetate and C: methanol

Note. Gallic acid (peak 1), catechin (peak 2), chlorogenic acid (peak 3), caffeic acid (peak 4), ellagic acid (peak 5), rutin (peak 6), quercetin (peak 7), luteolin (peak 8) and apigenin (peak 9).

Several functions have been attributed to flavonoids, including plant protection against incident ultraviolet and visible rays; protection against insects, fungi, viruses and bacteria; animal attraction for pollination; antioxidant action; plant hormone action control; allelopathic action and enzyme inhibition (Simões, Schenkel, Gosmann, Mello, & Mentzz, 2010).

The involvement of phenols including hydroquinone, ellagic acid and gallic acid esters in plant defenses and their involvement in interrelationships between animals and plants with different activities, including seed germination inhibition, fungal growth and plant growth in general has been highlighted in chemical ecology (Simões et al., 2010).

The gallic acid found in all *P. viridis* extracts and fractions is considered a special secondary metabolite, widespread in the plant kingdom, that shows various biological activities including an allelopathic effect on other plants (Woodson, Ames, Selassie, Hansch, & Weinshilboum, 1983; Souza Filho et al., 2006; Li, Wang, Ruan, Pan, & Jiang, 2010). Similarly, caffeic acid has been detected with abundant occurrence in soil, and its inhibitory effect on the germination and growth of various plants has been proven under laboratory conditions (Inderjit, 1995).

This was the first chemical quantification of phenolic and flavonoid compounds in *P. viridis* because most studies performed using that species have investigated the occurrence of the alkaloid N, N-dimethyltryptamine in its composition. Given that the focus of the present study was to examine the allelopathic performance of *P. viridis,* these findings are of great value to knowledge regarding the species' ecology and its allelopathic potential, and they may support future research in pursuit of a bio-herbicide as a biological alternative with specific action that is less harmful to the environment.

4. Conclusions

The findings indicate that *P. viridis* has an allelopathic effect on seed germination and seedling growth of *L. sativa,* both positively and negatively, varying according to the concentration, and the phenolic and flavonoid compounds identified in extracts from the species under study, or even the alkaloid present in its leaves, which may act as allelochemicals together or alone, may account for these effects.

Acknowledgements

We thank CAPES for the financial support and CEBUDV for authorizing plant material collection and allowing this study to be conducted.

References

Aquila, M. E. A., Ungaretti, J. A. C., & Michelin, A. (1999). Preliminary observation on allelopathic activity in *Achyrocline satureoides* (Lam.) DC. *Acta Hort., 502*, 383-388. http://dx.doi.org/10.17660/ActaHortic.1999.502.63

Bagchi, G. D., Jain, D. C., & Kumar, S. (1997). Arteether: A potent plant growth inhibitor from *Artemisia annua*. *Phytochemistry, 45*, 1131-1133. http://dx.doi.org/10.1016/S0031-9422(97)00126-X

Barbosa Filho, V. M., Waczuk, E. P., Kamdem, J. P., Abolaji, A. O., Lacerda, S. R., Costa, J. G. M., ... Posser, T. (2014). Phytochemical constituents, antioxidant activity, cytotoxicity andosmotic fragility effects of Caju (*Anacardium microcarpum*). *Industrial Crops and Products, 55*, 280-288. http://dx.doi.org/10.1016/j.indcrop.2014.02.021

Chon, S., Coutts, J. H., & Nelson, C. J. (2000). Effects of light, growth media, and seedling orientation on bioassays of alfalfa autotoxicity. *Agron. J., 92*, 715-720. http://dx.doi.org/10.2134/agronj2000.924715x

Chung, I. M., Ahn, J. K., & Yun, S. J. (2001). Assessment of allelopathic potential of barnyard grass (*Echinochloa crus-galli*) on rice (*Oryza sativa* L.) cultivars. *Crop Protect., 20*, 921-928. http://dx.doi.org/10.1016/S0261-2194(01)00046-1

Colpo, E., Vilanova, C. D. D. A., Reetz, L. G. B., Duarte, M. M. M. F., Farias, I. L. G., Meinerz, D. F., ... Rocha, J. B. T. (2014). Brazilian nut consumption by healthy volunteers improves inflammatory parameters. *Nutrition, 30*, 459-465. http://dx.doi.org/10.1016/j.nut.2013.10.005

Eberlein, C. V. (1987). Germination of *Sorghum almum* seeds and longevity in soil. *Weed Sci., 35*, 796-801. Retrieved from http://www.jstor.org/stable/4044573

Ferreira, A. G., & Áquila, M. E. A. (2000). Alelopatia: Uma área emergente da ecofisiologia. *Revista Brasileira de Fisiologia Vegetal, Edição Especial, 12*, 175-204.

Ferreira, A., & Borghetti, F. (2004). *Germinação: Do Básico Ao Aplicado* (p. 323). Porto Alegre: Artmed.

Frescura, V. D. S. (2012). *Avaliação do potencial antiproliferativo, Genotóxico e antimutagênico das espécies Psychotria Brachypoda (Müll. Arg.) Briton e Psychotria birotula Smith & Downs (Rubiaceae)* (Santa Maria, p. 73, MSc. Thesis, UFSM).

Fujii, Y., & Hiradate, S. (2007). *Allelopathy: New Concepts & Methodology* (p. 398). Enfield: Science and Publishing House Publishers.

Gatti, A., Perez, S., & Lima, M. (2004). Atividade alelopatica de extratos aquosos de *aristolochia esperanze* O. Kuntze na germinação e crescimento de *Lactuca sativa* L. e Raphanus *sativus* L. *Acta. Botanica Brasilica, 18*, 459-472. http://dx.doi.org/10.1590/S0102-33062004000300006

Goldfarb, M., Pimentel, L. W., & Pimentel, N. W. (2009). Alelopatia: Relações nos agroecossistemas. *Tecnologia & Ciência Agropecuária, 3*(1), 23-28.

Hong, N. H., Xuan, T. D., Eiji, T., & Khanh, T. D. (2004). Paddy weed control by higher plants from Southeast Asia. *Crop Protect., 23*, 255-261. http://dx.doi.org/10.1016/j.cropro.2003.08.008

Inderjit, & Dakshini, K. M. M. (1995). On laboratory bioassays in allelopathy. *Bot. Rev., 61*, 28-44. http://dx.doi.org/10.1007/BF02897150

Klimaczewski, C. V., Saraiva, R. A., Roos, D. H., Boligon, A. A., Athayde, M. L., Kamdem, J. P., ... Rocha, J. B. T. (2014). Antioxidant activity of *Peumus boldus* extract and alkaloid boldine against damage induced by Fe(II)-citrate in rat liver mitochondria *in vitro*. *Industrial Crops and Products, 54*, 240-247, http://dx.doi.org/10.1016/j.indcrop.2013.11.051

Li, Z. H., Wang, Q., Ruan, X., Pan, C. D., & Jiang, D. A. M. (2010). Phenolics and plant allelopathy. *Molecules, 15*, 8933-8952. http://dx.doi.org/10.3390/molecules15128933

Macias, F. A., Gallindo, J. C. G., & Molinillo, J. M. G. (2000). Plant biocommunicators: aplication of allelopathic studies. In J. C. Teus & P. Luijendijk (Eds.), *2000 Years of Natural Products Research Past, Present and Future* (pp. 137-161). Leiden: Pytoconsult.

Mano, A. R. O. (2006). *Efeito alelopático do extrato aquoso de sementes de cumaru (Amburana cearensis S.) sobre a germinação de sementes desenvolvimento e crescimento de plântulas de alface, picão-preto e carrapicho* (Fortaleza, 102f. MSc Thesis, Universidade Federal do Ceará).

Maraschin-Silva, F., & Aqüila, M. E. A. (2006). Potencial alelopático de espécies nativas na germinação e Crescimento inicial de *Lactuca sativa* (Asteraceae). *Acta Bot. Brasílica, 20*, 61-69. http://dx.doi.org/10.1590/S0102-33062006000100007

Mazzafera, P. (2003). Efeito alelopático do extrato alcoólico do cravo-da-índia e eugenol. *Rev. Bras. Bot., 26*, 231-238. http://dx.doi.org/10.1590/S0100-84042003000200011

Medeiros, A. R. (1989). *Determinação de Potencialidade Alelopáticas em Agroecossistemas [Determining Allelopathic Potential in Agroecosystems]* (Piracicabap. 92f. PhD Thesis, Escola Superior de Agricultura Luiz de Queiroz).

Pepin, G., & Duffort, G. (2004). Ayahuasca: Liane de l´âme, chamanes et soumission chimique. *Annales de Toxicologie Analytique, XVI*(1), 76-84. http://dx.doi.org/10.1051/ata/2004027

Pires, R. M. O., França, A. C., Nery, M. C., Silva, L. H. M. C., Santos, S. R., Reis, R. R. F., & Reis, L. A. C. (2010). *Potencial alelopático de cascas de café no crescimento de plantas* (pp. 1082-1086). XXVII. Congresso Brasileiro da Ciência das Plantas Daninhas.

Quinteiro, M. M. C., Teixeira, D. C., Moraes, M. G., & Silva, J. G. (2006). Anatomia foliar de *Psychotria viridis* Ruiz & Pav. (Rubiaceae) [leaf anatomy of *Psychotria viridis* Ruiz & Pav. (Rubiaceae)]. *Revista Universidade Rural, 26*, 30-41.

Reigosa, M. J., Sánchez-Moreiras, A., & González, L. (1999). Ecophysiological Approach in Allelopathy. *Crit. Rev. Plant Sci., 18*, 577-608. http://dx.doi.org/10.1016/S0735-2689(99)00392-5

Rice, E. L. (1984). *Allelopathy* (2nd ed., p. 422). New York: Academic Press.

Roy, M. M. (1986). Effects of pH on germination of *Dichrostachys cinerea* (L.). *Wight and Arn. J. Tree Sci., 5*, 62-64.

Schultes, R. E., & Hofmann, A. (1993). *Plantas de los Dioses: Orígenes del uso de los alucinógenos* (p. 192). México: Fondo de Cultura Económica.

Simões, C. M. O., Schenkel, E. P., Gosmann, G., Mello, J. C. P., Mentzz, L. A., & Farmacogosia, P. P. R. (2010). *Da Planta Ao Medicamento* (6th ed., p. 1104). Porto Alegre: UFRGS.

Souza Filho, A. P. S., Rodrigues, L. R. A., & Rodrigues, T. J. D. (1997). Efeito do potencial alelopático de três leguminosas forrageiras sobre três invasoras de pastagens. *Pesquisa Agropecuária Brasileira, 32*, 165-170.

Souza Filho, A. P. S., Santos, R. A., Santos, L. S., Guilhon, G. M. P., Santos, A. S., Arruda, M. S. P., ... Arruda, A. C. (2006). Potencial alelopático de Myrcia guianensis. *Planta Daninha, 24*, 649-656. http://dx.doi.org/10.1590/S0100-83582006000400005

Souza, S. A. M., Cattelan, L. V., Vargas, D. P., Piana, C. F. de B., Bobrowski, V. L., & Rocha, B. H. G. (2005). Efeito de extratos aquosos de plantas medicinais nativas. do Rio Grande do Sul sobre a germinação de sementes de alface. *Publicatio UEPG Ciências Biologicas e da Saúde, 11*(3), 29-38. http://dx.doi.org/10.5212/publicatio%20uepg.v11i3.418

Taylor, C. M. (2007). *Psychotria, L. Wanderley, M. Das G. L. (Coord.). Flora Fanerogâmica do Estado de São Paulo* (Vol. 5, pp. 259-460). São Paulo, Rima: FAPESP.

Taylor, C., Gomes, M., & Zappi, D. (2014). *Psychotria in Lista de Espécies da Flora Do Brasil*. Jardim Botânico do Rio de Janeiro. Retrieved from September, 2014, from http://floradobrasil.jbrj.gov.br/jabot/floradobrasil/FB24581

Woodson, L. C., Ames, M. M., Selassie, C. D., Hansch, C., & Weinshilboum, R. M. (1983). Thiopurine methyltransferase. Aromatic thiol substrates and inhibition by benzoic acid derivatives. *Mol. Pharmacol., 24*, 471-478.

Abbreviations

AI, alternate furrow irrigation; FI, fixed furrow irrigation; CI, conventional furrow irrigation; AN, alternate nitrogen supply; FN, fixed nitrogen supply; CN, conventional nitrogen supply; UP, under the plant; SP, south of the plant; NP, north of the plant; V_6, V_{12}, VT, R_2 and R_6 represents 6 collars, 12 collars, tasseling, filling and maturity of maize development stage, respectively.

13

Influence of Management Practices on Selected Cowpea Growth Attributes and Soil Organic Carbon

E. T. Sebetha[1] & A. T. Modi[2]

[1] Crop Science Department, School of Agriculture, Science and Technology, North-West University, Mafikeng Campus, Mmabatho, South Africa

[2] Crop Science, School of Agriculture, Earth and Environmental Sciences, University of KwaZulu-Natal, Scottsville, South Africa

Correspondence: E. T. Sebetha, Crop Science Department, School of Agriculture, Science and Technology, North-West University, Mafikeng Campus, Private Bag x 2046, Mmabatho 2735, South Africa. E-mail: erick.sebetha@nwu.ac.za

Abstract

Cowpea is a multi-purpose nitrogen fixing crop that can be grown as a vegetable, grain legume and a fodder. The objectives of this study were to investigate the growth response of cowpea to different cropping systems at different locations and determine nitrogen fertilization on cowpea growth and soil organic carbon content. Three cropping systems were used, namely, maize-cowpea rotation, cowpea monocropping and maize-cowpea intercropping at three locations (Potchefstroom, Taung, and Rustenburg) in South Africa during 2011/12 and 2012/13 planting seasons. Nitrogen fertilizer was applied at two rates where no application was the control at all locations and application according to soil analysis recommendation for maize requirement was applied at each location. The variables measured for cowpea growth were days to 100% flowering and physiological maturity, number of leaves and nodules per cowpea plant. Soil organic carbon was determined for each treatment. The results showed that, maize-cowpea rotation and monocropping reached days to 100% flowering and maturity significantly earlier compared to intercropping. Cowpea planted at Potchefstroom and Rustenburg reached days to 100% flowering and physiological maturity significantly earlier than cowpea planted at Taung. Cowpea planted at Taung had significantly higher number of nodules per plant than cowpea planted at Potchefstroom and Rustenburg. There was also a positive correlation between soil organic carbon and cowpea growth. It is concluded that the positive effect of cowpea in agronomic systems is enhanced by the correct cropping system, although it is affected by location.

Keywords: flowering, monocropping, nodules, organic carbon, rotation

1. Introduction

Cowpea is grown traditionally by small scale farmers as mixed or relay crop in association with cereals. Cowpea is a crop that play diverse role in contributing to the food security, income generation and soil amelioration for small-scale farming conditions (Amajoyegbe & Elemo, 2013). Analysing growth help to monitor the independent and interactive effects of various factors affecting yield (Addo-Quaye et al., 2011). Ghanbari et al. (2009) reported that intercropped species might utilize the growth resources more efficiently than sole crops and resources may support a greater number of plants. It was further indicated that intercrops utilize plant growth resources such as light, water and nutrients more efficiently than the equivalent sole crops. In other studies, Cowpea growth parameters such as plant height and days to flowering were not significantly affected by intercropping (Alhaji, 2008). Cowpea is highly sensitive to high moisture condition because it enhances high vegetative growth with negative effect on final yield (Oyelade & Anwanane, 2013). Cowpeas that are planted in intercropping flowered later than those in sole crops (Moriri et al., 2010). Sole cowpea reaches physiological maturity earlier than those planted in intercropping. They indicated that shading effect causes by taller maize plants delays flowering and maturity of cowpeas. The competitive relationships between the non-legume and the legume affect the growth of the leguminous crops in close proximity (Tosti et al., 2010). Fertilizer application results in significant improvement of plant height, number of leaves per plant and reduces days to flowering (Abayomi et al., 2008). Legumes require nitrogen at early vegetative stage and phosphorus fertilizers to enhance the processes of nodulation in legumes (Abayomi et al., 2008). The high amount of nitrogen application has been reported to reduce nodulation in legumes but as little as 20-25 kg N/ha has been reported to enhance early vegetative growth and increases nodulation without compromising the process of nitrogen fixation in legumes (Amba et al., 2013). According to Liu et al. (2006) a productive soil should have an organic matter content of at least 4% (2.32% soil organic carbon). The correlation between soil organic carbon and cowpea growth has not been investigated extensively. In this study, the interaction effects of location, cropping system, and nitrogen fertilizer on cowpea growth and soil organic carbon were evaluated. The objective of this study therefore was to determine the effect of location, cropping system and nitrogen fertilization on cowpea physiological growth and soil organic carbon.

2. Materials and Methods

2.1 Experimental Sites

The study was conducted at three dryland locations in South Africa, namely Taung situated at 27°30′S and 24°30′E, Potchefstroom situated at 27°26′S and 27°26′E and Rustenburg situated at 25°43′S and 27°18′E. Taung experimental site is situated in grassland savannah with annual mean rainfall of 1061 mm that begins in October. Potchefstroom has clay percentage of 34 and receives annual mean rainfall of 622.2 mm, with daily temperature range of 9.1 to 25.2 °C during planting (Macvicar et al., 1977). Rustenburg has clay percentage of 49.5 and receives an annual mean rainfall of 661 mm. Potchefstroom has plinthic catena soil, eutrophic, red soil widespread (Pule-Meulenberg et al., 2010). The soil at Taung is described as Hutton, deep, fine sandy dominated red freely drained, eutrophic with parent material that originated from Aeolian deposits (Staff, 1999). The soil at Rustenburg has dark, olive grey and clay soil, bristle consistency, medium granular structure (Botha et al., 1968).

2.2 Experimental Design

The experiment was established in 2010/11 planting season and data for experiment was collected during 2011/12 and 2012/13 planting seasons. The experimental design was factorial experiment laid out in random complete block design (RCBD) with three replicates. The statistical method was based on the previously published study by Blade et al. (1997). This technique allows accurate randomisation and analysis of variance for a multivariate design.

The experiment consisted of three cropping systems (monocropping, rotational and intercropping), three locations (Potchefstroom, Taung, and Rustenburg) and two levels of nitrogen fertilizer (urea) at each location, i.e., the amount of 0 and 20; 0 and 17; 0 and 23 kg N ha^{-1} applied on cowpea plots at Potchefstroom, Rustenburg, and Taung respectively. Maize cultivar (PAN 6479) and cowpea (Bechuana white) were used as test crop.

2.3 Agronomic Practices

Cowpea and maize seeds were sown at the same time during planting in all cropping systems. Two seeds of cowpea were sown per hole and thinning was performed after emergence to maintain one plant on the intra-row spacing. Cowpea seeds were sown at inter-row and intra-row spacing of 0.9 and 0.3 meters respectively under monocropping and rotational systems. In intercropping plots of cowpea, seeds were sown at inter-row and intra-row spacing of 0.45 and 0.3 meters respectively. The previous crop planted at Potchefstroom before the establishment of the experiment was drybean while at Taung was maize. At Rustenburg, the previous crop planted before the establishment of the trial was cotton. The herbicide that was used before and during the experiment was Roundup.

2.4 Data Collection and Analysis

Days to 100% flowering were recorded during 2011/12 and 2012/13 planting seasons. Three plants (one per middle row) were dug by their roots to determine nodule per plant during five weeks after planting, before flowering. Inoculation was performed during the first planting season of 2010/11 and no inoculants were applied to cowpea seeds during the second and third season of 2011/12 and 2012/13 planting season. Number of leaves per plant was recorded from three plant harvested in the middle rows prior to flowering period. Days to physiological maturity were recorded when the cowpea pods were matured and brown in colour. The cowpea plant height was recorded prior to 100% flowering. The data was not considered since there were no significant interactions between treatment factors.

Soil samples were collected at the depth of 0-30 cm at each plot of cowpea for organic carbon analysis. Soil samples were air-dried and grinded using mortar and pestle (porcelain). Samples were weight at the quantity of 0.5 g into the glass beakers with capacity of 250 cm^3. The laboratory procedure used to determine organic carbon was Walkley Black method (Walkley, 1935).

$$\text{Organic C\%} = \frac{\text{cm}^3 \text{ Fe (NH}_4)_2(\text{SO}_4)_2 \text{ blank} - \text{cm}^3 \text{ Fe (NH}_4)_2(\text{SO}_4)_2 \text{ sample} \times M \times 0.3 \times f}{\text{Soil mass (g)}}$$

Where, M = Concentration of Fe $(\text{NH}_4)_2(\text{SO}_4)_2$ in mol dm^{-3}.

Analysis of variance was performed using GenStat 15th edition (2012). Least significant difference (LSD) was used to separate means. A probability level of less than 0.05 was considered as significant statistically (K. A. Gomez & A. A. Gomez, 1984). The first and second order interactions were considered on days to 100% flowering. Only second order interactions were considered on days to physiological maturity. The first and second order interactions were also considered on number of leaves, nodules per plant and soil organic matter.

3. Results

3.1 Days to 100% Flowering

Interaction of cropping system × nitrogen had significant effect (P = 0.033) on days to 100% flowering as indicated in Table 1. Rotational cowpea without nitrogen fertilizer flowered significantly early at 71.4 DAP than other cropping system. Interaction of cropping system × location had significant effect (P = 0.005) on days to 100% flowering. Cowpea monoculture and rotational cowpea planted at Potchefstroom significantly flowered earlier at 68.8 and 66.3 DAP respectively than other cropping system. Cowpea monoculture and rotational cowpea planted at Rustenburg significantly flowered earlier at 64.8 and 64.3 DAP respectively than other cropping system. Interaction of nitrogen × location had significant effect (P < 0.001) on days to 100% flowering. Cowpea supplemented with nitrogen fertilizer at Potchefstroom, Rustenburg and Taung had significantly flowered early at 67.3, 62.2 and 71.7 DAP respectively as compared to cowpea without nitrogen fertilizer. Interaction of nitrogen × season had significant effect (P = 0.003) on days to flowering. Cowpea supplemented with nitrogen fertilizer during 2011/12 and 2012/13 planting seasons flowered significantly earlier at 66.9 and 67.2 DAP respectively than cowpea without nitrogen fertilizer. Interaction of location × season had significant effect (P < 0.001) on days to flowering. Cowpea planted at Rustenburg during 2011/12 planting season flowered significantly earlier at 63.7 DAP as compared to other locations. Cowpea planted at Potchefstroom and Rustenburg during 2012/13 planting season flowered significantly early at 62.7 and 66.5 DAP respectively than other location.

Table 1. The interaction effect of cropping system × nitrogen, cropping system × location, nitrogen × location, nitrogen × season and location × season on cowpea days to 100% flowering

Cropping system	Nitrogen fertilizer		
	N-Fert	Zero-N	
Intercowpea	66.6	72.7	
Monocowpea	66.7	73.2	
Rotation	65.9	71.4	
LSD$_{(0.05)}$	1.26		
Cropping system	Location		
	Potch	Rust	Taung
Intercowpea	70.5	66.2	75.3
Monocowpea	68.8	64.8	76.1
Rotation	66.3	64.3	75.3
LSD$_{(0.05)}$	1.54		
Nitrogen	Location		
	Potch	Rust	Taung
N-Fert	67.3	62.2	71.7
Zero-N	69.8	68.1	79.4
LSD$_{(0.05)}$	1.26		
Nitrogen	Season		
	2011/12	2012/13	
N-Fert	66.9	67.2	
Zero-N	73.4	71.4	
LSD$_{(0.05)}$	1.03		

Location	Season	
	2011/12	2012/13
Potch	74.4	62.7
Rust	63.7	66.5
Taung	72.4	78.8
LSD$_{(0.05)}$	1.26	

Note. N-Fert = Nitrogen fertilizer; Zero-N = Zero nitrogen fertilizer; Potch = Potchefstroom; Rust = Rustenburg; Intercowpea = Intercropped cowpea; Monocowpea = Monocropped cowpea.

3.2 Days to Physiological Maturity

The interaction of cropping system × nitrogen × location had significant effect (P < 0.001) on days to physiological maturity as indicated in Table 2. Cowpea monoculture and rotational cowpea planted at Potchefstroom supplemented with nitrogen fertilizer had reached physiological maturity significantly early at 91.3 and 91.7 days respectively than other cropping system. At Rustenburg, cowpea monoculture and rotational cowpea supplemented with nitrogen fertilizer reached physiological maturity significantly early at 89.7 and 89.8 days respectively. At Taung, cowpea monoculture and rotational cowpea supplemented with nitrogen fertilizer reached physiological maturity significantly early at 111.7 and 111.5 days respectively than other cropping system. Cowpea monoculture and rotational cowpea planted at Potchefstroom without supplement of nitrogen fertilizer reached physiological maturity significantly early at 98.0 and 98.2 days respectively than intercropping. At Rustenburg, cowpea monoculture and rotational cowpea without supplement of nitrogen fertilizer reached physiological maturity significantly early at 96.8 and 96.0 days than other cropping system. At Taung, cowpea monoculture and rotational cowpea without supplement of nitrogen fertilizer reached physiological maturity significantly early at 120.3 and 120.2 days respectively than intercropping. Interaction of cropping system × location × season had significant effect (P < 0.001) on days to physiological maturity. Cowpea monoculture and rotational cowpea planted at Potchefstroom during 2011/12 planting season, reached physiological maturity significantly early at 101.7 and 101.8 days respectively than intercropping. During 2012/13 planting season, cowpea monoculture and rotational cowpea planted at Potchefstroom reached physiological maturity significantly early at 87.7 and 88.0 days respectively. At Rustenburg, cowpea monoculture and rotational cowpea planted during 2012/13 planting season reached physiological maturity significantly early at 81.3 and 80.7 days. At Taung, rotational cowpea planted during 2011/12 planting season reached physiological maturity significantly early than other cropping systems. During 2012/13 planting season, cowpea monoculture and rotational cowpea planted at Taung reached physiological maturity significantly early at 104.3 and 104.5 days respectively. The interaction of nitrogen × location × season had significant effect (P < 0.001) on days to physiological maturity. Cowpea planted at Potchefstroom during 2011/12 and 2012/13 planting seasons and supplemented with nitrogen fertilizer significantly reached physiological maturity significantly early at 99.2 and 88.6 days respectively than cowpea without nitrogen supplement. Cowpea planted at Rustenburg during 2011/12 and 2012/13 planting seasons and supplemented with nitrogen fertilizer significantly reached physiological maturity significantly early at 101.9 and 82.8 days respectively. Cowpea planted at Taung during 2011/12 and 2012/13 planting seasons and supplemented with nitrogen fertilizer also significantly reached physiological maturity significantly early at 122.2 and 105.0 days respectively than cowpea without nitrogen fertilizer.

Table 2. The interaction effects of cropping system × nitrogen × location, cropping system × location × season and nitrogen × location × season on cowpea days to physiological maturity

Cropping system	N-Fert			Zero-N		
	Potch	Rust	Taung	Potch	Rust	Taung
Intercowpea	98.7	97.5	117.7	99.5	102.7	122.8
Monocowpea	91.3	89.7	111.7	98.0	96.8	120.3
Rotation	91.7	89.8	111.5	98.2	96.0	120.2
LSD$_{(0.05)}$	1.13					
Cropping system	Potch		Rust		Taung	
Intercowpea	103.2	95.0	105.3	94.8	128.5	112.0
Monocowpea	101.7	87.7	105.2	81.3	127.7	104.3
Rotation	101.8	88.0	105.2	80.7	127.2	104.5
LSD$_{(0.05)}$	1.13					
Nitrogen	Potch		Rust		Taung	
	2011/12	2012/13	2011/12	2012/13	2011/12	2012/13
N-Fert	99.2	88.6	101.9	82.8	122.2	105.0
Zero-N	105.2	91.9	108.6	88.4	133.3	108.9
LSD$_{(0.05)}$	0.92					

Note. N-Fert = Nitrogen fertilizer; Zero-N = Zero nitrogen fertilizer; Potch = Potchefstroom; Rust = Rustenburg; Intercowpea = Intercropped cowpea; Monocowpea = Monocropped cowpea.

3.3 Number of Leaves per Cowpea Plant

Number of leaves per cowpea plant was significantly affected (P < 0.001) by the interaction of location × season as indicated in Table 3. Cowpea planted at Rustenburg and Potchefstroom during 2011/12 planting season had significantly higher number of leaves per plant of 50.9 and 49.0 respectively than cowpea planted at Taung. Cowpea planted at Potchefstroom and Taung during 2012/13 planting season had significantly higher number of leaves per plant of 50.0 and 56.6 respectively than cowpea planted at Rustenburg.

Table 3. The interaction effect of location × season on cowpea number of leaves per plant

Location	Season	
	2011/12	2012/13
Potch	49.0	50.0
Rust	50.9	38.3
Taung	43.7	56.6
LSD $_{(0.05)}$	7.65	

Note. Potch = Potchefstroom; Rust = Rustenburg.

3.4 Number of Nodules per Cowpea Plant

Number of nodules per cowpea plant was significantly affected by the interaction of nitrogen × location (P = 0.017) as indicated on Table 4. Cowpea planted at Potchefstroom and not supplemented with nitrogen fertilizer had significantly higher number of 8.0 than cowpea supplemented with nitrogen fertilizer. There were no significantly difference between cowpea supplemented with nitrogen fertilizer and cowpea without nitrogen fertilizer at both Rustenburg and Taung. Interaction of location × season had significant effect (P < 0.001) on number of nodules per plant. Cowpea planted at Taung during 2011/12 planting season had significantly higher number of nodules of 7.1 per plant than cowpea planted of other locations. During 2012/13 planting season, cowpea planted at Taung and Potchefstroom had significantly higher number of nodules per plant of 17.2 and 10.9 as compared to other location.

Table 4. The interaction effects of nitrogen × location and location × season on number of nodules per cowpea plant

Nitrogen	Location		
	Potch	Rust	Taung
N-Fert	4.8	5.7	12.7
Zero-N	8.0	6.3	11.6
LSD$_{(0.05)}$	2.07		
Location	Season		
	2011/12	2012/13	
Potch	1.8	10.9	
Rust	3.7	8.3	
Taung	7.1	17.2	
LSD$_{(0.05)}$	2.07		

Note. N-Fert = Nitrogen fertilizer; Zero-N = Zero nitrogen fertilizer; Potch = Potchefstroom; Rust = Rustenburg; Intercowpea = Intercropped cowpea; Monocowpea = Monocropped cowpea.

3.5 Soil Organic Carbon Content at Harvest

The interaction of location × season ($P < 0.001$) had significantly affected soil organic carbon as indicated in Table 5. Soil collected at Potchefstroom and Rustenburg during 2011/12 planting season had significantly higher organic carbon of 0.75 and 0.57 % respectively than Taung. During 2012/13 planting season, Potchefstroom and Rustenburg also had significantly higher soil organic carbon of 0.66 and 0.58% respectively than Taung.

Table 5. The interaction effect of location × season on soil organic carbon content during harvest

Location	Season	
	2011/12	2012/13
Potch	0.75	0.66
Rust	0.57	0.58
Taung	0.29	-0.08
LSD$_{(0.05)}$	0.109	

Potch = Potchefstroom; Rust = Rustenburg.

3.6 Correlation between Soil Organic Carbon and Cowpea Growth

The correlations between soil organic carbon and cowpea days to 100% flowering was weak ($R^2 = 0.064$) during 2011/12 planting season as indicated in Figures 1-4. The correlation between soil organic carbon and cowpea days to 100% flowering was strong ($R^2 = 0.814$) during 2012/13 planting season. The correlations between soil organic carbon and cowpea number of days to physiological maturity were low positive, $R^2 = 0.470$ and $R^2 = 0.624$ during 2011/12 and 2012/13 planting seasons, respectively. The correlations between soil organic carbon and cowpea number of leaves per plant were weak, $R^2 = 0129$ and $R^2 = 0.164$ during 2011/12 and 2012/13 respectively. The correlations between soil organic carbon and cowpea number of nodules per plant were low positive, $R^2 = 0.688$ and $R^2 = 0.483$ during 2011/12 and 2012/13 planting seasons, respectively.

Figure 1A: Correlation of soil organic carbon and days to 100% flowering of cowpea during 2011/12 planting season.

Figure 1B: Correlation of soil organic carbon and days to 100% flowering of cowpea during 2012/13 planting season.

Figure 2A: Correlation of soil organic carbon and cowpea days to physiological maturity during 2011/12 planting season.

Figure 2B: Correlation of soil organic carbon and cowpea days to physiological maturity during 2012/13 planting season.

Figure 3A: Correlation of soil organic carbon and cowpea leaves per plant during 2011/12 planting season.

Figure 3B: Correlation of soil organic carbon and cowpea leaves per plant during 2012/13 planting season.

Figure 4A: Correlation of soil organic carbon and cowpea nodules per plant during 2011/12 planting season.

Figure 4B: Correlation of soil organic carbon and cowpea nodules per plant during 2012/13 planting season.

Figures 1-4. The correlation between cowpea growth and soil organic carbon

4. Discussion

The earlier days to 100% flowering under cowpea planted on rotational system may have been attributed to improvement of soil structure caused by previous crops. The shading by maize under intercropping plots caused delay in days to 100% flowering. This contradicts the findings by Njouku and Muoneke (2008) who reported that there was no effect on cowpea intercropping on days to 50% flowering. Moraditochaee et al. (2012) reported that nitrogen deficiency lead to premature flowering. Cowpea planted on Monocropping and rotational systems had reduced competition for resources such as sunlight and soil nutrients, and these resulted in earlier days to physiological maturity (Sarkar et al., 2013).

The earlier physiological maturity of cowpea planted on monocropping system confirms the statements by Moriri et al. (2010) that sole cowpea reached physiological maturity earlier than those planted in intercropping. According to Amujoyegbe and Elemo (2013) site and time of introduction of cowpea affected growth of cowpea. Higher number of leaves under monocropping and rotational cowpea may have been attributed to fertility of soil (Pereira vaz Ferreirra et al., 2015) that led to increase in growth of cowpea (Abraha, 2013). The production of more leaves under monocropping and rotational cowpea means higher light interception and more photo-assimilate production (Babaji et al., 2011; Kouyate et al., 2012). Blade et al. (1992) reported that cowpea growth was severely depressed by competition with other plants. The higher number of nodules per plant on cowpea planted at Taung may have been attributed to sandy soil on that site. The differences of soil organic carbon by sites corroborate the findings by Fu et al. (2004) who reported that soil organic carbon is affected by environmental factors such as topography, parent material, soil depth and land use. Topography influences precipitation and temperature, both of which will affect the soil carbon (Tsui et al., 2004). The differences in soil organic carbon by seasons may have been attributed to soil temperatures and rainfall. This supports the statement by Fang et al. (2008) who reported higher soil microbial biomass carbon in rainy season than in dry season. It was also revealed that soil carbon was significantly positively correlated with soil temperature. The higher soil organic carbon at Potchefstroom and Rustenburg may have been attributed to clay content on those sites. The higher soil organic carbon in monocropping cowpea plots was due to improved soil structure and fertility, which led to high carbon content. Soil carbon increases was found to be generally greater with higher level of soil fertility. Alvarez (2005) reported that carbon sequestration increases as nitrogen fertilizer was applied to the system, and this contradicted the findings of this study. Nitrogen fertilization had no effect on soil organic carbon. This corroborates the findings by Russell et al. (2009) who reported that N fertilization offset gains in carbon inputs to the soil in such a way that soil carbon sequestration was virtually nil despite up to 48 years of N addition.

5. Conclusions

It has been shown that cowpea growth and soil organic carbon were higher under monocropping and rotational systems. The application of nitrogen fertilizer played a significant role on improvement of cowpea growth. Cowpea nodulation tends to be higher under location with higher percentage of sandy soil. Intercropping system suppresses the growth of cowpea and also results with reduced soil organic carbon. In this study, it was found that, the application of nitrogen fertilizer has no influence on soil organic carbon. Organic carbon tends to increase on site with higher soil clay percentage. Interaction of location × season plays a vital role on improvement of cowpea growth and soil organic carbon. There is positive correlation between soil organic carbon and cowpea days to flowering, physiological maturity, and number of nodules per plant.

References

Abayomi, Y. A., Ajibade, T. V., Sammuel, O. F., & Saadudeen, B. F. (2008). Growth and yield responses of cowpea (*Vigna unguiculata* (L.) Walp) genotypes to nitrogen fertilizer (NPK) application in the Southern Guinea Savanna zone of Nigeria. *Asian Journal of Plant Sciences, 7*, 170-176. http://dx.doi.org/10.3923/ajps.2008.170.176

Abraha, L. (2013). The effect of intercropping maize with cowpea and lablab on crop yield. *Herald Journal of Agricultural and Food Science Research, 2*(5), 156-170.

Addo-Quaye, A. A., Darkwa, A. A., & Ampiah, M. K. P. (2011). Performance of three cowpea (*Vigna unguiculata* (L.) Walp) varieties in two agro-ecological zones of the central region of Ghana I: Dry matter production and growth analysis. *ARPN Journal of Agricultural and Biological Science, 6*(2), 1-9.

Alhaji, I. H. (2008). Performance of some cowpea varieties under sole and intercropping with maize. *African Research Review, 2*(3), 278-291. http://dx.doi.org/10.4314/afrrev.v2i3.41073

Alvarez, R. A. (2005). Review of nitrogen fertilizer and conservation tillage effects on soil organic carbon storage. *Soil Use and Management, 21*(1), 38-52. http://dx.doi.org/10.1079/SUM2005291

Amba, A. A., Agbo, E. B., & Garba, A. (2013). Effect of nitrogen and phosphorus fertilizers on nodulation of some selected grain legumes at Bauchi, Northern Guinea Savanna of Nigeria. *International Journal of Biosciences, 3*(10), 1-7. http://dx.doi.org/10.12692/ijb/3.10.1-7

Amujoyegbe, B. J., & Elemo, K. A. (2013). Growth performance of maize/cowpea in intercrop as influenced by time of introducing cowpea and nitrogen fertilizer. *International Research Journal Plant Science, 4*(1), 1-11.

Babaji, B. A., Yahaya, R. A., Mahadi, M. A., Jaliya, M. M., Sharifi, A. I., Kura, H. N., ... Ajeigbe, H. (2011). Growth attributes and pod yield of four cowpea (*Vigna unguiculata* (L.) Walp.) varieties as influenced by residual effect of different application rates of farmyard manure. *Journal of Agricultural Sciences, 3*(2), 165-171. http://dx.doi.org/10.5539/jas.v3n2p165

Blade, S. F., Mather, D. E., & Singth, B. B. (1992). Evaluation of yield stability of cowpea under sole and intercrop management in Nigeria. *Euphytica, 61*, 193-201. http://dx.doi.org/10.1007/BF00039658

Blade, S. F., Shetty, S. V. R., Terao, T., & Singth, B. B. (1997). Recent developments in cowpea cropping systems research. In B. B. Singh, D. R. Mohan Raj, K. E. Dashiell, & L. E. N. Jackai (Eds.), *Advances in Cowpea Research*. International Institute of Tropical Agriculture and Japan International Research Center for Agricultural Sciences.

Botha, A. D. P., Snyman, H. G., Hahne, H. C. H., Prinsloo, A. L., Steenkamp, C. J., & Duplessis, D. P. (1968). Eienskappe van die gronde van die navorsings institute vir Tabak. *Tegniese Mededeling 74*. Rustenburg: Department van Landbou-Tegniese Dienste.

Fang, L. N., Yang, X. D., & Du, J. (2008). Effects of land use pattern on soil microbial biomass carbon in Xishuangbanna. *Ying Yong Sheng Tai Xue Bao, 22*(4), 837-844.

Fu, B. J., Liu, S. L., Ma, K. M., & Zhu, Y. G. (2004). Relationships between soil characteristics, topography and plant diversity in a heterogeneous deciduous broad-leaved forest near Beijing, China. *Plant and Soil, 261*, 47-54. http://dx.doi.org/10.1023/B:PLSO.0000035567.97093.48

Ghanbari, A., Dahmardeh, M., Siahsar, B. A., & Ramroudi, M. (2009). Effect of maize (*Zea mays* L.) cowpea (*Vigna unguiculata* L.) intercropping on light distribution, soil temperature and soil moisture in arid environment. *Food, Agriculture & Environment, 8*(1), 102-108.

Gomez, K. A., & Gomez, A. A. (1984). *Statistical Procedures for Agricultural Research*. John Wiley and Sons. New York.

Kouyate, Z., Krasova-Wade, T., Yattara, I. I., & Neyra, M. (2012). Effects of Cropping System and Cowpea Variety on Symbiotic Potential and Yields of Cowpea (*Vigna unguiculata* L. Walp) and Pearl Millet (*Pennisetum glaucum* L.) in the Sudano-Sahelian Zone of Mali. *International Journal of Agronomy, 12*, 1-8. http://dx.doi.org/10.1155/2012/761391

Liu, X., Herbert, S. J., Hashemi, A. M., Zhang, X., & Ding, G. (2006). Effects of agricultural management on soil organic matter and carbon transformation- a review. *Plant Soil and Environment, 52*(12), 531-543.

Macvicar, C. N., De Villiers, J. M., Loxton, R. F., Vester, E., Lambrechts, J. J. N., Merryweather, F. R., ... Harmse, H. J. (1977). Soil classification. A binomial system for South Africa. *Science Bull., 390*, ARC-Institute for Soil Climate and Water, Pretoria.

Moraditochaee, M., Bidarigh, S., Azarpour, E., Danesh, R. K., & Bozorgi, H. R. (2012). Effects of nitrogen fertilizer management and foliar spraying with amino acid on yield of cowpea (*Vigna unguiculata* L.). *International Journal of Agriculture and Crop Sciences, 4*(20), 1489-1491.

Moriri, S., Owoeye, L. G., & Mariga, I. K. (2010). Influence of component crop densities and planting patterns on maize production in dryland maize/cowpea intercropping systems. *African Journal of Agricultural Research, 5*(11), 1200-1207.

Njoku, D. N., & Muoneke, C. O. (2008). Effect of cowpea planting density on growth, yield and productivity of component crops in cowpea/cassava intercropping system. *Journal of Tropical Agriculture, Food, Environment and Extension, 72*(2), 106-113. http://dx.doi.org/10.4314/as.v7i2.1591

Oyelade, O. A., & Anwanane, N. B. (2013). Maize and cowpea production in Nigeria. *Advances in Agriculture, Sciences and Engineering Research, 3*(11), 1343-1358.

Pereira vaz Ferreira, I. C., de Lima Pereira Sales, N., & de Araujo, A. V. (2015). Effect of temperature and photoperiod on the development of *fusariosis* in pineapples. *African Journal of Agriculture Research, 10*(2), 76-83. http://dx.doi.org/10.5897/AJAR2014.8577

Pule-Meulenberg, F., Belane, A. K., Krasova-Wadet, T., & Dakora, F. D. (2010). Symbiotic functioning and bradyrhizobial biodiversity of cowpea (*Vigna unguiculata* L. Walp.) in Africa. BMC Microbiology. http://dx.doi.org/10.1186/1471-2180-10-89

Russell, A. E., Cambardella, C. A., Laird, D. A., Jaynes, D. B., & Meek, D. W. (2009). Nitrogen fertilizer effects on soil carbon balances in Midwestern U.S. agricultural systems. *Ecological Applications, 19*, 1102-1113. http://dx.doi.org/10.1890/07-1919.1

Sarkar, S., Majumdar, B., & Kundu, D. K. (2013). Strip-cropping of legumes with jute (*Corchorus olitorius*) in jute-paddy-lentil cropping system. *Journal of Crop and Weed, 9*(1), 207-209.

Soil Survey Staff. (1999). *Keys to soil taxonomy* (8th ed.). Poca-hontas Press Inc., Blacksburg. Virginia.

Tosti, G., Benincasa, P., & Guiducci, M. (2009). Competition and facilitation in hairy vetch-barley intercrops. *Italian Journal of Agronomy, 3*, 239-247.

Tsui, C. C., Chen, Z. S., & Hsien, C. F. (2004). Relationships between soil properties and landscape position in a lowland rain forest of Southern Taiwan. *Geoderma, 123*, 131-142. http://dx.doi.org/10.1016/j.geoderma.2004.01.031

Walkley, A. (1935). An examination of methods for determining organic carbon and nitrogen in soils. *Journal of Agricultural Science, 25*, 598-609. http://dx.doi.org/10.1017/S0021859600019687

14

Evaluation of Spatial Variability of Soil Physico-Chemical Characteristics on Rhodic Ferralsol at the Syferkuil Experimental Farm of University of Limpopo, South Africa

Kopano Conferance Phefadu[1] & Funso Raphael Kutu[2]

[1] Department of Plant Production, Soil Science and Agricultural Engineering, University of Limpopo, Sovenga, South Africa

[2] Food Security & Safety Niche Area Research Group, Department of Crop Science, School of Agricultural Sciences, North-West University, Mafikeng Campus, Mmabatho, South Africa

Correspondence: Funso Raphael Kutu, Department of Crop Science, School of Agricultural Sciences, North-West University, Mafikeng Campus P/Bag X2046, Mmabatho 2735, South Africa. E-mail: funso.kutu@nwu.ac.za

The research was partly financed through the VLIR-UOS project (Grant number: ZIUS2016AP21) of University of Limpopo in support of the first author as part of a Master of Science (Soil Science) degree work. Assistance provide by Dr. M. G. Zerizghy, Mr. M. Mpati and Mr. Richard Tswai (ARC-ISCW, Pretoria) in the production of variability maps is also acknowledged.

Abstract

Spatial variability among selected soil physical and chemical properties in twelve profiles dug across the research block of the University of Limpopo experimental farm was investigated. The soils were moderately shallow to deep, contain variable textural classes and classified as Rhodic ferralsol. Over 90% of the samples were considered as slightly alkaline based on the water-measured pH values but decreased to marginally over 27% when measured in KCl. The electrical conductivity of the soils revealed a generally non-saline field. Bray P1, EC, exchangeable cations, extractable Zn and effective cation exchange capacity contents differed significantly ($p < 0.05$) with depth while K, Mg, Ca, Mn, organic carbon and ECEC differed significantly ($p < 0.05$) across profiles. Semi-variograms for the measured variables had low values indicating the existence of considerable level of spatial variability. Spatial dependence among top and subsoil pH, EC, organic carbon, sand, silt clay and bulk density ranged between weak and strong. Results revealed a significant spatial variability of the characterized parameters across the research block because to differences in tillage, cropping pattern and nutrient specific application over the years.

Keywords: spatial variability, soil physico-chemical properties, geostatistics, university research farm

1 Introduction

The provision of information about spatial variability of soil attributes is essential to achieve a better understanding of the complex relations between soil properties (Goovaerts, 1998), establish appropriate management practices for soil resources use (Bouma et al., 1999) and better management of spatially variable soils (Mohammadi, 2002). Spatial variability of soil physical and chemical properties within or among agricultural fields represents inherent attributes. However, the variability may either be attributed to geological and pedological soil forming factors or induced and exacerbated by tillage and soil management practices such as fertilizer use (Iqbal et al., 2005). Therefore, an ideal experimental field is one in which soil variability has been minimised for a specific crop or soil physical/chemical treatments (Cerri et al., 2004). Over the past 20 years, soils on the University of Limpopo experimental farm have continually being used for conducting various experiments ranging from cereal through legume crops production to horticultural crops. Cereal crops by their nature are heavy feeders requiring large amount of nutrients, particularly nitrogen, N (Nsanzabaganwa et al., 2014) while legumes are able to fix N into the soil. Many of the crop evaluation trials carried out on the field are often accompanied by variable fertiliser use that imposes a high degree of nutrient variability on the field. Thus, the farm is often subjected to various extensive tillage operations particularly during land preparation in each planting season. Despite the long term history of intensive and continuous use and various management operations, the farm has no reliable detailed spatial variability information.

Many researchers have applied geostatistics to provide description and distribution of the spatial variability of soil physico-chemical properties (Mohammadi, 2002; Lin et al., 2005; Vaezi et al., 2010; Staugaitis & Sumskis, 2011; Akbas, 2014; Reza et al., 2015). Characterizing the spatial variation of soil variables can provide important implications on water and nutrient management as well as fertilizer use during agricultural production (Saglam et al., 2011). Agricultural sustainability depends to a large extent on improvements in soil physical and chemical properties that are largely controlled by several factors including mineral nutrition that has been largely described as the most important (Jat et al., 2006). Cerri et al. (2004) indicated that understanding the distribution and nature of soil properties in the field is essential in refining agricultural management practices while minimizing environmental damage. Information on the spatial variability of soil properties could therefore lead to better management decisions aimed at correcting problems, maintaining productivity, fertility and sustainability of the soils (Özgöz, 2009). Detailed soil characterization particularly on a research farm where high degree of accuracy and precision is required for prescribing recommendation will allow researchers to follow crop and soil management practices aligned with the soil conditions (Castrignanò et al., 2000). The study objectives of this paper therefore include, to: (i) evaluate the spatial distribution of soil physical and chemical characteristics in the research block, study the correlation between soil physical and chemical characteristics, and (ii) identify the trends in variability across the research block.

2 Method

2.1 Description of the Study Location

This study was conducted at the University of Limpopo Experimental Farm, Syferkuil (23°50′36.86″S; 29°40′54.99″E; 1324 meters above sea level), which is located in the Mankweng area within Capricorn District of Limpopo Province, South Africa. The area experiences hot summers with an annual rainfall of 350-500 mm. The research block is regularly used for agronomic and plant nutrition studies by students and researchers from various national and international institutions through research project collaboration by local researchers within the University. Soils at this farm are formed *in situ* on basalt, sandstone and biotic gneiss, possess inherent poor fertility status (FAO, 2009); and locally classified as Hutton according to South Africa classification system or Rhodic Ferralsol (WRB, 2006). The 1 650 ha farm size serves as the University's students' demonstration, agronomic and plant nutrition research as well as animal production studies. Currently on the farm, about 50 ha are allocated for rainfed crops, 80 ha for irrigated crops and 40 ha are used for rotation of winter and summer crops.

2.2 Sampling Points Selection and Digging of the Soil Profiles

Twelve soil profile pits were dug across the research block. The areas where the profile pits were dug were randomly selected for even distribution across the entire block. The coordinates of each profile pit were measured using a GPS device (Trimble Juno 3D) and the map showing the distribution of the pits across the study location and the total depth of each profile pit are shown in Figure 1.

2.3 Horizon Demarcation, Physical Parameters Characterization and Soil Sampling

All profile pit horizons were demarcated based on soil colour (moist and dry state) using the Munsell soil colour chart according to Schoeneberger et al. (1998). Soil structure was characterized based on the soil structure types while soil samples taken from each soil profile horizon were analysed for selected soil chemical and physical parameters using standard laboratory procedures. Some of the physical properties namely: depth, structure and consistency were documented in the field.

2.4 Analyses of Physical and Chemical Properties of Soil Samples

Soil samples collected were air-dried, ground to pass through a 2-mm sieve and used for the various determinations. Soil physical properties namely soil texture and bulk density (BD) were determined using the hydrometer method (Sheldrick & Hand Wang, 1993) and the cylindrical core method (Campbell & Henshall, 1991) respectively. Electrical conductivity (EC) was measured in a 1:5 ratio of soil/water suspension using a digital conductivity meter while pH was measured in water as well as in 1mol dm^{-3} potassium chloride (KCl) at a ratio of 1:2.5 using a digital electronic pH meter. Organic carbon (OC) was determined by Walkley-Black chromic acid wet oxidation method (Nelson & Sommers, 1996), available phosphorus (P) was determined by Bray-1 extraction followed by molybdenum blue colorimetry (Okalebo et al., 2002) and exchangeable potassium (K), sodium (Na), calcium (Ca) and magnesium (Mg) were extracted using 1M NH_4OAC, pH7 solution and concentration of each nutrient determined on atomic absorption spectrophotometer (Okalebo et al., 2002). Effective cation exchange capacity (ECEC) was estimated by summation of exchangeable cations and exchangeable acidity (Okalebo et al., 2002). Extractable iron (Fe), copper (Cu), zinc (Zn) and manganese (Mn) in the soil samples were determined following Ambic-1 procedure (The Non-Affiliated Soil Analysis Work Committee, 1990).

2.5 Statistical analyses and creation of semi-variograms for the measured soil parameters

The collected data were subjected to classical statistical methods to obtain the minimum, maximum, mean, median, skewness (Shapiro & Wilk, 1965), and standard deviation for each horizon (n = 22). A one way analysis of variance was also performed using Statistix 8.1 to compare each variable across the soil profiles using LSD test at 5%. A Pearson-correlation analysis was performed to establish the significances of the linear relations between all measured variables. Semi-variograms of selected soil parameters were created using ArcMap10.2 software while the raw data were interpolated using Simple Kriging method (Santra et al., 2008).

3 Results

3.1 Distribution of Selected Soil Physical Parameters in the Research Block

Soil physical parameters measured revealed that the profiles were generally moderately shallow to deep (Table 1). Profiles 10 and 11 located on the eastern side of the field were the shallowest while profile 7 located at the central part of the field represented the deepest. The soil depth variability map revealed that the soils are deeper in the central part of the field towards the western part but shallowest in the eastern part of the field (Figure 1). The proportion of sand, silt and clay in all soil samples collected from the profiles ranged from 61-87%, 1-15% and 7-27%, respectively; broadly categorised as sandy loam, loamy sand and sandy clay loam. The BD values were generally relatively high and ranged from 1.20 g/cm^3 to 1.80 g/cm^3 with obvious variation across and within the profile pits. According to Lal (2006), normal bulk density for clay ranged from 0.90 to 1.40 g/cm^3 while that for sand ranged from 1.40 to 1.90 g/cm^3 with potential root restriction occurring at $\geq$ 1.40 g/cm^3 for clay and $\geq$ 1.60 g/cm^3 for sand. Other soil physical properties (colour, structure and consistency) and shown in Table 2. Soil colour (dry and moist) is highly variable and ranged from reddish brown to very dark greyish brown depending on the sampling depth while the predominant soil structural type was blocky. The consistencies of the soil samples determined dry were mainly firm and friable.

Table 1. Textural and bulk density variations across the twelve soil profiles

Profile ID	Profile depth (cm)	% Sand	% Silt	% Clay	Texture class	BD (g/cm^3)
RBP1T	80	71	12	17	Sandy loam	1.48
RBP1S		74	12	14	Sandy loam	1.35
RBP2T	60	71	15	14	Sandy loam	1.47
RBP2S		84	9	7	Loamy sand	1.27
RBP3T	61	67	9	24	Sandy clay loam	1.55
RBP3S		61	12	27	Sandy clay loam	1.58
RBP4T	79	81	2	17	Sandy loam	1.75
RBP4S		67	9	24	Sandy clay loam	1.54
RBP5T	80	77	2	21	Sandy clay loam	1.74
RBP5S		84	9	7	Loamy sand	1.50
RBP6T	85	74	2	24	Sandy clay loam	1.56
RBP6S		80	9	11	Loamy sand	1.46
RBP7T	100	77	2	21	Sandy clay loam	1.78
RBP7S		74	9	17	Sandy loam	1.57
RBP8T	98	84	2	14	Sandy loam	1.69
RBP8S		68	7	25	Sandy clay loam	1.57
RBP9T	45	87	2	11	Loamy sand	1.65
RBP9S		84	2	14	Sandy loam	1.80
RBP10T	30	84	1	15	Loamy sand	1.78
RBP11T	28	87	2	11	Loamy sand	1.72
RBP12T	94	87	2	11	Loamy sand	1.72
RBP12S		74	2	24	Sandy clay loam	1.60
CV %	36	10	75	36		9

Note. RBP1T = Research block profile 1 topsoil; RBP1S = Research block profile 1 subsoil; BD = bulk density; CV = Coefficient of variation.

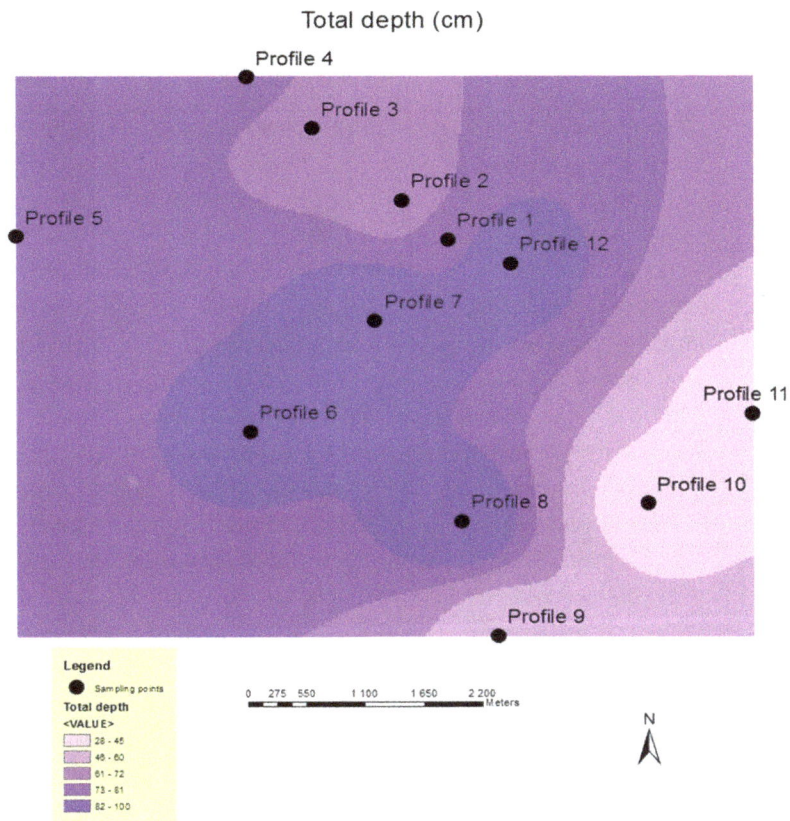

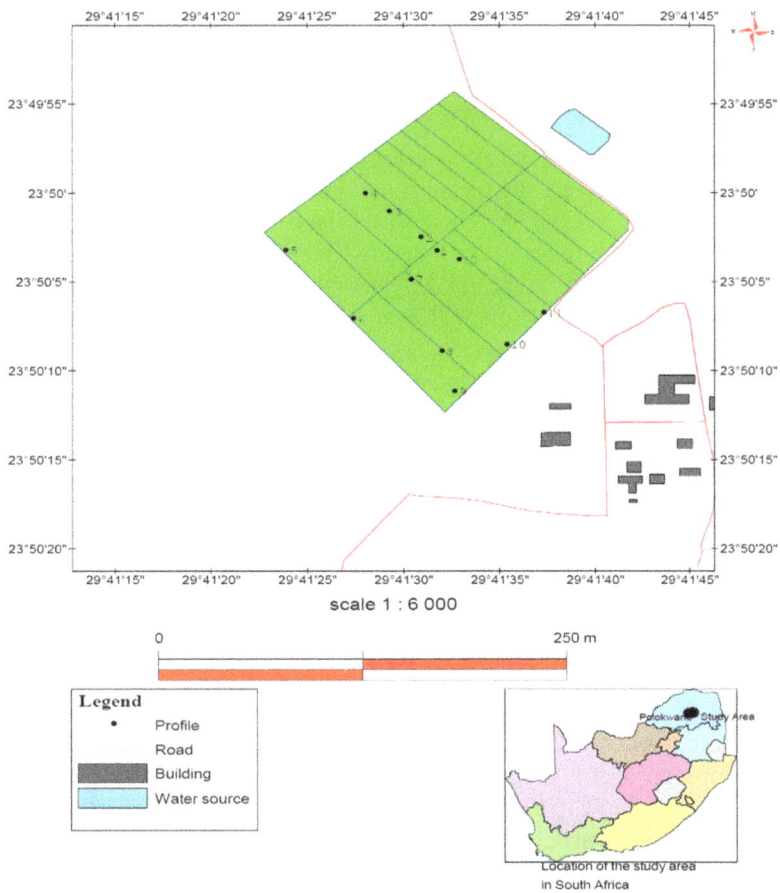

Figure 1. Maps of the study location showing the twelve soil profiles and the total soil depths of each profile

Table 2. Physical parameters of the twelve soil profiles dug across the research block within the experimental farm

Profile ID	Horizon thickness (cm)	Soil colour (dry)	Soil colour (moist)	Soil structure	Soil consistence (dry)
RBP1T	0-48	-5YR 4/4 (Reddish brown)	5YR 3/3 (Dark reddish brown)	Blocky	Firm
RBP1S	48-80	7.5YR 4/6 (Strong brown)	7.5YR 3/4 (Dark brown)	Blocky	Firm
RBP2T	0-22	7.5YR 4/6 (Strong brown)	7.5YR 3/4 (Dark brown)	Granular	Friable
RBP2S	22-60	7.5YR 6/8 (Reddish yellow)	7.5YR 4/6 (Strong brown)	Platy	Extremely firm
RBP3T	0-34	7.5YR 4/4 (Dark brown)	7.5YR 3/4 (Dark brown)	Blocky	Firm
RBP3S	34-61	7.5YR 5/6 (Strong brown)	7.5YR 4/4 (Dark brown)	Blocky	Friable
RBP4T	0-37	5YR 4/4 (Reddish brown)	5YR 3/3 (Dark reddish brown)	Blocky	Friable
RBP4S	37-79	5YR 5/8 (Yellowish red)	5YR 4/4 (Reddish brown)	Blocky	Friable
RBP5T	0-24	10YR 3/2 (Very dark greyish brown)	10YR 2/2 (Very dark brown)	Blocky	Firm
RBP5S	24-80	10YR 5/2 (Greyish brown)	10YR 3/3 (Dark brown)	Blocky	Friable
RBP6T	0-32	5YR 4/4 (Reddish brown)	5YR 3/3 (Dark reddish brown)	Blocky	Firm
RBP6S	32-85	5YR 5/4 (Reddish brown)	5YR 4/4 (Reddish brown)	Blocky	Friable
RBP7T	0-46	5YR 4/3 (Reddish brown)	5YR 3/4 (Dark reddish brown)	Blocky	Firm
RBP7S	46-100	7.5YR 5/4 (Brown)	7.5YR 4/3 (Dark brown)	Blocky	Friable
RBP8T	0-30	7.5YR 4/4 (Dark brown)	7.5YR 3/3 (Dark brown)	Blocky	Friable
RBP8S	30-98	5YR 4/6 (Yellowish red)	5YR 3/4 (Dark reddish brown)	Blocky	Friable
RBP9T	0-20	5YR 4/6 (Yellowish red)	5YR 3/4 (Dark reddish brown)	Blocky	Firm
RBP9S	20-45	5YR 5/8 (Yellowish red)	5YR 3/4 (Dark reddish brown)	Blocky	Firm
RBP10T	0-30	7.5YR 4/6 (Strong brown)	7.5YR 3/4 (Dark brown)	Blocky	Firm
RBP11T	0-28	7.5YR 4/6 (Strong brown)	7.5YR 3/3 (Dark brown)	Blocky	Firm
RBP12T	0-30	5YR 4/4 (Reddish brown)	5YR 3/3 (Dark reddish brown)	Blocky	Firm
RBP12S	30-94	5YR 4/6 (Yellowish red)	5YR 3/4 (Dark reddish brown)	Blocky	Friable

3.2 Distribution of Selected Soil Chemical Parameters in the Research Block

The measured chemical parameters for the soil samples are contained in Table 3. The pH value measured in water showed that over 90% of the samples were slightly alkaline while pH in 1 M potassium chloride solution revealed that about 73% of the soil samples were acidic. There was a significant ($p < 0.05$) variation in the measured pH values across and down the profiles. Soil pH values measured in both water and KCl had similar pattern of spatial variation; with generally lower values at the surface soil than subsurface depth. Virtually all the measured soil pH values in water were outside the desired pH range (6.5 to 7.2). The measured EC value of all soil samples was though high but generally non-saline; and revealed a non-significant variation both across and with the profiles. Organic carbon content in this field was low with marginal and non-significant variation across the field.

About 36% of the soil samples collected mainly from the topsoil were at or about the critical level of 10-16 mg kg^{-1} for available Bray-1 P for grain crops and were more than nine times higher than in soil samples obtained from the subsoil depth. None of the exchangeable bases in the soil samples from this field was below the prescribed critical level. Bray-1 P, EC, exchangeable K, Ca, Mg and Na, extractable Zn and ECEC contents differed significantly ($p < 0.05$) with increasing soil depth while K, Ca, Mg, Mn, OC and ECEC differed significantly across soil profiles (data not shown). Based on Waskom et al. (2014) classification standard of EC level of > 2 dS/m or 2000 mS/cm for salt affected soils, the measured EC values are low and therefore the field has no incidence of salt or salinity problem despite the seeming high pH values. However, the coefficient of variation (CV) for the measured soil chemical parameters across the different profiles ranged from 5.8-45.8%. Among the soil chemical parameters, EC, P, K and Na recorded the most variable (CV > 35%) while measured pH, Cu and Mn values recorded the low variation (CV < 15%) on the field (Table 4). For all soil chemical parameters, the mean values are close to the median values. The distribution of pH, OC, available P, Fe, Cu and Zn as determined from the coefficient of skewness was normal (< 0.5) while EC, ECEC, Mn, Ca, Mg, Na and K did not follow a normal distribution with the coefficient of skewness greater than 0.5.

3.3 Spatial Distribution of Selected Measured Soil Physical and Chemical Parameters

Spatial dependence of soil properties may be attributed to either intrinsic factors, extrinsic factors or both (Behera et al., 2011). Semi-variogram parameters (nugget, range, sill and nugget to sill ratio) measured in soil samples collected to describe the spatial distribution across the different profiles on the field are presented in Table 5. Large nugget values were obtained for topsoil and subsoil EC and clay and topsoil sand suggesting that additional soil samplings at shorter distances are needed to detect spatial dependence and more accurate maps (Mousaviard et al., 2012). According to Cambardella et al. (1994), spatial dependence was categorized using nugget/sill ratio with values $\geq 25\%$ implying strong, between 25 and 75% were considered as moderate while values $> 75\%$ were considered as weak. Hence, the spatial dependency of total depth, subsoil pH, EC and sand based on nugget/sill ratio was weak. However, the observed spatial dependence topsoil EC, OC, sand content and subsoil BD was strong while topsoil clay content and pH showed moderate dependence. The semi-variogram graphs of soil physical and chemical parameters (Figure 2) revealed considerable variability across the field while the spatial variability maps (Figure 3) revealed distinct textural (sand and clay) distribution pattern between the topsoil and subsoil horizons. Topsoil horizons with high sand content were found in the south eastern part of the field while high clay content found in the north and western parts. On the other hand, subsoil horizons with high sand content were found in the north, west and southern parts of the field while high clay content were found in the north and south eastern parts. Topsoil horizons containing high BD were found in the south eastern and north western part of the field while soils in the subsoil horizons with high BD were found in the southern part of the field. The maps revealed that the topsoil horizons were dominated with high BD while the subsoil horizons were characterized by low BD.

According to Wilding (1985), variability described in terms of the range of coefficient of variation can be grouped as least ($< 15\%$), moderate (15-35%) and most ($> 35\%$). Among the measured soil chemical parameters (Table 4), the content of EC, P, K and Na represented the most variable. On the other hand OC, Ca, Mg, Fe, Zn and ECEC were moderately variable while pH, Cu and Mn were the least variable. The values of range for pH, EC and OC measured from the semi-variogram for both topsoil and subsoil were low indicating a great amount of variability within the field. Majority of these chemical parameters were slightly skewed with a coefficient of skewness ≥ 0.5 (Table 5). Variables that were normally distributed included pH, OC, P, Fe, Cu and Zn while EC, ECEC, K, Ca, Mg, Na and Mn were not normally distributed. High topsoil pH were found in the north, east and south eastern parts of the field whereas high pH levels appeared to be dominated in the subsoil horizons except in the north western and southern parts. The spatial distribution pH on the field showed highly varied topsoil that increased from west to the east but with partly uniform subsoil (Figure 3).

Table 3. Selected chemical parameters of the twelve soil profiles dug across the research block within the experimental farm

Profile ID	pH$_W$	pH$_{KCl}$	EC mS/cm	Bray-1 P mg/kg	K mg/kg	Ca mg/kg	Mg mg/kg	Na mg/kg	Fe mg/kg	Cu mg/kg	Zn mg/kg	Mn mg/kg	ECEC Cmol$_{(-)}$/kg	OC %
RBP1T	7.52	6.78	56	2	70	810	508	103	8.56	1.64	0.54	33	8.88	0.30
RBP1S	8.38	6.96	94	1	70	913	710	173	16	1.60	0.52	23	11.36	0.35
RBP2T	8.04	7.21	117	7	173	1068	668	88	14	1.84	1.08	39	11.69	0.38
RBP2S	8.49	7.25	139	1	60	1360	890	178	12	1.88	0.72	31	15.08	0.80
RBP3T	7.86	6.20	89	12	238	788	505	60	11	1.80	1.64	47	8.98	0.88
RBP3S	7.44	6.15	74	1	70	1055	683	115	17	1.84	0.52	32	11.60	0.83
RBP4T	7.23	6.39	66	14	228	718	433	40	23	1.76	1.80	42	7.93	0.49
RBP4S	7.99	6.10	80	1	98	958	708	128	18	1.56	0.48	26	11.45	0.35
RBP5T	6.92	6.23	70	9	150	1053	538	20	17	1.72	1.44	33	10.18	0.88
RBP5S	8.50	7.49	230	1	70	1910	920	138	13	1.56	0.76	27	17.93	0.75
RBP6T	7.08	5.95	44	15	298	805	420	23	15	1.80	0.72	23	8.36	0.71
RBP6S	8.18	7.22	88	3	150	1050	620	68	20	1.92	0.24	21	11.05	0.65
RBP7T	7.32	6.54	73	10	128	803	495	25	12	1.44	1.24	29	8.54	0.73
RBP7S	8.55	7.36	163	1	73	853	910	200	17	1.68	0.36	20	12.84	0.67
RBP8T	7.90	6.54	43	17	90	613	373	20	10	1.48	1.28	35	6.47	0.58
RBP8S	7.85	6.45	76	1	33	645	443	93	11	1.92	0.40	20	7.38	0.49
RBP9T	6.85	5.34	35	16	88	385	203	15	8.12	1.36	1.88	28	3.89	0.57
RBP9S	7.17	6.09	49	2	33	493	308	10	4.88	1.72	0.44	16	5.14	0.66
RBP10T	7.52	6.39	40	15	70	393	265	8	9.16	1.16	1.68	31	4.37	0.27
RBP11T	7.91	6.90	59	19	78	600	390	40	13	1.28	1.60	26	6.60	0.35
RBP12T	8.10	7.00	51	18	93	563	338	30	9.52	1.40	1.72	30	5.98	0.83
RBP12S	8.05	6.30	52	2	43	655	465	95	13	1.76	0.44	19	7.64	1.04

Note. RBP1T = Research block profile 1 topsoil; RBP1S = Research block profile 1 subsoi; EC = electrical conductivity; OC = organic carbon; ECEC = effective cation exchange capacity.

Table 4. Summary of statistical analysis of measured chemical parameters of soil samples (n = 22) across the twelve soil profiles

Parameter	Minimum	Maximum	Mean	Median	Skewness	CV%
pH_W	6.85	8.55	7.77	7.88	-0.213	5.8
pH_{KCl}	5.34	7.49	6.58	6.49	-0.145	6.4
EC (mS/cm)	34.57	229.67	81.35	71.75	1.946	45.6
OC%	0.27	1.04	0.62	0.65	-0.028	19.9
Bray-1 P (mg/kg)	1	19	8	5	0.395	41.7
Exch. K (mg/kg)	33	298	109	83	1.393	39.6
Exch. Ca (mg/kg)	385	1910	841	804	1.487	21.7
Exch. Mg (mg/kg)	203	920	536	500	0.497	15.4
Exch. Na (mg/kg)	8	200	76	64	0.677	45.8
Extr. Fe (mg/kg)	4.88	23	13.28	13	0.315	26.2
Extr. Cu (mg/kg)	1.16	1.92	1.64	1.7	-0.665	9.7
Extr. Zn (mg/kg)	0.24	1.88	0.98	0.74	0.339	33.6
Extr. Mn (mg/kg)	16	47	29	29	0.555	12.0
ECEC (Cmol/kg)	3.89	17.93	9.24	8.71	0.670	15.7

Table 5. Semi-variogram parameters of the measured soil variables

Soil properties	Nugget	Range	Partial sill	Nugget/Sill ratio
Total depth	0	0.002	761.524	0
Topsoil				
pH	0.14	0.01	0.21	0.68
EC	449.63	0.00	239.30	1.88
OC	0.05	0.01	0.02	2.71
Clay	16.31	0.01	32.05	0.51
Sand	38.81	0.01	30.97	1.25
BD	0.01	0.01	0	
Subsoil				
pH	0	0.00	0.27	
EC	491.32	0.01	6938.28	0.07
OC	0.05	0.01	0	0
Clay	57.33	0.01	0	0
Sand	0	0.00	70.86	0
BD	0.02	0.01	0.02	1.06

Note. EC = electrical conductivity, OC = organic carbon, BD = bulk density.

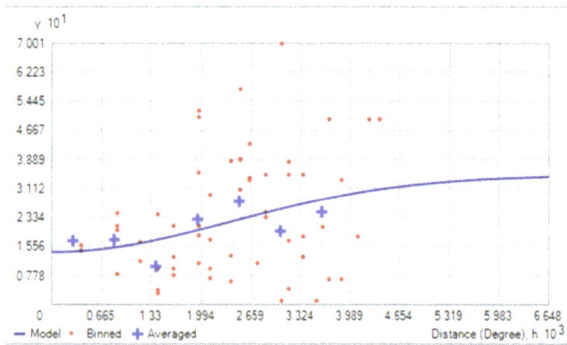

A1

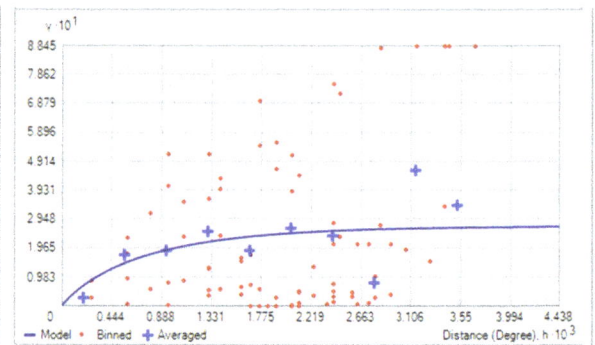

A2

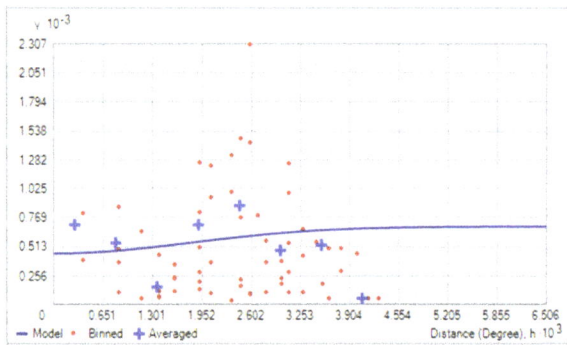

A3

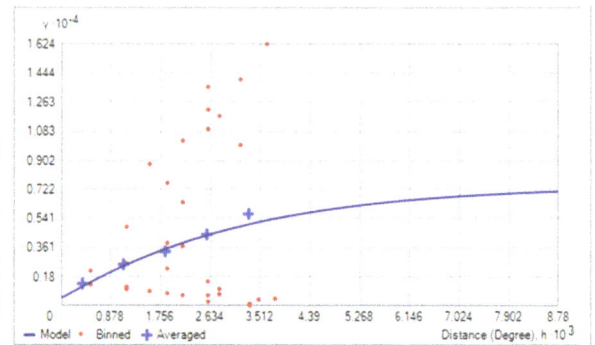

A4

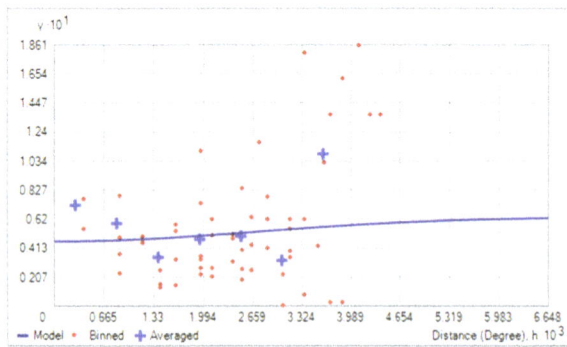

B1

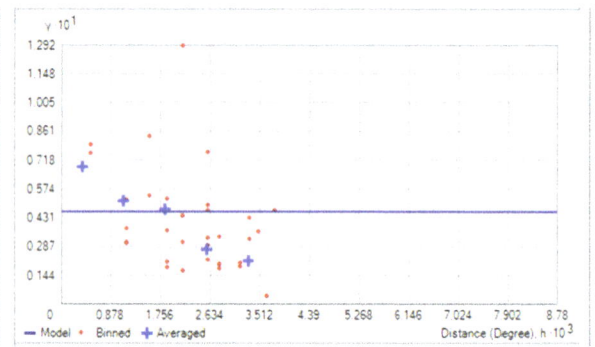

B2

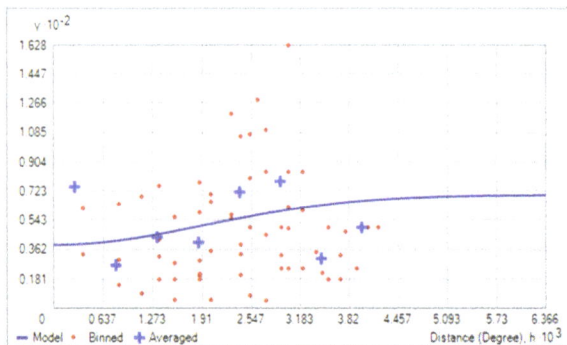

B3

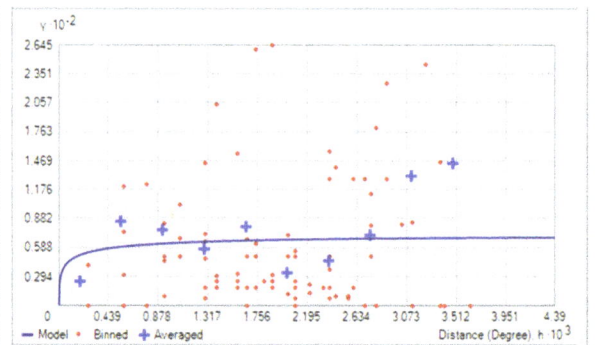

B4

C1

C2

C3

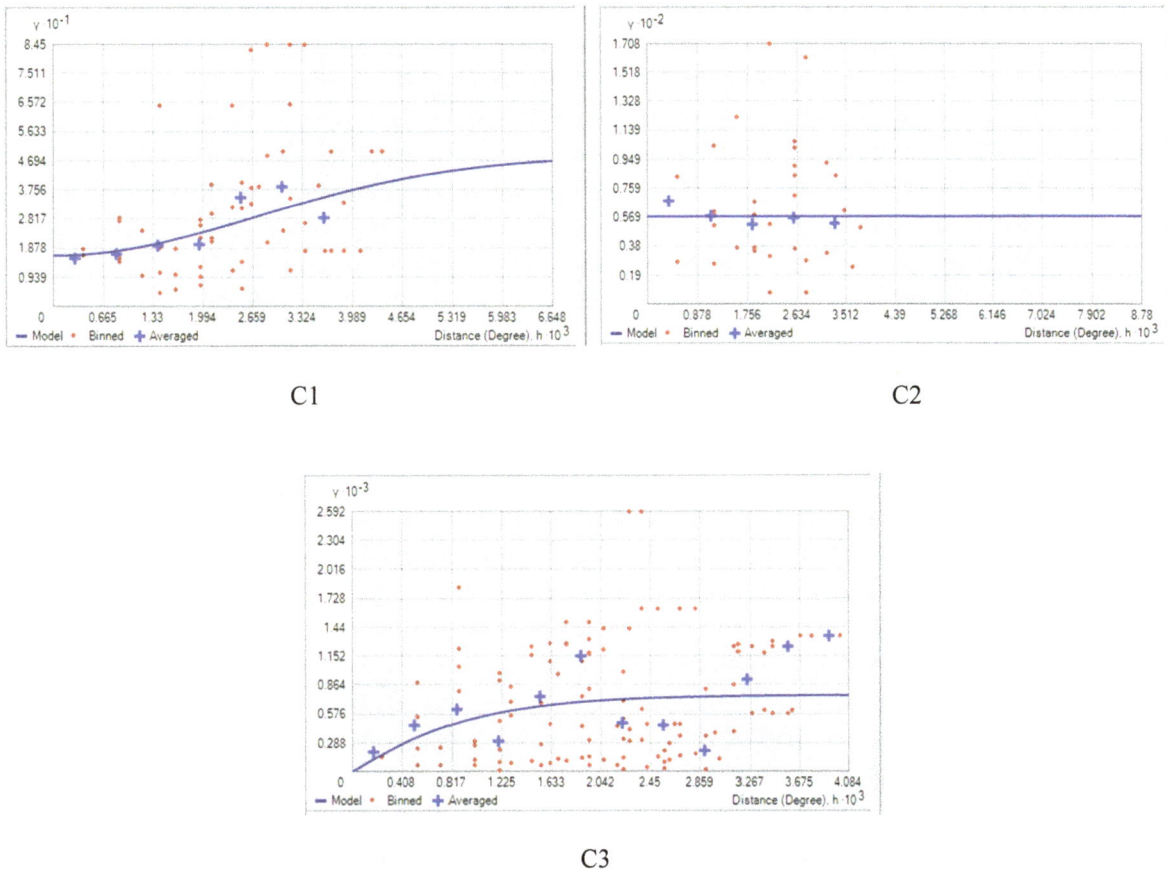

Figure 2. Semi-variograms for selected top and subsoil physical and chemical parameters measured from the field

Note. A1 & A2 = top and subsoil pH; A3 & A4 = top and subsoil EC; B1& B2 = content of top and subsoil organic carbon; B3 & B4 = contents of topsoil and subsoil sand; C1 & C2 = top and subsoil clay; C3 = total soil depth.

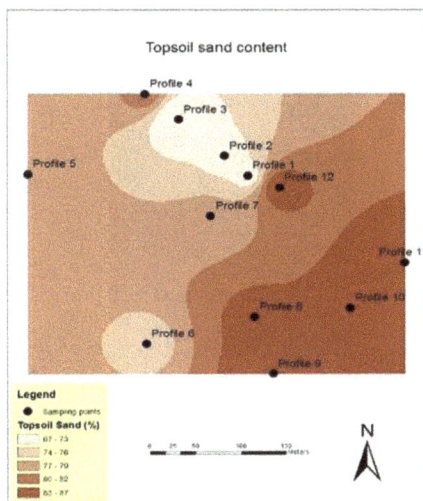

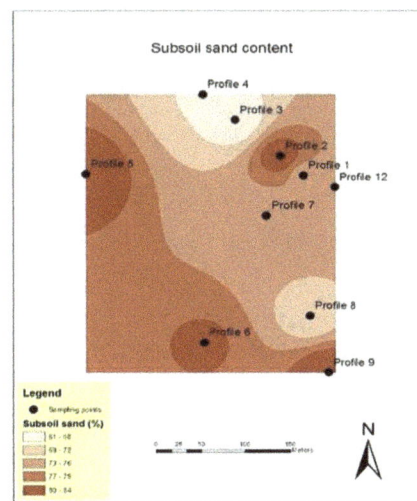

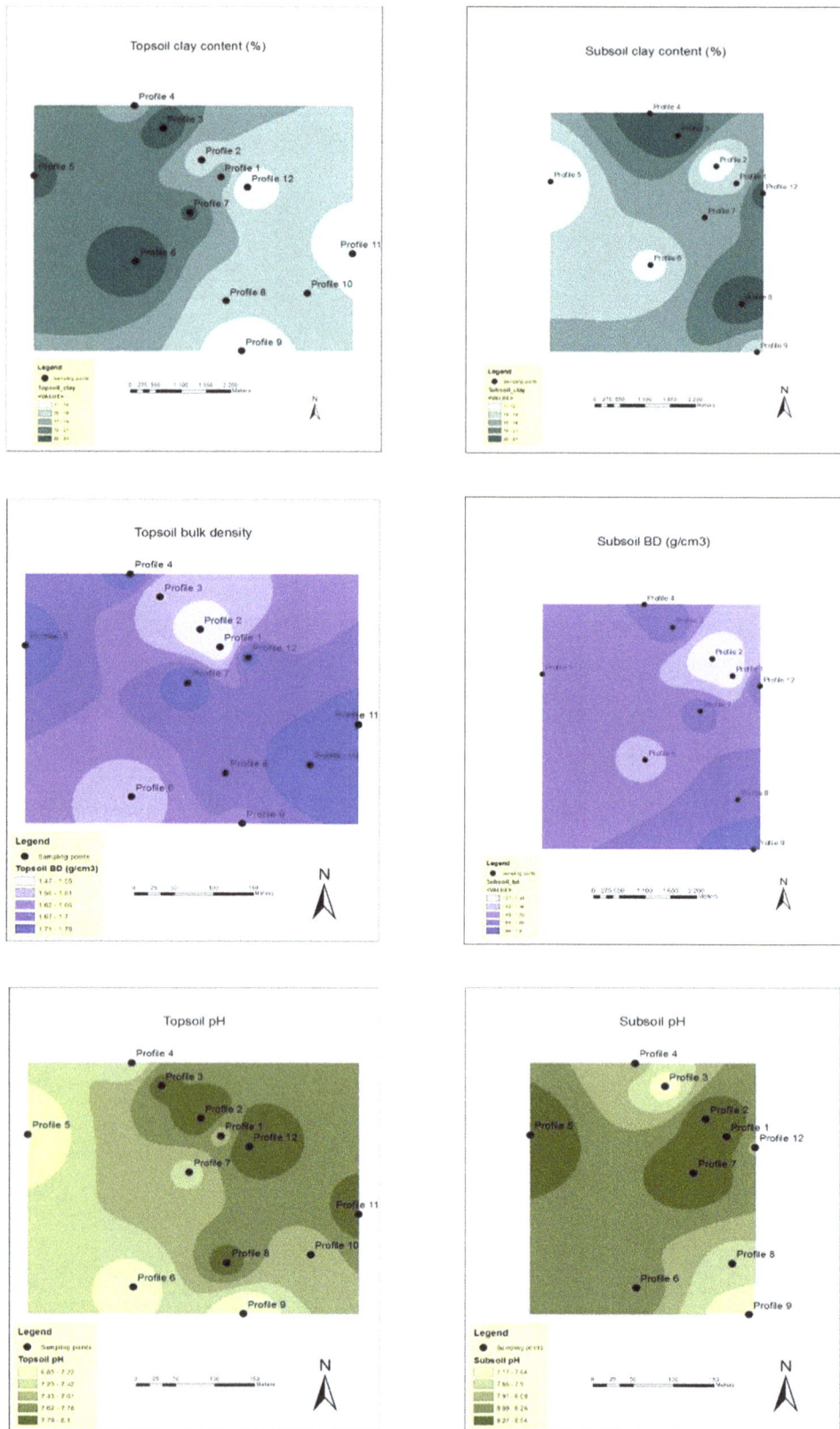

Figure 3. Spatial variability maps of topsoil and subsoil sand, clay, bulk density and pH values measured from the field

3.4 Correlation between Measured Soil Parameters

Linear correlation between measured parameters on the field (Table 6) indicated a significant ($p < 0.01$) positive relationship of EC with silt, pH, ECEC, Ca, Mg and Na content. BD had significant ($p \leq 0.05$) negative relationship with Na (-0.78), ECEC (-0.69) as well as pH value measured in water (-0.61). The ECEC content had significant ($p < 0.001$) positive correlation with Ca, Mg and Na but negative significant correlation with available P. There was significant positive correlation between pH and ECEC while a non-significant negative correlation was observed between clay content and ECEC. A significant negative correlation of BD was found with silt (-0.78) and poor non-significant correlation with organic carbon content (0.07).

Table 6. Pearson correlation matrix (r-value) between measured soil properties from the 12 profiles across different sampling depths

Parameters	Sand	Silt	Clay	BD	pH$_{KCl}$	pH$_W$	EC	OC	Bray P1	K	Ca	Mg	Na	ECEC	Fe	Mn	Cu	Zn
Sand	1																	
Silt	-0.60**	1																
Clay	-0.80***	0.01	1															
BD	0.33	-0.78***	0.17	1														
pH$_{KCl}$	0.15	0.46*	-0.53	-0.45	1													
pH$_W$	-0.02	0.52	-0.36	-0.61*	0.81***	1												
EC	-0.06	0.55*	-0.33	-0.52	0.70**	0.66**	1											
OC	-0.04	-0.20*	0.20	0.04	-0.08	0.00	0.14	1										
Bray P1	0.51	-0.01	-0.15	0.58*	-0.30	-0.41	-0.52	-0.09	1									
K	-0.16	-0.07	0.25	0.04	-0.19	-0.33	-0.11	0.11	0.43	1								
Ca	-0.16	0.56*	-0.22	-0.59*	0.56*	0.47	0.85***	0.23	-0.53	0.04	1							
Mg	-0.34	0.71***	-0.11	-0.70**	0.65**	0.68**	0.87***	0.14	-0.00**	-0.10	0.85***	1						
Na	-0.39	0.74***	-0.06	-0.78***	0.56*	0.74***	0.71**	-0.01	0.75***	-0.27	0.59*	0.88***	1					
ECEC	-0.28	0.68**	-0.15	-0.69**	0.63*	0.60*	0.89***	0.19	-0.63*	0.00	0.95***	0.97***	0.78***	1				
Fe	-0.32	0.24	0.22	-0.21	0.20	0.15	0.26	0.02	-0.18	0.42	0.375	0.468	0.32	0.46	1			
Mn	-0.12	0.15	0.04	0.04	-0.05	-0.16	-0.02	-0.03	0.41	0.50	0.08	-0.05	-0.19	0.03	0.11	1		
Cu	-0.55	0.48*	0.33	-0.51	0.11	0.13	0.26	0.37	-0.58*	0.26	0.39	0.43	0.37	0.44	0.38	-0.02	1	
Zn	0.47	-0.52*	-0.21	0.54*	-0.28	-0.40	-0.33	-0.80	0.86***	0.33	-0.37	-0.55*	-0.62*	-0.48	-0.16	0.64*	-0.57*	1

Note. CEC = effective cation exchange capacity; OC = organic carbon; BD = bulk density; * implies significant at $P < 0.05$; ** implies significant at $P < 0.01$; *** implies significant at $P < 0.00-1$.

4. Discussion

The observed spatial variations across the field have serious implications on soil volume, root growth, and crop management practices such as irrigation and fertilizer use. The textural types obtained on the field are well suited for crop production as they influence water holding capacity, CEC, soil workability, soil fertility and crop productivity (Heil & Schmidhalter, 2011). The observed high BD may be attributed to soil compaction following repeated use of tractor implements for tillage operations over the years. Soil compaction results in high bulk density (Afzalinia et al., 2011) causing restricted root growth, poor air and water movement within soil. The generally blocky structure of soils observed in the profiles might have resulted from the presence of high content of shrinking and swelling clay mineral types (Horn & Smucker, 2005). The observed soil colour variation could be related to the content of OC, soil texture, the presence of iron-containing minerals and the larger ecosystem processes (Viscarra-Rossel et al., 2006). Soils with firm and friable consistencies similar to those found in present study are reportedly favourable to workability and trafficability depending on their moisture content level (Huang et al., 2011).

The significant difference and high variability in most of the measured chemical parameters on the field may be related to land-use systems (Agoume & Birang, 2009). The slightly alkaline nature observed in this field could be related to the presence of high amount of exchangeable cations. Chik and Islam (2011) reported that soils containing high amounts of K, Ca, and Mg are likely to be alkaline while Kilic et al. (2012) attributed high soil pH in cultivated lands to high salt concentration of irrigation water. Frequent supplementary irrigation was performed on this field to support crop growth during winter and summer growing seasons. Balanced soil pH has an important influence on soil nutrient availability, solubility of toxic nutrient elements and CEC (Arain et al., 2000). Notwithstanding the high pH values on the field, the low EC values may be due to the low sodium levels in the soils, which were well below the critical level of 2000 mS/cm prescribed by Waskom et al. (2014). Soil variables that revealed normal distribution contributed to the observed results accuracy while the observed variation in soil OC level on the field may be due to differences in over the years cropping systems, diverse crop residues type and their decomposition rate (Tsui et al., 2004).

The significant distribution of micronutrients (Zn, Fe, Cu and Mn) observed across the profiles is in agreement with earlier findings (Verma et al., 2005). This may have been influenced by such characteristics as organic matter, clay and pH contents (Singh et al., 1989), and the cation activity ratios in soil solution on the field (Mayland &Wilkinson, 1989). The highly variable spatial dependence among measured variables may be attributed to soil intrinsic properties as influenced by the parent materials and extraneous factors like intensive and diverse cropping, soil erosion and fertilization (Alvares et al., 2011). The observed variation of the measured parameters over long distances on the field is probably due to increase of semivariance to points where the locations are considered independent of each other (Karl & Maurer, 2010).

In conclusion, the present study revealed the existence of fairly high level of spatial variability of soil physico-chemical properties across this intensively utilized research block associated with land use and management practices. Of all the measured parameters, electrical conductivity, Bray-1 P, exchangeable K, Ca and Na as well as extractable Fe and Zn showed a huge percent variation across the different depths and locations within the field. The correlation analyses indicated that there is negative and positive inverse or direct relationship among the measured soil physical and chemical properties. The findings explain the eventual anomalies/variabilities inherent in the results of current and similar research fields including soil that could be used for future experiments thus allowing researchers to implement management practices that are aligned with crop and soil requirements.

References

Afzalinia, S., Solhjou, A. A., & Eskandari, I. (2011). Effects of subsoiling on some soil physical properties and wheat yield in a dry land ecological condition. *Journal of Agricultural Science & Technology, 1*, 842-847.

Agoume, V., & Birang, A. M. (2009). Impact of land-use systems on some physical and chemical soil properties of an oxisol in the humid forest zone of Southern Cameroon. *Tropicultura, Yaoundé, 27*, 15-20.

Alvares, C. A., De Moraes-Gonçalves, J. L., Vieira, S. R., Da Silva, C. R., & Franciscatte, W. (2011). Spatial variability of physical and chemical attributes of some forest soils in south-eastern of Brazil. *Scientia Agricola, Piracicaba, 68*, 697-705. http://dx.doi.org/10.1590/S0103-90162011000600015

Arain, M. A., Ahmed, M., & Khan, M. A. (2000). Some physic-chemical characteristics of soil in sugarcane cultivated areas of Nawabshah, Sindh, Pakistan. *Pakistan Journal of Botany, Nawabshah, 32*, 93-100.

Behera, S. K., Singh, M. V., Singh, K. N., & Todwal, S. (2011). Distribution variability of total and extractable zinc in cultivated acid soils of India and their relationship with some selected soil properties. *Geoderma, 162*, 242-250. http://dx.doi.org/10.1016/j.geoderma.2011.01.016

Bouma, J., Stoorvogel, J., Van Alphen, B. J., & Booltink, H. W. G. (1999). Pedology, precision agriculture, and the changing paradigm of agricultural research. *Soil Science Society of America Journal, 63*, 1763-1768. http://dx.doi.org/10.2136/sssaj1999.6361763x

Cambardella, C. A., Moorman, T. B., Nocak, J. M., Parkin, T. B., Karlen, D. L., Turco, R. F., & Konopka, A. E. (1994). Field-scale variability of soil properties in central Iowa soils. *Soil Science Society of America Journal, 58*, 1501-1511. http://dx.doi.org/10.2136/sssaj1994.03615995005800050033x

Campbell, D. J., & Henshall, J. K. (1991). Bulk Density. In K. A. Smith & C. E. Mullins (Eds.), *Soil Analysis-Physical methods* (pp. 329-366). New York: Mareel Dekker.

Castrignanò, A., Giugliarini, L., Risaliti, R., & Martinelli, N. (2000). Study of spatial relationships among some soil physico-chemical properties of a field in central Italy using multivariate geostatistics. *Geoderma, 97*, 39-60. http://dx.doi.org/10.1016/S0016-7061(00)00025-2

Cerri, C. E. P., Bernoux, M., Chaplot, V., Volkoff, B., Victoria, R. L., Melillo, J. M., ... Cerri, C. C. (2004). Assessment of soil property spatial variation in an Amazon pasture: Basis for selecting an agronomic experimental area. *Geoderma, 123*, 51-68. http://dx.doi.org/10.1016/j.geoderma.2004.01.027

Chik, Z., & Islam, T. (2011). Study of chemical effects on soil compaction characterizations through electrical conductivity. *International Journal of Electrochemical Science, 6*, 6733-6740.

FAO. (2009). *Climate and Rainfall*. Retrieved September 28, 2015, from http://wwww.fao.org/wairdocs/ilri/x5524e/x5524e03.htm

Goovaerts, P. (1998). Geostatistical tools for characterizing the spatial variability of microbiological and physico-chemical soil properties. *Biology and Fertility of Soils, 27*, 315-334. http://dx.doi.org/10.1007/s003740050439

Heil, K., & Schmidhalter, U. (2011). Characterization of soil texture variability using the apparent soil electrical conductivity at a highly variable site. *Computers and Geosciences, 39*, 98-110. http://dx.doi.org/10.1016/j.cageo.2011.06.017

Horn, R., & Smucker, A. (2005). Structure formation and its consequences for gas and water transport in unsaturated arable and forest soils. *Soil and Tillage Research, 82*, 5-14. http://dx.doi.org/10.1016/j.still.2005.01.002

Huang, P. M., Li, Y., & Sumner, M. E. (Eds.). (2011). *Handbook of Soil Sciences: Properties and Processes*. Florida: CRC Press.

Iqbal, J., Thomasson, J. A., Jenkins, J. N., Owens, P. R. & Whisler, F. D. (2005). Spatial variability analysis of soil physical properties of alluvial soils. *Soil Science Society of America Journal, 69*, 1338-1350. http://dx.doi.org/10.2136/sssaj2004.0154

Jat, N. L., Jatin, N. K., & Choudary, G. R. (2006). Integrated nutrient management in fenugreek (*Trigonellafoenum graecum*). *Indian Journal of Agronomy, 51*, 331-333.

Karl, J. W., & Maurer, B. A. (2010). Spatial dependency of predictions from image segmentation: A variogram-based method to determine appropriate scales for producing land-management information. *Ecological Informatics, 5*, 194-202. http://dx.doi.org/10.1016/j.ecoinf.2010.02.004

Kilic, K., Kilic, S., & Kocyigit, R. (2012). Assessment of spatial variability of soil properties in areas under different land uses. *Bulgarian Journal of Agricultural Science, 18*, 722-732.

Lal, R. (2006). *Encyclopedia of Soil Science*. Florida: Taylor and Francis.

Lin, H., Wheeler, D., Bell, J., & Wilding, L. (2005). Assessment of spatial variability at Multiple Scales. *Ecological Modelling, 182*, 271-290. http://dx.doi.org/10.1016/j.ecolmodel.2004.04.006

Mayland, H. F., & Wilkinson, S. R. (1989). Soil factors affecting magnesium availability in plant-animal systems: A review. *Journal of Animal Science, 6*, 3437-3444.

Mohammadi, J. (2002). Spatial variability of soil fertility, wheat yield and weed density in a One-hectare field in Shahre Kord. *Journal of Agricultural Science and Technology, 4*, 83-92.

Mousavifard, S. M., Momtaz, H., Sepehr, E., Davatgar, N., & Sadaghiani, M. H. R. (2012). Determining and mapping some soil physico-chemical properties using geostatistical and GIS techniques in the Naqade region, Iran. *Archives of Agronomy and Soil Science, 59*, 1573-1589. http://dx.doi.org/10.1080/0365034 0.2012.740556

Nelson, D. W., & Sommers, L. E. (1996). Total carbon, organic carbon and organic matter. In D. L. Sparks (Eds.), *Methods of soil analysis* (Part 3, 2nd ed., SSSA Book Series No. 5, pp. 961-1010). Madison (WI): American Society of Agronomy and Soil Science of America.

Nsanzabaganwa, E., Das, T. K., Rana, D. S., & Kumar, S. N. (2014). Nitrogen and phosphorus effects on winter maize in an irrigated agroecosystem in western Indo-Gangetic plains of India. *Maydica, 59*, 152-160.

Okalebo, J. R., Gathua, K. W., & Woomer, P. L. (2002). *Laboratory methods of soil and water analysis: A working manual*. Nairobi: TSBF-CIAT and SACRED-Africa.

Özgöz, E. (2009). Long term conventional tillage effect on spatial variability of some soil physical properties. *Journal Sustainable Agriculture, 33*, 142-160. http://dx.doi.org/10.1080/10440040802395056

Reza, S. K., Nayak, D. C., Chattopadhyay, T., Mukhopadhyay, S., Singh, S. K., & Srinivasan, R. (2015). Spatial distribution of soil physical properties of alluvial soils: a geostatistical approach. *Archives of Agronomy and Soil Science, 62*, 972-981. http://dx.doi.org/10.1080/03650340.2015.1107678

Saglam, M., Ozturk, H. S., Ersahin, S., & Ozkan, A. I. (2011). Spatial variation of soil physical properties in adjacent alluvial and colluvial soils under Ustic moisture regime. *Hydrology and Earth System Sciences Discussions, 8*, 4261-4280. http://dx.doi.org/10.5194/hessd-8-4261-2011

Santra, P., Chopra, U. K., & Chakraborty, D. (2008). Spatial variability of soil properties and its application in predicting surface map of hydraulic properties in an agricultural farm. *Current Science, 95*, 937-945.

Schoeneberger, P. J., Wysocki, D. A., Benham, E. C., & Broderson, W. D. (1998). *Field book for describing and sampling soils*. Lincoln, NE: NRCS, USDA, National Soil Survey Centre.

Shapiro, S. S., & Wilk, M. D. (1965). An analysis of variance test for normality (complete samples). *Biometrika, Great Britain, 52*, 591-611. http://dx.doi.org/10.1093/biomet/52.3-4.591

Sheldrick, B., & Hand Wang, C. (1993). Particle-size distribution. In M. R. Carter (Ed.), *Soil sampling and methods of analysis* (pp. 499-511). Ann Arbor (MI): Canadian Society of Soil Science/Lewis Publishers.

Singh, K., Kuhad, M. S., & Dhankar, S. S. (1989). Influence of soil characteristics on profile distribution of DTPA-extractable micronutrient cations. *Indian Journal of Agricultural Sciences, 59*, 310-312.

Staugaitis, G., & Sumskis, D. (2011). Spatial variability of soil pH as influenced by various sampling methods and geostatistical technoiques. *Zemdirbyste Agriculture, 98*, 323-332.

The Non-Affiliated Soil Analyses Work Committee. (1990). *Handbook of Standard Soil Testing Methods for Advisory Purposes*. Pretoria: Soil Science Society of South Africa,

Tsui, C. C., Chen, Z. S., & Hsieh, C. F. (2004). Relationships between soil properties and slope position in a lowland rain forest of southern Taiwan. *Geoderma, 123*, 131-142. http://dx.doi.org/10.1016/j.geoderma.2004.01.031

Vaezi, A. R., Bahrami, H. A., Sadeghi, S. H. R., & Mahdian, M. H. (2010). Spatial variability of soil erodibility factor (K) of the USLE in North West of Iran. *Journal of Agricultural Science and Technology, 12*, 241-252.

Verma, Y. K., Setia, R. K., Sharma, P. K., Singh, C., & Ashok, K. (2005). Pedospheric variations in distribution of DTPA-extractable micronutrients in soils developed on different physiographic units in central parts of Punjab, India. *International Journal of Agriculture Biology, 7*, 243-246.

Viscarra-Rossel, R. A., Minasny, B., Roudier, P., & Mcbratney, A. B. (2006). Colour space model for soil science. *Geoderma, 133*, 320-337. http://dx.doi.org/10.1016/j.geoderma.2005.07.017

Waskom, R. M., Bauder, J. G., & Andales, A. A. (2014). *Diagnosing saline and sodic soil problems*. Retrieved March 24, 2014, from http://www.ext.colostate.edu

Wilding, L. P. (1985). Spatial Variability: It's Documentation, Accommodation and Implication to Soil Surveys. In D. R. Nielsen & J. Bouma (Eds.), *Soil Spatial Variability* (pp. 166-194). The Netherlands: Pudoc, Wageningen.

World Reference Base for Soil Resources (WRB). (2006). A Framework for International Classification, Correlation and Communication. *World Soil Resources Rep.103*. IUSS/ISRIC/FAO.

15

Improving Soil Fertility and Crops Yield through Maize-Legumes (*Common bean* and *Dolichos lablab*) Intercropping Systems

Prosper I. Massawe[1], Kelvin M. Mtei[2], Linus K. Munishi[1] & Patrick A. Ndakidemi[1]

[1] Department of Sustainable Agriculture and Biodiversity Management, The Nelson Mandela African Institution of Science and Technology, Arusha, Tanzania

[2] Department of Water and Environmental Sciences, The Nelson Mandela African Institution of Science and Technology, Arusha, Tanzania

Correspondence: Prosper I. Massawe, Department of Sustainable Agriculture and Biodiversity Management, The Nelson Mandela African Institution of Science and Technology, P.O. Box 447, Arusha, Tanzania. E-mail: massawep@nm-aist.ac.tz

Abstract

Declining crops yield in the smallholder farmers cropping systems of sub-Saharan African (SSA) present the need to develop more sustainable production systems. Depletion of essential plant nutrients from the soils have been cited as the main contributing factors due to continues cultivation of cereal crops without application of organic/ inorganic fertilizers. Of all the plant nutrients, reports showed that nitrogen is among the most limiting plant nutrient as it plays crucial roles in the plant growth and physiological processes. The most efficient way of adding nitrogen to the soils is through inorganic amendments. However, this is an expensive method and creates bottleneck to smallholder farmers in most countries of sub-Saharan Africa. Legumes are potential sources of plant nutrients that complement/supplement inorganic fertilizers for cereal crops because of their ability to fix biological nitrogen (N) when included to the cropping systems. By fixing atmospheric N_2, legumes offer the most effective way of increasing the productivity of poor soils either in monoculture, intercropping, crop rotations, or mixed cropping systems. This review paper discuses the role of cereal legume intercropping systems on soil fertility improvement, its impact on weeds, pests, diseases and water use efficiency, the biological nitrogen fixation, the amounts of N transferred to associated cereal crops, nutrients uptake and partition, legume biomass decomposition and mineralization, grain yields, land equivalent ratio and economic benefits.

Keywords: cropping system, BNF, N-transfer, biomass decomposition, nutrient uptake, grain yields

1. Introduction

In traditional agriculture, arable land is left fallow for some years to allow soil to acquire self-rejuvenation, but due to increased population pressure, fallow periods are shorter and are not sufficient to restore the soil nutrient pools sufficient to support economic crop yields (TASDS, 2001). The agricultural systems based on high external inputs are not sustainable and threatens food security in Tanzania, particularly at the smallholder levels (Birech & Freyer, 2007). A sustainable cropping system as an integral part of a complete farming system of soil, water, air, plant, animal, and human resources have to be endorsed (Arshad & Martin, 2002; Rahman, 2013). Intercropping is the common cropping system which involved the cultivation of two or more crops at the same time in the same field (Balthazar, 2014). The common crops combination in intercropping systems in Tanzania and Africa at large is maize and a variety of legumes such as beans, cowpeas, dolichos lablab, pigeon peas, green gram and bambara nuts (Waddington et al., 1989; Balthazar, 2014). Maize-legumes are important components of intercropping systems in improving soil fertility, controlling weeds, diseases and insects, conserving soil moisture, reducing soil erosion and improving soil microbiology (Fageria et al., 2005; Delin et al., 2008). Maize is the most cereal crop produced by about 82% of all Tanzanian farmers (NBS, 2007). In sub-Saharan Africa, maize is a staple food for an estimated 50% of the population and provides 50% of the basic calories. Maize production estimates in Tanzania in 2005-2007 were 3.4 million tonnes, which were grown on an area of two million hectare or about 45% of the cultivated area (NBS, 2007). This low maize productivity is associated with high level of dependency of agriculture to exogenous factors such as drought, flooding, pests and high costs of agro-inputs.

Bean (*Phaseolus vulgaris* L.) is one of the most important legumes in the world because of its commercial value, extensive production, consumer use and nutrient value (Xavery et al., 2006; CABI, 2007). For example, beans consumption per capita in Eastern and Southern Africa is 40-50 kg year^{-1} (Blair et al., 2010). These crops are good sources of proteins, vitamins, and minerals such as Fe, Zn, P, Ca, Cu, K, and Mg, and are excellent sources of complex carbohydrates (Camacho & Gonzalez de Mejia, 1998). It has positive impact on soil fertility improvement through BNF process and upon incorporation of residues into the soils (Blair et al., 2010). In East Africa, Tanzania is a major beans producing country where it is estimated that over 75% of rural households in Tanzania depend on beans for home consumption as well as cash crop income (Hillocks et al., 2006).

Dolichos lablab (*Lablab purpureus*) is popular as a nitrogen-fixing green manure to contribute to soil N and improve soil quality. Lablab is a popular choice as a cover crop on infertile, acidic soils, and it is drought tolerant once established (Hector & Jody, 2002). When in symbiotic association with *Bradyrhizobium japonicum*, Dolichos lablab plants can fix up to 400 kg N ha^{-1}yr^{-1} (Benselama et al., 2014), reducing the need for expensive and environmentally damaging nitrogen fertilizer. Maass et al. (2010) observed that *D. lablab* may suffer from low yields when grown as a main cash crop, and suggest that it is more popular in home gardens and mixed-cropping schemes. Moreover, lablab is a traditional food and fodder crop in Africa, including Tanzania (Maass et al., 2010), and offers great potential for smallholder farming systems. However, nowadays, lablab's utilization by farmers is in steady decline, being outperformed by other leguminous species such as common bean (*Phaseolus vulgaris*) and cowpea (*Vigna unguiculata*) (Bourgault & Smith, 2010).

The Soil fertility depletion is a widespread limitation to yield improvement in maize based intercropping systems throughout Eastern and Southern Africa (Mekuria & Waddington, 2002; Keston et al., 2013). It is widely considered as a major factor contributing to low productivity and non-sustainability of existing production systems and a major source of low returns to other inputs and management committed to smallholder farmers (Sanchez & Jama, 2002; Mekuria & Waddington, 2002). As the case with most other SSA countries, Tanzania faces a challenge of declining soil fertility with nitrogen considered the main limiting factor to crops growth (Keston et al., 2013). Nitrogen is an essential element for plant growth and development and a key issue of agriculture because it is an important component of plant cells at the structural, genetic and metabolic levels, getting involved in many processes of plant growth and development which finally lead to yield as well as the quality of harvested organs such as seeds or shoot biomass (Salon et al., 2011). A study by Unkovich et al. (2008) indicates that nitrogen fertilizers contribute to resolving the challenge the world is facing, feeding the human population, although urea which is the most commonly used nitrogenous fertilizer, has now become a costly input for farmers. The solutions to smallholder farmers' soil fertility problems may be found in the strategic combination of organic resources and nitrogen-fixing legumes that work symbiotically with special bacteria, *rhizobia*, which live in the root nodules. The symbiosis is manifested by the development of nodules on the roots of legumes acting as factories of nitrogen fixation (Collins, 2004). *Rhizobium* inoculants may be used as a cheaper substitute for urea in the production of food legume crops (Karim et al., 2001). The beneficial effect of *rhizobial* inoculates in increasing yield of leguminous crops results from the activity of its root nodule bacteria, which fix atmospheric nitrogen making it available for the plants. Legumes can meet most of their N needs and contribute to soil N through symbiotic nitrogen fixation (Maobe et al., 1998). Estimates indicate that legumes can fix up to 200 kg N /ha/ year under optimal field conditions (Giller, 2001).

Decomposition and mineralization of legumes organic residues is another key process in ecosystem's carbon cycle, releasing carbon to the atmosphere and is determined by climate and litter quality (Zhang, 2009). It involves insect and microbial decomposers, organically-bound nutrients are released as free ions to the soil solution which are then available for uptake by plants (Mureithi et al., 2005). The use of high quality plant residues could ensure timely nutrients release for enhanced crop uptake. Legumes produce the high quality residues and therefore, offer a low cost opportunity for maintaining soil fertility by contributing nutrient during decomposition (Ibewiro et al., 2000; Baijukya et al., 2004) and improving soil organic matter and soil physical properties (Mureithi et al., 2005).

Plants acquire nutrients from two principal sources which are the soil, (through commercial fertilizer, manure and/or mineralization of organic matter); and the atmosphere (through symbiotic N fixation) (Vance, 2001; Rahman, 2013). Total mineral nutrient uptake is the sum of nutrient content in the stover and grain and estimates the total quantity of a mineral nutrient required to produce a crop (Bender, 2012).

Land equivalent ratio (LER) is an important tool used to evaluate the advantages of intercropping systems; it measures the yield advantage obtained by growing two or more crops or varieties as an intercrop compared to growing the same crops as a collection of separate monoculture (Mazaheri et al., 2006). It further estimates the levels of intercrop interference going on in the intercropping systems which indicates the resources utilization by the component crops (Yancey & Cecil, 1994).

Biological nitrogen fixation by grain legume crops has received a lot of attention because it is a significant N source in agricultural ecosystems (Peoples et al., 2002; Ndakidemi, 2006; Rahman, 2013). The N balance of a cropping system can be improved using legumes but the magnitude of biological N fixation of legumes is highly variable and depends on several factors, such as plant species, inoculation, soil nitrate and water contents (Bender, 2012). However, studies on N_2 fixation in the complex cereal-legume cropping systems are few and therefore there is a need to identify and develop cereal-legumes intercropping systems that would influence N_2 fixation in agricultural systems and complement fertilizer-N use as a sustainable means for soil fertility improvement and consequently crops yields.

2. Cereal-Legume Cropping Systems

Cropping system is an effective ecological farming system that manages and organizes crops so that they best utilize the available resources such as sunlight, water, air, soil, farm labour and equipments (Steiner, 1982). Intercropping system is a type of mixed cropping and defined as the agricultural practice of cultivating two or more crops in the same space at the same time (Andrews & Kassam, 1976; Sanchez, 1976). This is a common practice in SSA, and it is mostly practiced by smallholder farmers. Mixed cropping of cereals and legumes is widespread in the tropics (Molatudi & Mariga, 2012) because legumes used in crop production have traditionally enabled farmers to cope with erosion and with declining levels of soil organic matter and available N (Scott et al., 1987). The common crop combinations in intercropping systems of Tanzania and Africa are cereal-legume, particularly maize-cowpea, maize-soybean, maize-pigeon pea, maize-groundnuts, maize-beans, sorghum-cowpea, millet-groundnuts, and rice-pulses (Beets, 1982; Rees, 1986). It is popular because of its nutritional complimentarity (Edje, 1990). The features of an intercropping system differ with soil, local climate, economic situation and preferences of the local community (Steiner, 1982). In crop rotation, legumes contribute to a diversification of cropping systems and as N_2-fixing plant, it can reduce the mineral N fertilizer demand.

According to Huang et al. (2006) systems that intercrop maize with a legume are able to reduce the amount of nutrients taken from the soil as compared to a maize monocrop. When nitrogen fertilizer is added to the field, intercropped legumes use the inorganic nitrogen instead of fixing atmospheric nitrogen and thus compete with maize for nitrogen (Kutu & Asiwe, 2010). However, when nitrogen fertilizer is not applied, intercropped legumes will fix most of their nitrogen requirements from the atmosphere and not compete with maize for nitrogen resources (Adu-Gyamfi et al., 2007). Improving performance of maize/legume intercrops can significantly benefit the smallholder (SH) farmers by increasing yield on a limited amount of land, reducing risk of total crop failure, and maximizing the efficiency of labour utilization (Huang et al., 2006). In addition, some of intercropping systems help to stabilize soil nutrient levels, which will keep yields sustainable into the future (Kutu & Asiwe, 2010). Intercropping has negative impact especially harvesting two crops from within one field may be more challenging than harvesting the different crops from separate fields (Thobatsi, 2009). Further, Farmers who traditionally use herbicides to protect their maize plants from competition may face problems if their intercrops are susceptible to the herbicides (Scholl & Nieuwenhuis, 2004). In this case, farmers may have fewer options for herbicide-based weed control, or may have to completely abandon this strategy (Maqbool et al., 2006). Generally, intercropping may not help farmers with very low soil fertility problems because does not rehabilitate poor land successfully (Thobatsi, 2009). Therefore, intercropping systems are deliberately designed to optimize the use of spatial, temporal, and physical resources both above- and belowground, by maximizing positive interactions (facilitation) and minimizing negative ones (competition) among the components (Ndakidemi, 2006). The understanding of the efficient cereal-legume cropping system would be one of the solutions of maximizing land use, spreading economic risk and improving soil productivity through nitrogen fixation.

3. Impact of Cereal-Legume Cropping Systems on Weeds, Diseases, Insects and Water Use Efficiency

Cereal-legume cropping system is one of the traditional farming which control weeds in the small scale farms. This occurs when the competitive ability of the component crops population is higher than the dominant weeds (Dimitrios et al., 2010). For example a study by Belel et al. (2014) indicated that cereals and cowpea reduced striga infestation significantly due to unfavourable cover conditions created by the intercropped crops. Intercropping system helps to reduce weeds population once the crops are established due to increased leaf cover (Dimitrios et al., 2010). Thayamini et al. (2010) reported weed suppression in maize-groundnut and maize-bean intercropping as an integrated weed management tool. Intercropping show weeds control advantages over sole crops because the chemical control is difficult when the crops have emerged especially when a dicotyledonous crop species is combined with a monocotyledonous crop species. Some intercropped species produce toxic chemicals (allelopathy) which suppress growth of weeds (Belel et al., 2014). On other hands, intercrops may provide yield advantages without suppressing the growth of weeds below levels observed in sole crops if intercrops use resources that are not exploitable by weeds or convert resources into harvestable materials more efficiently than sole crops (Thayamini et al., 2010). Intercropping such as wheat-canola-pea has great suppressive effect on weeds compared to sole crop, indicating some type of synergism among crops within intercrops with respect to weed suppression (S. S. Rana & M. C. Rana, 2011). Generally, the weed suppressing ability of intercrop is dependent upon the component crops selected, genotype used, plant density adopted, proportion of component crops, their spatial arrangement and fertility moisture status of the soil (Belel et al., 2014; S. S. Rana & M. C. Rana, 2011; Dimitrios et al., 2010).

Cereal-legume cropping systems also control insects and diseases by provision of barrier that prevent the spread between the host and parasite. This has been reported by Seran and Brintha (2010) on bud worm and corn borer infestation in sole maize being greater than in maize intercropped with soybean. Another study by Thayamini et al. (2010) indicated that the average percentage of maize stalk borer infestation was significantly greater in monocropped (70 percent) than in intercropped maize-soybean. Cereal-legumes intercropping enhance the abundance of predators and parasites, which in turn prevent the build-up of insects and disease, thus minimizing the need of using expensive and dangerous chemical insecticides and fungicide (Sekamatte et al., 2003). Mixed crop species can also delay the introduction of diseases by reducing the spread of disease carrying spores and by modifying environmental conditions so that they are less favorable to the spread of certain pathogens (S. S. Rana & M. C. Rana, 2011). Therefore the simplification of cereal-legume cropping systems can affect the abundance and efficiency of the natural enemies or predators, which depend on habitat complexity for resources. Changes in environment and host plant quality lead to direct effects on the host plant searching behavior of herbivorous insects as well as indirect effects on their developmental rates and on interactions with natural enemies (S. S. Rana & M. C. Rana, 2011).

Cereal-legume cropping systems improve water use efficiency which leads to increases the use of other resources because of the early high leaf area index and higher leaf area which conserve water (Ogindo & Walker, 2005). Thayamini et al. (2010) indicated high water use efficiency in intercrops than sole crops under water limiting conditions hence leading to higher grain yields. This is because intercropping increased light interception, reduce water evaporation, and improve conservation of the soil moisture compared with maize alone (Ghanbari et al., 2010). The total water requirement of intercrop does not increase much compared to sole cropping. For example, a study by Rana and Rana (2011) showed that the water requirement of sole sorghum and intercropping with red gram was almost similar (584 and 585 mm, respectively). Therefore, the total water used in intercropping system is almost same as in sole crops, but yields are increased hence water use efficiency of intercropping is higher than sole crops.

4. Biological Nitrogen Fixation (BNF) in Cereal-Egume Cropping Systems

Dinitrogen (N_2) gas represents almost 80% of the earth's atmosphere and it is not directly available to plants (Davidson et al., 2007). BNF is the process whereby a number of species of bacteria use the enzyme nitrogenase to convert atmospheric N_2 into ammonia (NH_3), a form of nitrogen (N) that can then be incorporated into organic components, e.g. protein and nucleic acids, of the bacteria and associated plants (Davidson et al., 2007; Postgate, 1998). The process is coupled to the hydrolysis of 16 equivalents of ATP and is accompanied by the co-formation of one molecule of H_2 (Chi Chung et al., 2014). The overall reaction for BNF is:

$$N_2 + 8 H^+ + 8 e^- \rightarrow 2 NH_3 + H_2 \qquad (1)$$

The conversion of N_2 into ammonia occurs at a cluster called FeMoco, an abbreviation for the iron-molybdenum cofactor. The mechanism proceeds via a series of protonation and reduction steps wherein the FeMoco active site hydrogenates the N_2 substrate (Hoffman et al., 2013). The natural process of BNF offers an economic means of reducing environmental problems and improving the internal resources compared with the production of nitrogen fertilizer by industrial fixation (Aydinalp & Cresser, 2008). Intercropping legumes with non-leguminous crops can result in competition for water and nutrients (People et al., 1989). However, it has been shown that when mineral N is depleted in the root zone of the legume component by the non-leguminous intercrops, N_2 fixation of legumes will be promoted. Legume-*Rhizobium* symbiotic system is the most important biological nitrogen fixation (BNF) process in nature (Peoples et al., 1995), providing about 65% biosphere's available nitrogen for use including the agricultural system (Lodwig et al., 2003). N_2-fixing systems can thrive in soils poor in N and that they are a source of proteins and provides N for soil fertility. Typical environmental stresses faced by the legume nodules and their symbiotic partner (*Rhizobium*) may include photosynthate deprivation, water stress, salinity, soil nitrate, temperature, heavy metals, and biocides (Walsh, 1995). For such constraints to be controlled, legume crops can contribute (or fix) substantial quantities of N into the soil. BNF is important in legume-based cropping systems when fertilizer-N is limited (Fujita & Ofosu-Budu, 1996), particularly in SSA where nitrogen annual depletion was recorded at all levels at rates of 22 kg ha^{-1} (Smaling et al., 1997) and mineral-N fertilization is neither available nor affordable to smallholder farmers (Mugwe et al., 2009). Therefore, the rates of N_2 fixation tend to be highest when plant-available mineral N in the soil is limiting but water and other nutrients are plentiful. Also mineral nutrients may influence N_2 fixation in legumes and non legumes at various levels of the symbiotic interactions: infection and nodule development, nodule function, and host plant growth (O'Hara, 2001). Robson (1978) summarized the nature of the interaction between nutrient supply and combined nitrogen on legume growth as a means for estimating symbiotic sensitivity to their supply or concentration. He highlighted that Co and Mo are required in high amounts for symbiotic nitrogen fixation for host-plant growth than Cu, Ca and P. Although there is an experimental evidence for specific requirements for 11 nutrients (B, Ca, Co, Cu, Fe, K, Mo, Ni, P, Se and Zn) for symbiotic development in some species of legume, only four of these elements (Ca, P, Fe and Mo) appear to cause significant limitations on the productivity of symbiotic legumes in some agricultural soils (O'Hara, 2001). High rates of N_2 fixation in some agricultural soils are commonly achieved because most cropping systems are dominated by cereals that utilise large quantities of soil mineral N (Peoples et al., 2002). Keeping in mind the importance of biological nitrogen fixation process in cereal/legume cropping systems, this will help to utilize the fixed nitrogen to its full potential.

5. Nitrogen Transfer from Legumes to Cereal Crops

The movement of biologically fixed nitrogen from legume to the cereal crop is not well known although; evidence suggests that associated non-legumes may benefit through N-transfer from legumes (Fujita et al., 1992). This N-transfer is considered to occur through root excretion, N leached from leaves, leaf fall, and animal excreta if present in the system (Fujita et al., 1992). The limited studies carried out within SSA suggested that N_2-fixed by a leguminous component may be available to the associated cereal in the current growing season (Eaglesham et al., 1981), known as direct N transfer (Stern, 1993). Further, Eaglesham et al. (1981) showed that 24.9 percent of N fixed by cowpea was transferred to maize crop.

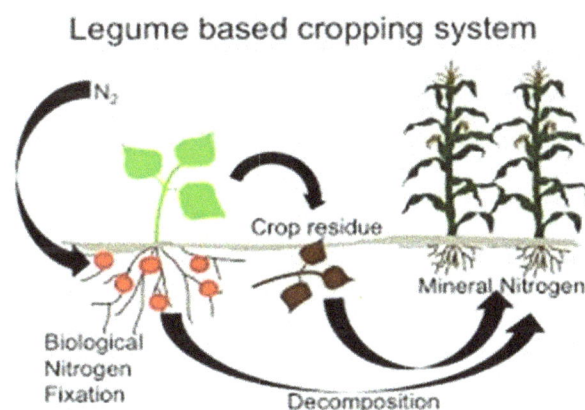

Figure 1. Legume-based cropping systems

Source: https://www.google.co.tz/search?q=cereal-legume+cropping+systems

However, Ofori and Stern (1987) and Danso et al. (1993) reported that there is little or no current N transfer in cereal-legume intercropping system. In addition, Fujita et al. (1992) reported that benefits to the associated non-leguminous crop in intercropping systems is influenced by component crop densities, which determine the closeness of legume and non-legume crops, and legume growth stages. In crop mixtures, any species utilizing the same combination of resources will be in direct competition because most annual crop mixtures such as those involving cereals and legumes are grown almost at the same period, and develop root systems that explore the same soil zone for resources (Reddy et al., 1994; Jensen et al., 2003). Controlled studies showed a significant direct transfer of fixed-N to the associated non-legume species (Eaglesham et al., 1981; Stern, 1993; Ndakidemi, 2006). In mixed cultures, where row arrangements and the distance of the legume from the cereal are far, nitrogen transfer could decrease (Giller et al., 1991). Research has shown that competition between cereals and legumes for nitrogen may in turn stimulate N_2 fixation activity in the legumes (Hardar-son & Atkins, 2003). The cereal component effectively drains the soil N, forcing the legume to fix more N_2. Despite claims for substantial N-transfer from grain legumes to the associated cereal crops, the evidence indicate that benefits are limited (Giller et al., 1991). Benefits are more likely to occur to subsequent crops as the main transfer path-way is due to root and nodule senescence and fallen leaves (Ledgard & Giller, 1995). Therefore, it is important to develop clear understanding on how the fixed nitrogen becomes available to non-leguminous crops in the cereal-legumes intercropping systems.

6. Nutrients Uptake and Partitioning in Plants

Mineral nutrients are usually obtained from the soil through plant roots, but many factors can affect the efficiency of nutrient acquisition (Hell & Hillebrand, 2001). First, the chemistry and composition of certain soils can make it harder for plants to absorb nutrients. The rate of nutrient uptake by roots depends on the concentration of the particular nutrient at the root surface, root properties or plant species, and requirement of the plants (Paul et al., 2005). The nutrients may not be available in certain soils, or may be present in forms that the plants cannot use (Hell & Hillebrand, 2001). Soil properties like water content, pH, and compaction may exacerbate these problems. Figure 2 indicates the maximum availability for the majority of nutrients are at pH = 6.5 i.e. under slightly acidic conditions while the availability of metal cations (mostly microelements) increases with acidity, with the exception of Molybdenum (Fageria & Baligar, 2008).

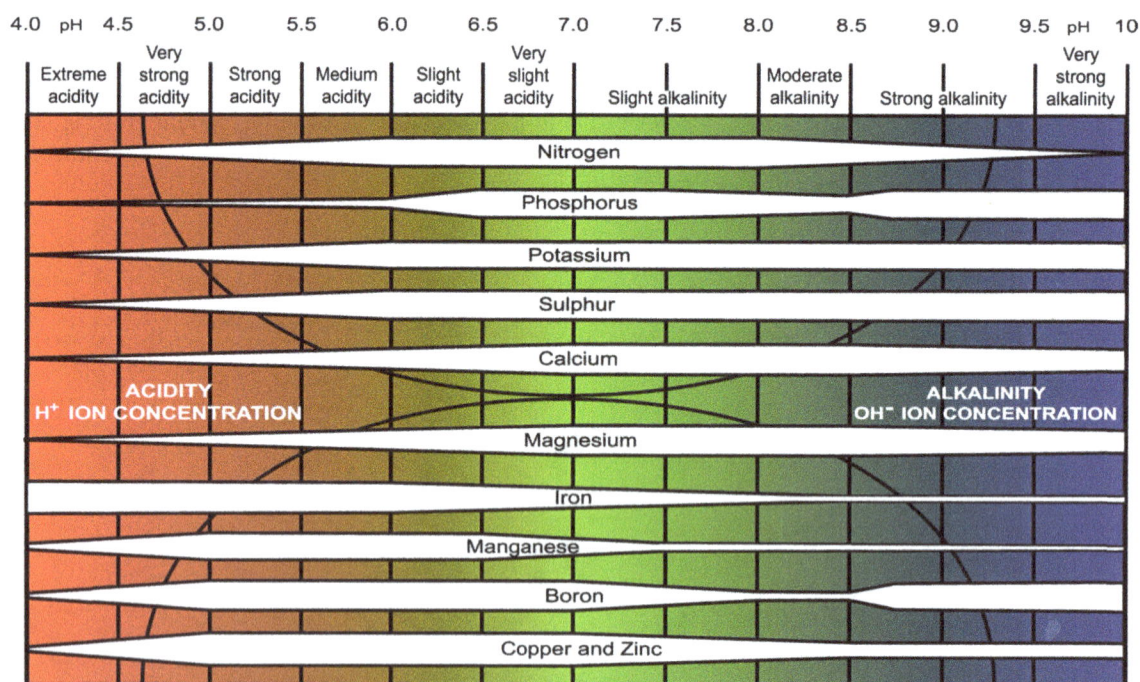

Figure 2. The influence of soil pH on nutrients availability

Soil pH has direct effect on nutrients availability; maximum availability and growth are achieved near neutral pH range (Figure 2). Second, some plants possess mechanisms or structural features that provide advantages when growing in certain types of nutrient limited soils (Jones & Ljung, 2012). In fact, most plants have evolved nutrient uptake mechanisms that are adapted to their native soils and are initiated in an attempt to overcome nutrient limitations. One of the most universal adaptations to nutrient-limited soils is a change in root structure that may increase the overall surface area of the root to increase nutrient acquisition or may increase elongation of the root system to access new nutrient sources (Jones & Ljung, 2012). These changes can lead to an increase in the allocation of resources to overall root growth, thus resulting in greater root to shoot ratios in nutrient-limited plants (Lopez-Bucio, 2003). The plant species are assumed to have different rooting and uptake patterns, such as cereal/legume intercropping system, more efficient use of available nutrients may occur and higher N-uptake in the intercrop have been reported, compared to monocrops (Fujita & Ofosu-Budu, 1996). Dahmardeh et al. (2010) reported that maize-cowpea intercropping increases the amount of nitrogen, phosphorus and potassium contents compared to monocrops of maize. Despite the beneficial effects of the intercropping to the cereal crops, it may also accelerate soil nutrient depletion, particularly for phosphorous, due to more efficient use of soil nutrients and higher removal through the harvested crops (Mucheru-Muna et al., 2010). Recent efforts on replenishment of soil fertility in Africa have been through the introduction of legumes as intercrop and/or in rotation to minimize external inputs (Sanginga & Woomer, 2009). Since plants are non-motile and often face nutrient shortages in their environment; they use mechanisms in an attempt to acquire sufficient amounts of the macro- and micronutrients required for proper growth, development and reproduction (Vance, 2001). These mechanisms include changes in the developmental program and root structure to better "mine" the soil for limiting nutrients, induction of high affinity transport systems and the establishment of symbioses and associations that facilitate nutrient uptake (Beyer, 2010). Together, these mechanisms allow plants to maximize their nutrient acquisition abilities while protecting against the accumulation of excess nutrients, which can be toxic to the plant (Vance, 2001). It is clear that the ability of plants to utilize such mechanisms exerts significant influence over crop yields as well as plant community structure, soil ecology, ecosystem health, and biodiversity. Therefore, the plant tissues analysis can be useful to diagnose plant nutritional problems and monitor effectiveness of a soil fertility improvement through maize-legume cropping systems and fertilization.

7. Legume Biomass Decomposition and Mineralization

Low soil fertility and nutrients mining are the main causes of decline in crop yield and productivity in Africa. Despite the prevalence of poor soils, fertilizers application rate in sub-Saharan Africa is only 9 kg ha^{-1} compared with global mean of 101 kg ha^{-1} (Camara & Heinemann, 2006) with high cost and limited access being cited as main contributing factors. The organic system favours the use of renewable resources and emphasizes the use of techniques that integrate natural processes such as nutrient cycling, biological nitrogen fixation and soil regeneration (Birech & Freyer, 2007). The magnitude of nutrient cycling is a function of (i) the biomass production, (ii) the nutrient contents and (iii) the decomposition rate (Lehmann et al., 2000). Decomposition is a complex process regulated by the interactions between organisms (fauna and microorganisms), physical environmental factors (particularly temperature and moisture) and resources quality (defined here by lignin, nitrogen and condensed and soluble polyphenol concentrations) (Swift et al., 1989). Fujita and Ofosu-Budu, (1996) reported that, the soil N may be replenished through decomposition of legume residues/ biomass. The nutrient accumulation within the biomass alone, does not give adequate information about the nutrient turnover, which is the relevant parameter for the magnitude of nutrient cycling. For this purpose, the biomass production has to be measured, which is more difficult for creeping legumes like Dolichos lablab in comparison with cereal crops (Lehmann et al., 2000). Further, there is evidence that the mineralization of decomposing legume roots in the soil can increase N availability to the associated crop (Dubach & Russelle, 1994; Schroth et al., 1995; Evans et al., 2001). However, litter decomposition is one of the key biogeochemical processes in forest ecosystems and it is estimated that the nutrients released during litter decomposition can account for 69-87% of the total annual requirement of essential elements for forest plants (Swift et al., 1989). The leaf litter decomposition in cropping systems is obviously easier than on the decomposition of any other parts of the plant because of the two reasons (Anderson, 1993). First, leaf litter decomposition is faster, and thus less time-consuming, especially in the tropics and subtropics (Harmon et al., 1999; Anderson, 1993). The decomposition of twig and stem litter takes significantly longer due to the higher content of more recalcitrant compounds contained in these highly lignified plant parts, and therefore long-term observations need to be considered (Harmon et al., 1999). Second, the concentration of nutrients contained in leaf litter is generally higher than in any other parts of a plant. Decomposition of legume crops is mainly determined by its chemical composition (Carbon to Nitrogen (C:N) ratio of the material) and on climatic conditions (Jama & Nair, 1996). Decomposition of residues takes place in two phases; the first phase is rapid which is controlled by C:N ratio and the second phase are slower controlled by lignin and polyphenol content of the residue (Jama & Nair, 1996). Marandu (2005) reported that, the rate of first phase decomposition is accelerated by high contents of soluble nitrogen in the residue. The critical levels of nitrogen below which the rate of decomposition is retarded are 18 to 22 mg/g (Mellilo et al., 1982). High lignin and polyphenol content of the residue retard the decomposition reducing nitrogen net mineralization while enhancing immobilization in the second phase. Determining the ratio of carbon to nitrogen (C:N) in the legume crop biomass is the most common way to estimate how quickly biomass N will be mineralized and released nutrients for use by the succeeding crop (Zhi-an et al., 2001). As a general rule, legume crop residues with C:N ratios lower than 25:1 will release N quickly. Values exceeding 30 parts carbon to one part nitrogen (C:N ratio of 30:1) are generally expected to immobilize N during the early stages of the decomposition process. Legume cover crops such as hairy vetch and crimson clover, when killed at flowering immediately before maize planting; generally will have C:N ratios of 10:1 to 20:1 (Ranells & Wagger, 1992). During decomposition, some materials will decompose fairly fast losing about 50% of their dry matter in 4-6 weeks (Wangari & Msumali, 2000). Residues with C:N ratios greater than

25:1 decompose more slowly and their N is more slowly released. N release from green manure occurs slowly at or below 5 °C increases to a maximum at 30-35 °C and declines at a higher temperature, the release is also favoured by soil conditions that are neither exceptionally dry nor flooded (Seiter & Horwath, 2004). The intercrop legume may accrue N to the soil and this may not become available until after the growing season, improving soil fertility to benefit a subsequent crop (Ofori & Stern, 1987; Ledgard & Giller, 1995). According to Peoples and Herridge (1990) to maximize the contribution of legume N to a following crop, it is necessary to maximize total amount of N in legume crop, the proportion of N derived from N_2 fixation, the proportion legume N mineralized and the efficiency of utilization of this mineral N. Ammonia volatilization from legume residues may be high when they are left on the soil surface, but the losses do not appear to match those measured in some fertilized systems. Larsson et al. (1998) estimated that 17% of the N in alfalfa mulch was lost as ammonia within 30 days of placement. Janzen and McGinn (1991) measured 15% volatilization losses from a lentil (*Lens culinaris*) green manure when left on the soil surface, and Venkatakrishnan (1980) measured ammonia losses of 23% from a sesbania (*Sesbania rostra*) green manure after 63 days. Losses of N by ammonia volatilization from legume residues as well as fertilizers can be greatly reduced or eliminated by incorporating amendments into the topsoil (Janzen & McGinn, 1991; Larsson et al., 1998). It is worth noting that incorporating legume residues may reduce N losses through ammonia volatilization at the expense of increasing potential denitrification losses of N (Peoples et al., 1995). While there are not sufficient data to state conclusively that legume-N is less susceptible to ammonia volatilization than fertilizer N, there are a couple of mechanisms that would help explain such a difference. The temporary immobilization of N and the production of acidic products during cover-crop decomposition might reduce NH_3 volatilization losses. Moreover, the addition of green manures to flooded rice systems may have the effect of increasing the CO_2 in the floodwater, and reducing the pH and thus ammonia loss (Peoples et al., 1995). Data by Diekmann et al. (1993) are consistent with such a mechanism, where flooded rice fields experienced greater percent losses when receiving N from urea compared to N from green manures. The effect of plant tissue quality on decomposition and N mineralization rates has been well documented (Myers et al., 1994; Peoples et al., 1995; Fillery, 2001), but a great work remains to test and optimize different crop combinations appropriate to particular regions. Therefore, it is important to explore the most efficient legume biomass decomposition that would have great effects on soil fertility improvement in cereal/legumes cropping systems.

8. Grain Yields in Cereal-Legume Cropping Systems

One of the most important reasons for intercropping is to ensure that an increased and diverse productivity per unit area is obtained compared to sole cropping (Sullivan, 2003). Intercropping provides an efficient utilization of environmental resources, decreases the cost of production, provides higher financial stability for farmers, decreases pest damages, inhibits weeds growth more than monocultures, and improves soil fertility through nitrogen increasing to the system and increase yield and quality (Deveikyte et al., 2009). Land productivity measured by land equivalent ratio and monetary gain showed advantages of mixed cropping of cereals and legumes (Molatudi & Mariga, 2012). However, plants planted close together usually have an increased risk of diseases, high percentage of lodging and increased interplant competition for light, water and nutrients (Molatudi & Mariga, 2012). The number of plants that can be supported per unit area is largely dependent upon soil water availability. Grain yield is the most important outcome of cereal-legume crops and it depends on a number of yield components of cereal and legume. Several studies have shown that over time, average dry matter (DM) and grains yields are higher with intercropping than when each of the plant species in the mixture is grown as a monoculture (Vandermeer, 1989). In most parts of SSA, legumes are usually inter-cropped with cereals to improve land productivity through soil amelioration (Adeleke & Haruna, 2012). This is because the cereal-legume cropping systems help to minimize excessive loss of N while maximizing N use efficiency and meeting cereal-legume grain yields. However, there is no sufficient information on yields of using different cereal-legume cropping systems in Tanzania. Therefore, it is expected that, the success of cereal-legume cropping systems can have economic impact to smallholder farmers.

9. Land Equivalent Ratio (LER)

Land equivalent ratio is defined as the relative land area under sole crops that is required to produce the yields achieved by intercropping (Waddington et al., 1989). Therefore it shows the efficiency of intercropping for using the environmental resources compared with monocropping. Land equivalent ratio (LER) can be used to assess land returns from the pure stand yields and from each separate crop within the mixture (Dariush et al., 2006). Muoneke et al. (2007) found that the productivity of the intercropping system indicated yield advantage of 2-63 percent as depicted by the LER of 1.02-1.63 showing efficient utilization of land resource by growing the crops together than separate planting. Further studies by Kipkemoi et al. (2001), Addo-Quaye et al. (2011), Osman et al. (2011), Raji (2007) and Samba et al. (2007) found that LER was greater than unity, implying that it will be more productive to intercrop cereal-legumes than grow them in monoculture. It is usually stipulated that the level of management must be the same for intercropping and sole cropping (Osman et al., 2011). The LER is interpreted based on the following criteria (Dariush et al., 2006).

LER > 1, indicates Intercropping to be more efficient than the Monocropping;

LER < 1, indicates a loss in efficiency due to intercropping;

LER = 1, indicates no difference in yield between the intercrop and monocrop.

Lithourgidis et al. (2011) proposed the simplest equation for computing LER; that is,

$$LER = \Sigma(Ypi/Ymi) \qquad (2)$$

Where, Yp is the yield of each crop in the intercrop; Ym is the yield of each crop in the monocrop.

Generally, monoculture legumes have higher yields than those intercropped (Waddington et al., 1989).

However, LER cannot address the problem of assessment of yield advantage/disadvantage if the objective is to assess the yield advantage on the basis of the criterion, namely yield per plant. Mead and Willey (1980) reported that land equivalent ratio does not give the exact value of yields, instead, it represents the yield advantages or disadvantages of intercrops compared with sole crops and the time factor is less considered for crop maturity. A lot of maize-legume cropping systems have been done focusing on intercropping and monocropping but no research when maize-legume seeds inoculations have been used in intercropping. Identifying appropriate LER will help to indicate the efficient cropping system among the cropping systems which involves maize-legume (*Dolichos lablab* and *Phaseolus vulgaris*).

10. The Economic Benefits of Cereal-Legumes Intercropping over Sole Cropping

One of the advantages of cereal-legumes intercropping includes potential for increased profitability and low fixed costs for land as a result of a second crop in the same field (Thobatsi, 2009). Cereal-legumes intercropping system has higher cash return to smallholder farmers than sole cropping (Seran & Brintha, 2010). A study by Vijay et al. (2014) reported the maximum economic benefits or the highest net return were obtained when cereal crops and legumes were planted at the same time under intercrops system. Further studies by Ashish et al. (2015) and Osman et al. (2010) using benefit cost ratio (BCR) and monetary advantages index (MAI) revealed that maize-cowpea intercropping was found to be profitable than their sole crops and increase the income for smallholder farmers, and compensate losses due to uneven condition. Cereal-legumes intercropping could enhance total productivity of the system with low input investment by changing planting population and configuration (Ashish et al., 2015). Furthermore, intercrop can give higher yield than sole crop yields, greater yield stability, more efficient use of nutrients, better weed control, provision of insurance against total crop failure, improved quality by variety, also cereal as a sole crop requires a larger area to produce the same yield as cereal in an intercropping system (Matusso et al., 2014). Despite these multiple benefits, few studies have been done to highlight the effects of intercropping legumes with cereals on agricultural productivity and economic benefits. The study on legume-cereal intercropping system is needed to show the economic benefits and how it can meet the household food requirements.

11. Conclusions

Intercropping is an old practice used by subsistence farmers, especially under rain-fed conditions. Maize/legume intercropping system has become one of the solutions for food security among small scale maize producers due to unaffordability of chemical nitrogenous fertilizers and limited access to arable land. From the comparisons of typical cereal-legume systems and monoculture practices, this review suggests that sustainable agriculture involves the successful management of agriculture resources to satisfy changing human needs while maintaining or enhancing the environment quality and conserving natural resources. It relies greatly on renewable resources, and on-farm nitrogen contributions as achieved largely through biological nitrogen fixation. Intercropping provides a balanced diet, reduces labour peaks, and minimizes crop-failure risks. It has also been suggested that intercropping can reduce the adverse effects of pests (diseases, insects, and weeds), provide higher returns, ameliorate soil fertility and protect soil against erosion. Therefore, in order to feed the growing population steps toward greater agricultural sustainability through effective use and management of internal resources focusing on cereal-legume cropping systems should be given due consideration.

References

Addo-Quaye, A. A., Darkwa, A. A., & Ocloo, G. K. (2011). Yield and productivity of component crops in a maize-soybean intercropping system as affected by time of planting and spatial arrangement. *Journal of Agricultural and Biological Science, 6*(9), 50-57.

Adeleke, M. A., & Haruna, I. M. (2012). *Residual Nitrogen Contributions from Grain Legumes to the Growth and Development of Succeeding Maize Crop.*

Adu-gyamfi, J. J., Myaka, F. A., Sakala, W. D., Odgaard, R., Vesterager, J. M., & Hogh, J. (2007). Biological nitrogen fixation and nitrogen and phosphorus budgets in farmer managed intercrops of maize pigeon pea in semi-arid southern and eastern Africa. *Plant and Soil, 295,* 127-136. http://dx.doi.org/10.1007/s11104-007-9270-0

Anderson, J., & Ingram, J. (1993). *Tropical soil biology and fertility: A handbook of methods.* CAB International: Wallingford, UK.

Andrew, D. J., & Kassam, A. H. (1976). The importance of multiple cropping in increasing world food supplies. In R. I. Papendick, A. Sanchez, & G. B. Triplett (Eds.), Multiple cropping (pp. 1-10). American Society Agronomy, Madison, WI., USA.

Anonymous. (n.d.). *Legume-based cropping systems.* Retrieved June 27, 2016, from https://www.google.co.tz/search?q=cereal-legume+cropping+systems

Arshad, M. A., & Martin, S. (2002). Identifying critical limits for soil quality indicators in agro-ecosystems. *Agriculture, Ecosystems and Environment, 88,* 153-160. http://dx.doi.org/10.1016/S0167-8809(01)00252-3

Ashish, D., Ista, D., Vineet, K., Rajveer, S. Y., Mohit, Y., Dileep, G., … Tomar, S. S. (2015). Potential Role of Maize-Legume Intercropping Systems to Improve Soil Fertility Status under Smallholder Farming Systems for Sustainable Agriculture in India. *International Journal of Life Sciences Biotechnology and Pharma Research, 4,*

Aydinalp, C., & Cresser, M. C. (2008). The Effects of Global Climate Change on Agriculture. *American-Eurasian J. Agric. And Environ. Sci., 3*(5), 672-676.

Baijukya, F. P. (2004). *Adapting to change in banana-based farming systems of Northwest Tanzania: The potential role of herbaceous legumes* (PhD thesis. Wageningen University, Netherlands).

Baligar, V. C., & Fageria, N. K. (2007). Agronomy and Physiology of Tropical Cover Crops. *Journal of Plant Nutrition, 30,* 1287-1339. http://dx.doi.org/10.1080/01904160701554997

Baltazari, A. (2014). *Bean density suppression of weeds in maize bean intercropping under conventional and conservation tillage systems in Arusha* (p. 136). Tanzania Sokoine University of Agriculture.

Beets, W. C. (1982). *Multiple cropping and tropical farming systems* (p. 156). Westview Press, Boulder.

Belel, M. D., Halim, R. A., Rafii, M. Y., & Saud, H. M. (2014). Intercropping of Corn with Some Selected Legumes for Improved Forage Production. *Journal of Agricultural Science, 6,* 3. http://dx.doi.org/10.5539/jas.v6n3p48

Bender, R. (2012). *Nutrients uptake and partitioning in high yielding corn* (p. 79). Thesis submitted in partial fulfillment of the requirements for the degree of Master of Science in crop sciences in the graduate college of the University of Illinois at Arbana Champaign.

Benselama, A., Tallah, S., Ounane, S. M., & Bekki, A. (2014). Effects of inoculation by bradyrhizobium japonicum strains on nodulation, nitrogen fixation, and yield of lablab purpureus in Algeria. *Turkish Journal of Agricultural and Natural Sciences, 21,* 57-66.

Beyer, P. (2010). Golden Rice and "Golden" crops for human nutrition. *New Biotechnology, 27,* 478-481. http://dx.doi.org/10.1016/j.nbt.2010.05.010

Birech, R., & Freyer, B. (2007). *Effect of plant biomass and their incorporation depth on organic wheat production in Kenya.*

Blair, M. W., Gonzalez, L. F., Kimani, P. M., & Butare, L. (2010). Genetic diversity, inter-gene pool introgression and nutritional quality of common beans (*Phaseolus vulgaris* L.) from central Africa. *Theor Appl Genet., 121,* 237-248. http://dx.doi.org/10.1007/s00122-010-1305-x

Bourgault, M., & Smith, D. L. (2010). Comparative study of common bean (*Phaseolus vulgaris* L.) and mungbean (*Vigna radiata* L. Wilczek) response to seven watering regimes in a controlled environment. *Crop & Pasture Science, 61,* 918-928. http://dx.doi.org/10.1071/CP10141

CABI (CAB International). (2007). Retieved March 30, 2016, from http://www.cabi.org

Camacho, B. M., & Gonzalez de Mejia, E. (1998). Comparative study of enzymes related to proline metabolism in tepary bean (*Phaseolus acutifolius*) and common bean (*Phaseolus vulgaris*) under drought and irrigated conditions, and various urea concentrations. *Plant Food Hum. Nutr., 52*, 119-132. http://dx.doi.org/10.1023/A:1008011529258

Camara, O., & Heinemann, E. D. (2006). *Overview of fertilizer situation in Africa*. Background paper prepared for the African Fertilizer summit. Abuja, Nigeria.

Chi Chung, L., Markus, W. R., & Yilin, H. (2014). Cleaving the N-N Triple Bond: The Transformation of Dinitrogen to Ammonia by Nitrogenases. In P. M. H. Kroneck, & M. E. S. Torres (Eds.), *The Metal-Driven Biogeochemistry of Gaseous Compounds in the Environment. Metal Ions in Life Sciences* (Vol. 14, pp. 147-174). http://dx.doi.org/10.1007/978-94-017-9269-1_7

Collins, A. S. (2004). *Leguminous cover crop fallows for the suppression of weeds* (p. 117). Thesis Presented to the Graduate School of the University of Florida in Partial Fulfillment of the Requirements for the Degree of Master of Science.

Dahmardeh, M., Ghanbari, A., Syasar, B., & Ramrodi, M. (2009). Intercropping maize (*Zea mays* L.) and cow pea (*Vigna unguiculata* L.) as a whole-crop forage: Effects of planting ratio and harvest time on forage yield and quality. *J. Food Agri. Environ, 7*, 505-509.

Danso, S. K. A., Hardarson, G., & Zapata, F. (1993). Misconceptions and practical problems in the use of the 15N soil enrichment techniques for estimating N fixation. *Plant and Soil, 152*, 25-52. http://dx.doi.org/10.1007/BF00016331

Dariush, M., Ahad, M., & Meysam, O. (2006). Assessing the land equivalent ratio (LER) of two corn [*Zea mays* L.] Varieties intercropping at various nitrogen levels in Karaj, Iran. *Journal of Central European Agriculture, 7*(2), 359-364.

Davidson, E. A., de Carvalho, C. J. R., Figueira, A. M., Ishida, F. Y., Ometto, J. P. H. B., Nardoto, G. B., ... Martinelli, L. A. (2007). Recuperation of nitrogen cycling in Amazonian forests following agricultural abandonment. *Nature, 447*, 995-998. http://dx.doi.org/10.1038/nature05900

Delin, S., Nyberg, A., Lindén, B., Ferm, M., Torstensson, G., Lerenius, C., & Gruvaeus, I. (2008). Impact of crop protection on nitrogen utilisation and losses in winter wheat production. *Eur. J. Agron, 28*, 361-370. http://dx.doi.org/10.1016/j.eja.2007.11.002

Deveikyte, I., Kadziuliene, Z., & Sarunaite, L. (2009). Weed suppression ability of spring cereal crops and peas in pure and mixed stands. *Agronomy Research, 7*(1), 239-244.

Diekmann, F. H., DeDatta, S. K., & Ottow, J. C. G. (1993). Nitrogen uptake and recovery from urea and green manure in lowland rice measured by 15N and non-isotope techniques. *Plant Soil, 147*, 91-99. http://dx.doi.org/10.1007/BF02185388

Dimitrios, B., Panyiota, P., Aristidis, K., & Aspasia, E. (2010). Weed suppression effects of maize-vegetable inorganic farming. *Int. J. Pest Mang, 56*, 173-181. http://dx.doi.org/10.1080/09670870903304471

Dubach, M., & Russelle, M. P. (1994). Forage legume roots and nodules and their role in nitrogen transfer. *Agron. J, 86*, 259-266. http://dx.doi.org/10.2134/agronj1994.00021962008600020010x

Eaglesham, A. R. J., Ayanaba, A., Ranga-Rao, V., & Eskew, D. L. (1981). Improving the nitrogen nutrition of maize by intercropping with cowpea. *Soil Biology and Biochemistry, 13*, 169-171. http://dx.doi.org/10.1016/0038-0717(81)90014-6

Edje, O. T. (1990). Relevance of the Workshop to Farming in Easten and Southern Africa in Research Methods for Cereal Legume Intercropping. In S. R. Palmer, & O. T. Odje (Eds.), *Proceeding of a Workshop for Research Methods Intercropping in Eastern and Southern Africa Waddington* (p. 5). Mexico. D.F. CIMMYT.

Evans, J., Mcneill, A. M., Unkovich, M. J., Fettell, N. A., & Heenan, D. P. (2001). Net nitrogen balances for cool-season grain legume crops and contributions to wheat nitrogen uptake: A review. *Austr. J. Exp. Agric, 41*, 347-359. http://dx.doi.org/10.1071/EA00036

Fageria, N. K., & Baligar, V. C. (2008). Ameliorating soil acidity of tropical Oxisols by liming for sustainable crop production. *Adv. Agron, 99*, 345-431. http://dx.doi.org/10.1016/S0065-2113(08)00407-0

Fageria, N. K., Baligar, V. C., & Bailey, B. A. (2005). Role of Cover Crops in Improving Soil and Row Crop Productivity. *Communications in Soil and Plant Analysis, 36*, 2733-2757. http://dx.doi.org/10.1080/00103620500303939

Fillery, I. R. P. (2001). The fate of biologically fixed nitrogen in legume-based dryland farming systems: A review. *Aust. J. Exp. Agric, 41*, 361-381. http://dx.doi.org/10.1071/EA00126

Fujita, K., & Ofosu-Budu, K. G. (1996). Significance of Intercropping in Cropping Systems. In O. Ito, C. Johansen, J. J. Adu-Gyamfi, K. Katayama, J. V. D. K. Kumar Rao, & T. J. Rego (Eds.), *Dynamics of Roots and Nitrogen in Cropping Systems of the Semi-Arid Tropics* (pp. 19-40). Japan International Research Center for Agricultural Sciences. International Agricultural Series No. 3 Ohwashi, Tsukuba, Ibavaki 305, Japan.

Fujita, K., Ofosu-Budu, K. G., & Ogata, S. (1992). Biological nitrogen fixation in mixed legume-cereal cropping systems. *Plant and Soil, 141*, 155-176. http://dx.doi.org/10.1007/BF00011315

Ghanbari, A., Dahmardeh, M., Siahsar, B. A., & Ramroudi, M. (2010). Effect of maize (*Zea mays* L.)-cowpea (*Vigna unguiculata* L.) intercropping on light distribution, soil temperature and soil moisture in and environment. *J. Food Agr. Environ., 8*, 102-108.

Giller, K. E. (2001). *Nitrogen fixation in tropical cropping system* (2nd ed., p. 448). CABI Publishing. http://dx.doi.org/10.1079/9780851994178.0000

Giller, K. E., Ormesher, J., & Awah, F. M. (1991). Nitrogen transfer from Phaseolus bean to intercropped maize measured using 15N-enrichment and 15N-isotope dilution methods. *Soil Biol. Biochem., 23*, 339-346. http://dx.doi.org/10.1016/0038-0717(91)90189-Q

Hardason, G., & Atkins, G. (2003). Optimizing biological N2 fixation by legumes in farming systems. *Plant and Soil, 252*, 41-54. http://dx.doi.org/10.1023/A:1024103818971

Harmon, M. E., Nadelhoffer, K. J., & Blair, J. M. (1999). Measuring decomposition, nutrient turnover, and stores in plant litter. In G. P. Robertson, C. S. Bledsoe, D. C. Coleman, & P. Sollins (Eds.), *Standard soil methods for long-term ecological research* (pp. 202-240). Oxford University Press, New York.

Havlin, J. L., Beaton, J. D., Tisdale, S. L., & Nelson, W. L. (2005). *Soil fertility and fertilizers. An introduction to nutrient management* (7th ed.). Pearson Prentice Hall, Upper Saddle River, NJ.

Hector, V., & Jody, S. (2002). *Sustainable Agriculture, green manure crops*. Cooperative extension services.

Hell, R., & Hillebrand, H. (2001). Plant concepts for mineral acquisition and allocation. *Current Opinion in Biotechnology, 12*, 161-168. http://dx.doi.org/10.1016/S0958-1669(00)00193-2

Hillocks, R. J., Madata, C. S., Chirwa, R., Minja, E. M., & Msolla, S. (2006). Phaseolus bean improvement in Tanzania, 1959-2005. *Euphytica, 150*(2), 215-231. http://dx.doi.org/10.1007/s10681-006-9112-9

Hoffman, B. M., Lukoyanov, D., Dean, D. R., & Seefeldt, L. C. (2013). Nitrogenase: A Draft Mechanism. *Acc. Chem. Res., 46*, 587-595. http://dx.doi.org/10.1021/ar300267m

Huang, R., Birch, C. J., & George, D. L. (2006). *Water use efficiency in maize production – The challenge and improvement strategies*. Maize Association of Australia 6th Triennial Conference 2006.

Ibewiro, B., Sanginga, N., Vanlauwe, B., & Merky, R. (2000) Nitrogen contribution from decomposing cover crop residues to maize in tropical derived savanna. *Nutrient Cycling in Agroecosystems, 57*, 131-140. http://dx.doi.org/10.1023/A:1009846203062

Jama, B. A., & Nair, P. K. R. (1996). Decomposition and nitrogen-mineralization pattern of Leucaena leucocephala and Cassia siamea mulch under tropical semiarid conditions in Kenya. *Plant and Soil, 179*, 275-285. http://dx.doi.org/10.1007/BF00009338

Janzen, H. H., & McGinn, S. M. (1991). Volatile loss of nitrogen during decomposition of legume green manure. *Soil Biol. Biochem., 23*, 291-297. http://dx.doi.org/10.1016/0038-0717(91)90066-S

Jensen, J. R., Bernhard, R. H., Hansen, S., McDonagh, J., Møberg, J. P., Nielsen, N. E., & Nordbo, E. (2003). Productivity in maize based cropping systems under various soil-water-nutrient management strategies in a semiarid, alfisol environment in East Africa. *Agric. Water Management, 59*, 217-237. http://dx.doi.org/10.1016/S0378-3774(02)00151-8

Jones, B., & Ljung, K. (2012). Subterranean space exploration: The development of root system architecture. *Current Opinion in Plant Biology, 15*, 97-102. http://dx.doi.org/10.1016/j.pbi.2011.10.003

Karim, M. R., Islam, F., Akkas, A. M., & Haque, F. (2001). On-farm trail with Rhizobium inoculants on lentil. *Bangladesh J Agric Res, 26*, 93-94.

Keston, O. W. N., Patson, C. N., George, Y. K., & Max William, L. (2013). Effects of sole cropped, doubled-up legume residues and inorganic nitrogen fertilizer on maize yields in Kasungu, Central Malawi. *Agricultural Science Research Journals, 3*(3), 97-106.

Kipkemoi, P. L., Wasike, V. W., Ooro, P. A., Riungu, T. C., Bor, P. K., & Rogocho, L. M. (2001). *Effects of intercropping pattern on soybean and maize yield in central Rift Valley of Kenya.* Unpublished paper. Kenyan Agricultural Research Institute.

Kutu, F. R., & Asiwe, J. A. N. (2010). Assessment of maize and dry bean productivity under different intercrop systems and fertilization regimes. *African Journal of Agricultural Research, 5*, 1627-1631.

Larsson, L., Ferm, M., Kasimir-Klemedtsson, A., & Klemedtsson, L. (1998). Ammonia and nitrous oxide emissions from grass and alfalfa mulches. *Nutr. Cycl. Agroecosyst, 51*, 41-46. http://dx.doi.org/10.1023/A:1009799126377

Ledgard, S. J., & Giller, K. E. (1995). Atmospheric N2-fixation as alternative nitrogen source. In P. Bacon (Ed.), *Nitrogen Fertilization and the Environment* (pp. 443-486). Marcel Dekker, New York.

Lehmann, J., da Silva, J., & Trujillo, L. (2000). Legume cover crops and nutrient cycling in tropical fruit tree production. *Acta Horticulturae, 531*, 65-72. http://dx.doi.org/10.17660/ActaHortic.2000.531.8

Lithourgidis, A. S., Dordas, C. A., Damalas, C. A., & Vlachostergios, D. N. (2011). Annual intercrops: An alternative pathway for sustainable agriculture. *Australian Journal of Crop Science, 5*(4), 396-410.

Lodwig, E. M., Hosie, A. H. F., Bourdès, A., Findlay, K., Allaway, D., Karunakaran, R., Downie, J. A., & Poole, P. S. (2003). Amino-acid cycling drives nitrogen fixation in the legume–Rhizobium symbiosis. *Nature, 422*, 722-726. http://dx.doi.org/10.1038/nature01527

Lopez-Bucio, J. (2003). The role of nutrient availability in regulating root architecture. *Current Opinion in Plant Biology, 6*, 280-287. http://dx.doi.org/10.1016/S1369-5266(03)00035-9

Maass, B. L., Knox, M. R., Venkatesha, S. C., Angessa, T. T., Ramme, S., & Pengelly, B. C. (2010). Lablab purpureus—A Crop Lost for Africa? *Tropical Plant Biology, 3*(3), 123-135. http://dx.doi.org/10.1007/s12042-010-9046-1

Maobe, S. N., Dyek, E. A., & Mureithi, S. G. (1998). Screening of soil improving herbaceous legumes for inclusion into smallholder farming systems in Kenya. Soil fertility research for maize based farming systems in Malawi and Zimbabwe. In S. R. Waddington, et al. (Eds.), *Proceedings of the Soil Fertility Network Results and Planning Workshop* (pp. 105-112). July 7-11, 1997, Harare, Zimbabwe.

Maqbool, M. M., Tanveer, A., Ata, Z., & Ahmad, R. (2006).Growth and yield of maize (*Zea mays* L.) as affected by row spacing and weed competition durations. *Pakistan Journal of Botany, 38*(4), 1227-1236.

Marandu, A. E. T. (2005). *Contribution of cowpea, pigeon pea and green gram to the nitrogen requirement of maize under intercropping and rotation on ferralsols in Muheza, Tanzania* (PhD Thesis).

Matusso, J. M. M., Mugwe, J. N., & Mucheru-Muna, M. (2014). Potential role of cereal legume intercropping systems in integrated soil fertility management in smallholder farming systems of Sub-Saharan Africa. *Research Journal of Agriculture and Environmental Management, 3*(3), 162-174.

Mazaheri, D., Madan, A., & Oveysi, M. (2006). Assessing the Land Equivalent Ratio (LER) of two corn varieties intercropping at various nitrogen levels in Karaj, Iran. *Journal of Central European Agriculture, 7*(2), 359-364.

Mead, R., & Willey, R. W. (1980). The concept of a 'land equivalent ratio' and advantages in yields from intercropping. *Exp. Agric, 16*, 217-228. http://dx.doi.org/10.1017/S0014479700010978

Mekuria, M., & Waddington, S. R. (2002). Initiatives to encourage farmer adoption of soil fertility technologies for maize-based cropping systems in southern Africa. In C. B. Barrett, F. Place, & A. A. Aboud (Eds.), *Natural Resources Management in African Agriculture: Understanding and Improving Current Practices* (pp. 219-233). CAB International, Wallingford, UK. http://dx.doi.org/10.1079/9780851995847.0219

Mellilo, J. M., Aber, J. D., & Musatore, J. F. (1982). Nitrogen and lignin control of hardwood leaf litter decomposition dynamics. *Journal of Ecology, 63*, 621-626. http://dx.doi.org/10.2307/1936780

Molatudi, R. L., & Mariga, I. K. (2012). Grain yield and biomass response of a maize/dry bean intercrop to maize density and dry bean variety. *African Journal of Agricultural Research, 7*(20), 3139-3146. http://dx.doi.org/10.5897/AJAR10.170

Moorhead, D. L., Currie, W. S., Rastetter, E. B., Parton, W. J., & Harmon, M. E. (1999). Climate and litter quality controls on decomposition: An analysis of modeling approaches. *Global Biogeochemical Cycles, 13*(2), 575-589. http://dx.doi.org/10.1029/1998GB900014

Mucheru-Muna, M., Pypers, P., Mugendi, D., Kung'u, J., Mugwe, J., Merckx, R., & Vanlauwe, B. (2010). A staggered maize–legume intercrop arrangement robustly increases crop yields and economic returns in the highlands of Central Kenya. *Field Crops Res, 115*, 132-139. http://dx.doi.org/10.1016/j.fcr.2009.10.013

Mugwe, J., Mugendi, D., Mucheru-Muna, M., Merckx, R., Hianu, J., & Vanlauwe, B. (2009). Determinants of the decision to adopt integrated soil fertility management practice by smallholder farmers in the central highlands of Kenya. *Expl Agric, 45*, 61-75. http://dx.doi.org/10.1017/S0014479708007072

Muoneke, C. O., Ogwuche, M. A. O., & Kalu, B. A. (2007). Effect of maize planting density on the performance of maize/soybean intercropping system in a guinea savannah agro-ecosystem. *African Journal of Agricultural Research, 2*(12), 667-677.

Mureithi, J. G., Gachene, C. K. K., & Wamuongo, J. W. (2005). Participatory evaluation of residue management effect on green manure legumes on maize yield in the central Kenya highlands. *Journal of Sustainable Agriculture, 25*, 49-68. http://dx.doi.org/10.1300/J064v25n04_06

Myers, A. (1988). Cereal cropping. *Plant and Soil, 174*, 30-33.

NBS. (2007). *Results of the 2002-03 National Agricultural Sample Census: National Bureau of Statistics* (p. 72). Ministry of Agriculture and Food Security, Ministry of Cooperatives and Marketing, and Ministry of Livestock, Dar es Salaam.

Ndakidemi, P. A. (2006). Manipulating legume/cereal mixtures to optimize the above and below ground interactions in the traditional African cropping systems. *African Journal of Biotechnology, 5*(25), 2526-2533.

O'Hara, G. W. (2001). Nutritional constraints on root nodule bacteria affecting symbiotic nitrogen fixation: A review. *Australian Journal of Experimental Agriculture, 41*, 417-433. http://dx.doi.org/10.1071/EA00087

Ofori, F., & Stern, W. R. (1987). Cereal-legume intercropping systems. *Advances in Agronomy, 40*, 41-90. http://dx.doi.org/10.1016/S0065-2113(08)60802-0

Ogindo, H. O., & Walker. S., (2005). Comparison of measured changes in seasonal soil water content by rained maize-bean intercrop and component cropping in semi arid region in South Africa. *Phys. Chem. Earth, 30*, 799-808. http://dx.doi.org/10.1016/j.pce.2005.08.023

Osman, A. N., Ræbild, A., Christiansen, J. L., & Bayala, J. (2011). Performance of cowpea (*Vigna unguiculata*) and Pearl Millet (*Pennisetum glaucum*) Intercropped under Parkia biglobosa in an Agroforestry System in Burkina Faso. *African Journal of Agricultural Research, 6*(4), 882-891.

Paul, R. A., Jonathan, R. C., & Rajeev, A. (2005). Nature of mineral nutrient uptake by plants. *Journal of Agricultural Sciences, 1*, 1-5.

Peoples, Landha, J. K., & Herridge, D. F. (1995). Enhancing legume N_2 fixation through plant and soil management. *Plant and Soil, 174*, 83-101. http://dx.doi.org/10.1007/BF00032242

Peoples, M. B., & Herridge, D. F. (1990). Nitrogen fixation by legumes in tropical and sub-tropical agriculture. *Adv. Agron, 44*, 155-223. http://dx.doi.org/10.1016/S0065-2113(08)60822-6

Peoples, M. B., Faizah, A. W., Rerkasem, B., & Herridge, D. F. (1989). Methods for evaluating nitrogen fixation by nodulated legumes in the field. *Monograph No. 11* (p. 76). Australian Centre for International Agricultural Research Canberra (ACIAR).

Peoples, M. B., Giller, K. E., Herridge, D. F., & Vessey, J. K. (2002). Limitations to biological nitrogen fixation as a renewable source of nitrogen for agriculture. In T. Finan, M. O'Brian, D. Layzell, K. Vessey, & W. Newton (Eds.), *Nitrogen Fixation: Global Perspectives* (pp. 356-360). CAB International, UK. http://dx.doi.org/10.1079/9780851995915.0356

Postgate, J. (1998). *Nitrogen Fixation* (3rd ed.). Cambridge University Press, Cambridg. Retrieved May 26, 2016, from http://en.wikipedia.org/wiki/Root_nodule

Rahman, M. M., Sofian-Azirun, M., & Boyce, A. N. (2013). Response of nitrogen fertilizer and legumes residues on biomass production and utilization in rice-legumes rotation. *The Journal of Animal & Plant Sciences, 23*(2), 589-595.

Raji, J. A. (2007). Intercropping soybean and maize in a derived savanna ecology. *African Journal of Biotechnology, 6*(16), 1885-1887. http://dx.doi.org/10.5897/AJB2007.000-2283

Rana, S. S., & Rana, M. C. (2011). *Cropping System* (p. 80). Department of Agronomy, College of Agriculture, CSK Himachal Pradesh Krishi Vishvavidyalaya, Palampur.

Ranells, N. N., & Wagger, M. G. (1992). Nitrogen release from crimson clover in relation to plant growth stage and composition. *Agronomy Journal, 84*, 424-430. http://dx.doi.org/10.2134/agronj1992.00021962008 400030015x

Reddy, K. C., Visser, P. L., Klaij, M. C., & Renard, C. (1994). The effects of sole and traditional intercropping of millet and cowpea on soil and crop productivity. *Exp. Agric, 30*, 83-88. http://dx.doi.org/10.1017/ S0014479700023875

Rees, D. J. (1986). Crop growth, development and yield in semi-arid conditions in Botswana. II. The effects of intercropping *Sorghum bicolor* with *Vigna unguiculata. Experimental Agriculture, 22*, 169-177. http://dx.doi.org/10.1017/S0014479700014241

Robson, A. D. (1978). Mineral nutrients limiting nitrogen fixation in legumes. In C. S. Andrew & E. J. Kamprath (Eds.), *Mineral Nutrition of Legumes in Tropical and Subtropical Soils* (pp. 277-293). CSIRO, Melbourne. Australia.

Salon, C., Avice, J. C., Larmure, A., Ourry, A., Prudent, M., & Voisin, A. S. (2011). Plant N fluxes and modulation by nitrogen, heat and water stresses: A review based on comparison of legumes and non legume plants. In A. S. A. B. Venkateswarlu (Ed.), *Abiotic Stress in Plants-Mechanism and Adaptation.*

Samba, T., Coulibay B. S., Koné, A., Bagayoko, M., & Kouyaté, Z. (2007). Increasing the productivity and sustainability of millet based cropping systems in the Sahelian zones of West Africa. In A. Bationo (Ed.), *Advances in Integrated Soil fertility Management in Sub-Saharan Africa: Challenges and Opportunities* (pp. 567-574). http://dx.doi.org/10.1007/978-1-4020-5760-1_54

Sanchez, P. A. (1976). *Properties and Management of Soils in the Tropics* (p. 618). John Wiley and Sons, Inc., New York and Toronto.

Sanchez. P. A., & Jama, B. A. (2002). Soil fertility replenishment takes off in east and southern Africa. In B. Vanlauwe, J. Diels, N. Sanginga, & R. Merckx (Eds.), *Integrated Plant Nutrient Management in Sub-Saharan Africa: From Concept to Practice* (pp. 23-45). CAB International, Wallingford.

Sanginga, N., & Woomer, P. L. (2009). *Integrated soil fertility management in Africa: Principles, practices, and developmental process.* CIAT.

Schöll, L. V., & Nieuwenhuis, R. (2004). *Soil fertility management, Agrodok2* (pp. 22-23). Netherland: Agromisa Foundation.

Schroth, G., Kolbe, D., Balle, P., & Zech, W. (1995). Searching for criteria for the selection of efficient tree species for fallow improvement, with special reference to carbon and nitrogen. *Fertilizer Res, 42*, 297-314. http://dx.doi.org/10.1007/BF00750522

Scott, T. W., Pleasant, J., Burt, R. F., & Otis, D. J. (1987). Contributions of ground cover, dry matter and nitrogen from intercrops and cover crops in a corn polyculture system. *Agron. J, 79*, 792-798. http://dx.doi.org/10.2134/agronj1987.00021962007900050007x

Seiter, S., & Horwath, W. R. (2004). Strategies for managing soil organic matter to supply plant nutrients. In Magdoff & R. R. Weil (Eds.), *Soil Organic Matter in Sustainable Agriculture* (pp. 269-293). CRC Press, Boca Raton. FL. http://dx.doi.org/10.1201/9780203496374.ch9

Sekamatte, B. M., Ogenga-Latigo, M., & Russell-Smith, A. (2003). Effects of maize-legume intercrops on termite damage to maize, activity of predatory ants and maize yields in Uganda. *Crop Prot, 22*, 87-93. http://dx.doi.org/10.1016/S0261-2194(02)00115-1

Seran, T. H., & Brintha, I. (2010). Review on maize based intercropping. *Journal of Agronomy, 9*(3), 135-145. http://dx.doi.org/10.3923/ja.2010.135.145

Smaling, E. M. A., Nandwa, S. M., & Janssen, B. H. (1997). Soil fertility in Africa is at stake. In R. J. Buresh, P. A. Sanchez & F. G. Calhoun (Eds.), *Replenishing soil fertility in Africa* (pp. 47-61). SSSA Special Publication 51. SSSA (Soil Science Society of America), Madison, Wisconsin, USA.

Steiner, K. G. (1982). *Intercropping in tropical smallholder agriculture with special reference to West Africa.* Schriftenreihe der GT2, N. 137, Eischbon, Germany.

Stern, W. R. (1993). Nitrogen fixation and transfer in an intercropping systems. *Field Crop Res, 34*, 335-356. http://dx.doi.org/10.1016/0378-4290(93)90121-3

Sullivan, P. (2003). *Overview of Cover Crops and Green Manures* (p. 16). Fundamentals of Sustainable Agriculture.

Swift, M. J., Frost, P. G. H., Campbell, B. M., Hatton, J. C., & Wilson, K. B. (1989). Nitrogen cycling in farming systems derived from Savanna: Perspectives and challenges. In M. Clarkholm & L. Bergstrom (Eds.), *Ecology of arable land* (pp. 63-76). Kluwer, Academic Publ., Dordrecht, The Netherlands. http://dx.doi.org/10.1007/978-94-009-1021-8_7

TASDS. (2001). *Tanzania Agricultural Sector Development Strategy.*

Thayamini, H., Seran, T. H., & Brintha, I. (2010). Review on Maize Based Intercropping. *Journal of Agronomy, 9*, 135-145. http://dx.doi.org/10.3923/ja.2010.135.145

Thobatsi, T. (2009). *Growth and Yield Response of Maize (Zea mays) and Cowpea (Vigna unguiculata L.) in Intercropping System* (p. 149, Thesis for Award of Msc. Degree at the University of Pretoria).

Unkovich, M. J., Pate, J. S., Sanford, P., & Armstrong, E. L. (1994). Potential precision of the delta 15N natural abundance method in field estimates of nitrogen fixation by crop and pasture legumes in South-West Australia. *Australian Journal of Agricultural Research, 45*, 119-132. http://dx.doi.org/10.1071/AR9940119

Vance, C. P. (2001). Symbiotic Nitrogen Fixation and Phosphorus Acquisition. Plant Nutrition in a World of Declining Renewable Resources. *Plant Physiology, 127*, 390-397. http://dx.doi.org/10.1104/pp.010331

Vandermer, J. H. (1989). *The ecology of intercropping.* Cambridge: Cambridge University Press. http://dx.doi.org/10.1017/CBO9780511623523

Venkatakrishnan, S. (1980). Mineralization of green manure (Sesbania aculeate, Pers.) nitrogen in sodic and reclaimed soils under flooded conditions. *Plant Soil, 54*, 149-152. http://dx.doi.org/10.1007/BF02182007

Vijay, K. C., Anil, D., Paramasivam, S. K., & Bhagirath, S. C. (2014). Productivity, Weed Dynamics, Nutrient Mining, and Monetary Advantage of Maize-Legume Intercropping in the Eastern Himalayan Region of India. *Plant Production Science, 17*(4), 342-352. http://dx.doi.org/10.1626/pps.17.342

Waddington, S. R., Palmer, A. F. E., & Edje, O. T. (1989). Research Methods for Cereal/Legume Intercropping. *Proceedings of a Workshop on Research Methods for Cereal/Legume Intercropping* (p. 250). Eastern and Southern Africa Held in Lilongwe, Malawi, January, 23-27, 1989.

Walsh, K. B. (1995). Physiology of the legume nodule and its response to stress. *Soil Biology & Biochemistry, 27*, 637-655. http://dx.doi.org/10.1016/0038-0717(95)98644-4

Wangari, N., & Msumali, G. P. (2000). Decomposition of Sesbania seban and Lantana camara green manures: Effect of the type of green manure and rate of application. *Soil Technologies for Sustainable Smallholder Farming Systems in East Africa* (pp. 13-20). Soil Science Society of East Africa, Signal Press, Nairobi.

Xavery, P., Kalyebara, R., Kasambala, S., & Ngulu, F. (2006). The Impact of Improved Bean Production Technologies in Northern and North Western Tanzania. *Occasional Publication Series No. 43* (p. 89). Pan African Bean Research Alliance, CIAT Africa Region, Kampala, Uganda.

Yencey, C., & Cesil, J. (1994). *Covers challenge cotton chemicals* (pp. 20-23). The New Farm.

Zhang, D. (2009). Rates of litter decomposition in terrestrial ecosystems: Global patterns and controlling factors. *Journal of Plant Ecology, 11*(4), 481-484.

Zhi-an, L., Shao-lin, P., Debbie, J. R., & Guo-yi, Z. (2001). Litter decomposition and nitrogen mineralization of soils in subtropical plantation forests of southern China, with special attention to comparisons between legumes and non-legumes. *Plant and Soil, 229*, 105-116. http://dx.doi.org/10.1023/A:1004832013143

16

Mapping Soil Moisture as an Indicator of Wildfire Risk Using Landsat 8 Images in Sri Lanna National Park, Northern Thailand

Kansuma Burapapol[1] & Ryota Nagasawa[2]

[1] The United Graduate School of Agricultural Sciences, Tottori University, Tottori, Japan

[2] Faculty of Agriculture, Tottori University, Tottori, Japan

Correspondence: Kansuma Burapapol, The United Graduate School of Agricultural Sciences, Tottori University, 4-101 Koyama Minami, Tottori 680-8553, Japan. E-mail: kansuma.bu@gmail.com

Abstract

Severely dry climate plays an important role in the occurrence of wildfires in Thailand. Soil water deficits increase dry conditions, resulting in more intense and longer burning wildfires. The temperature vegetation dryness index (TVDI) and the normalized difference drought index (NDDI) were used to estimate soil moisture during the dry season to explore its use for wildfire risk assessment. The results reveal that the normalized difference wet index (NDWI) and land surface temperature (LST) can be used for TVDI calculation. Scatter plots of both NDWI/LST and the normalized difference vegetation index (NDVI)/LST exhibit the triangular shape typical for the theoretical TVDI. However, the NDWI is more significantly correlated to LST than the NDVI. Linear regression analysis, carried out to extract the maximum and minimum LSTs (LST_{max}, LST_{min}), indicate that LST_{max} and LST_{min} delineated by the NDWI better fulfill the collinearity requirement than those defined by the NDVI. Accordingly, the NDWI-LST relationship is better suited to calculate the TVDI. This modified index, called $TVDI_{NDWI-LST}$, was applied together with the NDDI to establish a regression model for soil moisture estimates. The soil moisture model fulfills statistical requirements by achieving 76.65% consistency with the actual soil moisture and estimated soil moisture generated by our model. The relationship between soil moisture estimated from our model and leaf fuel moisture indicates that soil moisture can be used as a complementary dataset to assess wildfire risk, because soil moisture and fuel moisture content (FMC) show the same or similar behavior under dry conditions.

Keywords: wildfire, soil moisture, fuel moisture content, vegetation index, Landsat 8, northern Thailand

1. Introduction

Severely dry climate plays an important role in the occurrence of wildfires. In Thailand, forest wildfires are particularly prevalent during the dry season and are especially damaging because of forest loss and degradation. During the dry season, the number of wildfires in Thai conserved forest areas were 4207, 4982, and 6685 in 2014, 2015, and 2016, respectively (Forest Fire Control Division, 2016). These numbers indicate that the number of wildfires appears to be increasing because Thailand has been experiencing longer dry seasons and under dry conditions, wildfires can ignite easily, as fuel sources are readily available. Fuel availability, which drives wildfire occurrences and directly affects wildfire behavior, depends on fuel characteristics, which are fuel load (influencing fire intensity) and fuel moisture content (influencing both fire ignition and spread). It appears that recurring dry seasons foster fuel availability and reduce fuel moisture content, resulting in potentially more damaging high-intensity fires, which may spread rapidly during extremely dry conditions.

Soil moisture, defined as the volumetric water content of soil (Eller & Denoth, 1996), is an important indicator of dry conditions and is linked to wildfire occurrence. The reduction of water in soil increases dry conditions, resulting in more intense and longer burning fires (Kozlowski & Pallardy, 2002; Chmura et al., 2011). Previous studies pointed out that soil moisture affects wildfire occurrence. For example, Krueger et al. (2015) showed that large growing-season wildfires occurred exclusively under conditions of low soil moisture. Yebra et al. (2013) suggested that improving wildfire assessments involves using soil moisture as a representative for fuel moisture, which is a key factor for ignition and spread of wildfires. Therefore, surface measurements of soil moisture may provide opportunities for improving estimates of fuel moisture (Qi, Dennison, Spencer, & Riano, 2012), because both are physically linked through soil-plant interactions (Hillel, 1998).

Remote sensing techniques have been extensively used for the analysis of soil moisture, and have provided alternative tools for obtaining rapid estimates of soil moisture on large spatial scales (Goward, Xue, & Czajkowski, 2002; Sandholt, Rasmussen, & Andersen, 2002; Ishimura, Shimizu, Rahimzadeh, & Omasa, 2011). Vegetation indices (VIs), which are mathematical combinations of different spectral bands from satellite remotely sensed data, have been utilized to estimate soil moisture (Z. Gao, W. Gao, & Chang, 2011; Chen et al., 2015). The normalized difference vegetation index (NDVI) is the normalized reflectance difference between the near-infrared (NIR) and visible red (R) bands (Rouse, Haas, Deering, Schell, & Harlan, 1974; Tucker, 1979), which measures changes in chlorophyll content. As a result, it is considered a function of vegetation strength, which changes as vegetation interacts with soil moisture. The normalized difference water index (NDWI) is a more recent satellite-derived index from the NIR and short-wave infrared (SWIR) channels that reflects changes in both water content and spongy mesophyll in vegetation canopies (Gao, 1996). This index has been employed for the determination of vegetation water content and stress (Ceccato, Gobron, Flasse, Pinty, & Tarantola, 2002), and is therefore expected to be linked to soil moisture due to its impact on vegetation water stress. Moreover, land surface temperature (LST) can rise rapidly with water stress (Goetz, 1997), which is directly related to soil moisture. Accordingly, LST is also widely used as a soil moisture indicator (Carlson, 2007).

The relationship between VI and LST has been investigated to evaluate evapotranspiration rates. The VI-LST relationship normally shows a negative correlation, resulting in triangular-shaped VI-LST plots at different spatial scales (Nemani, Pierce, Running, & Goward, 1993; Goetz, 1997). Based on the VI-LST correlation, the temperature vegetation dryness index (TVDI), computed from the NDVI-LST relationship has become a widely used dryness index to estimate surface soil moisture (Sandholt, Rasmussen, & Andersen, 2002; Mallick, Bhattacharya, & Patel, 2009; Patel, Anapashsha, Kumar, Saha, & Dadhwal, 2009). For example, Wang, Qu, Zhang, Hao, and Dasgupta (2007) applied NDVI-LST produced from moderate resolution imaging spectroradiometer (MODIS) data to investigate the correlation with soil moisture determined by field measurements. The results revealed that NDVI-LST is strongly correlated with soil moisture and can be used to generate soil moisture estimates. Chen et al. (2015) used the TVDI (NDVI-LST) derived from Landsat-5 TM data to estimate soil moisture and found that the TVDI can reflect the soil moisture status under different tree species. In this study, we propose a new application of the NDWI-LST relationship, which could enhance the efficiency of the TVDI calculation. Additionally, the normalized difference drought index (NDDI), which combines information about both greenness and water obtained from the NDVI and the NDWI (Gu, Brown, Verdin, & Wardlow, 2007), has been applied in numerous studies to evaluate drought and it was found that it is an appropriate indicator for the dryness of a particular area (Renza, Martinez, Arquero, & Sanchez, 2010; Gouveia, Bastos, Trigo, & DaCamara, 2012). The NDDI appears to respond to soil moisture based on drought conditions, and was used in this study to determine soil moisture.

The objectives of this study are to estimate the spatial distribution of soil moisture using VIs based on Landsat 8 OLI/TIRS data and to evaluate the use of soil moisture data for wildfire risk assessment. Specifically, this paper includes: (1) soil moisture estimates for mapping the spatial distribution of soil moisture by combining TVDI and NDDI based on a regression approach. We propose a possible adaptation and application of NDWI and LST for constructing a TVDI based on the similar design of the triangular NDVI-LST space. We then compare the efficiencies of NDVI-LST and NDWI-LST for calculating the TVDI. (2) An investigation of the relationship between estimated soil moisture and fuel moisture measured in the field to assess the suitability of the simulated soil moisture data for wildfire prediction. (3) We hypothesize that (i) the NDWI-LST relationship performs as well as or better than the NDVI-LST relationship and can be applied for calculating TVDI, and (ii) that estimated soil moisture derived from our model is directly related to fuel moisture, influencing wildfire occurrence. In this study, we used the Landsat 8 TIRS and MODIS products for calculating LST and the Landsat 8 OLI product for determining TVDI and NDDI.

This study could also be used as an approach to enhance the efficiency of wildfire assessment using soil moisture as a surrogate for fuel moisture, identifying areas prone to wildfire across different landscapes. Until now, Thailand has not widely applied remote sensing to wildfire management. Using soil moisture measured by remote sensing as a complementary dataset for wildfire management may have the unique potential to predict wildfire danger for Thailand's forest areas and enhance the effectiveness of planning and decision-making in the area of wildfire management.

2. Materials and Methods

2.1 Study Area

Sri Lanna National Park, located in Chiang Mai province in northern Thailand, is the field measurement area for soil moisture (Figure 1). The park consists of a mountain chain, running north to south, with elevations varying from 400 to 1718 meters above sea level. The study mainly focused on 63,965 ha of dipterocarp forest and 20,528 ha of deciduous forest (Department of National Parks, Wildlife and Plant Conservation [DNP], 2003). Wildfires mostly occur during the dry season (from December to April), when trees shed their leaves, and leaf litter quickly accumulates, serving as available fuel to drive wildfires. The mean annual temperature in the study area is 26.7 °C, while the minimum and maximum temperatures of the coldest (January) and hottest months (April) are 11.0 °C and 39.5 °C, respectively (Thai Meteorological Department, 2014). The area receives an average precipitation of 1,156 mm yr^{-1}. August is the month with the highest precipitation of 256.76 mm, which decreases in the dry season and reaches a minimum of 4.10 mm in February (DNP, 2003). Soil properties in the park are closely related to slope; 92.2% of the total forest area is classified as a soil slope complex series, which is found in areas with slopes that exceed 35%. Sandy and sandy loam soils are dominant (DNP, 2003).

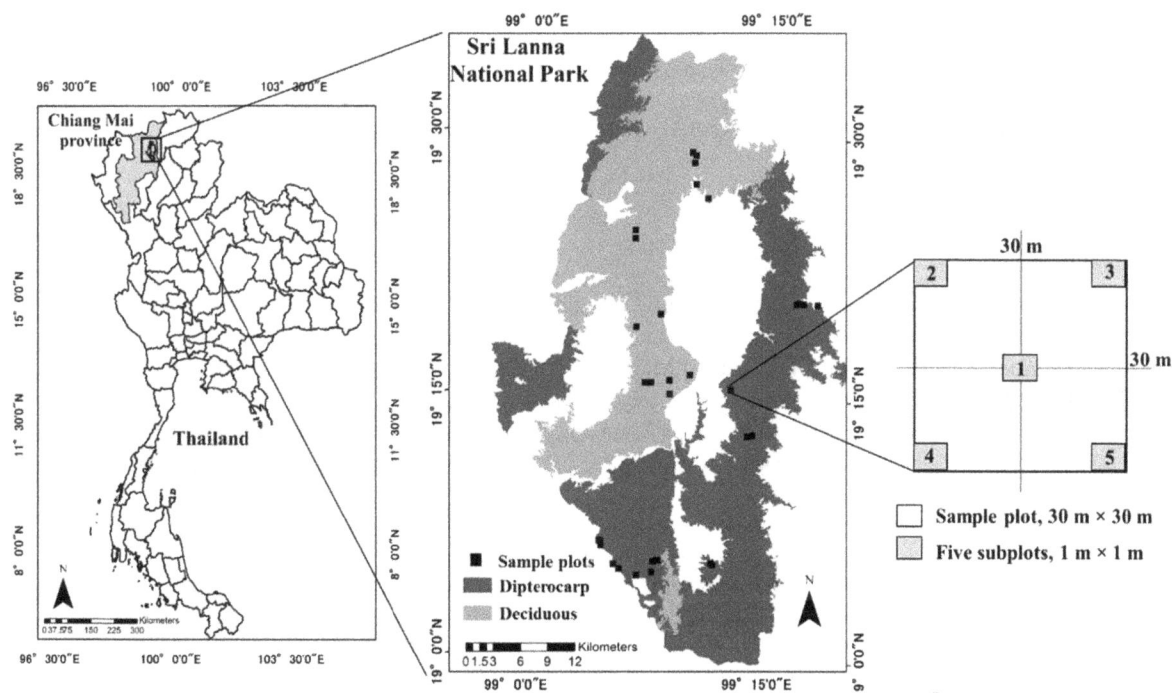

Figure 1. Location map and observation sites in Sri Lanna National Park

2.2 Field Measurements

Thirty-four sample plots with heterogeneous landscape and ecological conditions were selected using a topographic map. A Landsat 8 image provided radiometric and geometric corrections for different slopes, aspects, and forest types. The selected plots were evaluated during the dry season in March 2015. Larger 30 m × 30 m sample areas, corresponding to the spatial resolution of Landsat 8 images (30 m × 30 m pixel size) used for linear regression analysis, were divided into five subplots (1 m × 1 m) for collecting soil samples and fuel or litter from the ground surface.

2.2.1 Gravimetric Soil Moisture Measurements

Soil samples were collected from each of the five 1-m^2 subplots, which are representative of the soil within each sample plot. The soil samples were taken at a standard depth of 10 cm, because previous studies have indicated that it is feasible to estimate surface (0 to 0.76 cm) soil moisture from visible and NIR reflectance (Kaleita, Tian, & Hirschi, 2005). In addition, VIs show the highest correlation with surface soil moisture at 10 cm depth (Zhang,

Hong, Qin, & Zhu, 2013). Each soil sample was placed in a plastic container and sealed tightly for further laboratory analysis. For the gravimetric analysis of soil moisture, we first weighed the soil samples (wet weight in grams) using a standard laboratory scale and then placed them in a drying oven at 105 °C for 48 hours (Gardner, 1986). After drying, we weighed the dried soil samples (dry weight in grams). The percentage of gravimetric soil moisture was calculated using Equation 1:

$$Soil\ moisture = \frac{wet\ weight\ \text{-}\ dry\ weight}{dry\ weight} \times 100\% \qquad (1)$$

Five soil moisture measurements from each of the five subplots within each sample plot were averaged to obtain representative soil moisture for each 30-m^2 site, corresponding to the spatial resolution of the Landsat 8 images. The averaged soil moisture data from 34 sample plots were used for both training (80%) and validation (20%) data.

2.2.2 Leaf Fuel Moisture Measurements

Leaf fuel was collected for fuel moisture measurements, which were used for analyzing the relationship with simulated soil moisture. We specifically focused on dead leaves on the ground surface, because those represent the largest fuel component. A small sample of leaf litter was randomly collected from each 1-m^2 subplot and then placed into a sealed envelope for further laboratory analysis. In the laboratory, leaf litter samples were weighed and oven-dried at 80 °C for 48 hours, then weighed again to calculate the fuel moisture content (FMC) in percent following the procedure described by Desbois, Deshayes and Beudoin (1997). The most common FMC calculation is the ratio of water to dry weight as expressed by Equation 2. The FMC values for the five subplots were averaged to obtain a representative FMC for each 30-m^2 sample plot.

$$FMC = \frac{wet\ weight\ \text{-}\ dry\ weight}{dry\ weight} \times 100\% \qquad (2)$$

2.3 Remotely Sensed Data and Preprocessing

We used cloud-free Landsat 8 OLI/TIRS and MODIS eight-day composite LST datasets at a spatial resolution of 30 m and 1000 m, respectively, as primary data (Table 1). Estimates of soil moisture require: (i) Landsat 8 images to extract the TVDI and NDDI, and (ii) MODIS eight-day composite LST and Landsat 8 thermal infrared (TIR) data to produce the LST. Landsat 8 datasets used are the L1G level product and were geographically corrected and clipped based on the study area's boundary. The MODIS data were (i) projected to UTM Zone 47N with the WGS84 datum, (ii) clipped based on the study area's boundary, and (iii) co-registered to Landsat 8 images to reduce potential geometric errors.

Table 1. Selected Landsat 8 and MODIS images for dry season

Season	Parameter	Landsat 8		MODIS eight-day composite	
		Acquisition date	Spectral band	Acquisition date	Product
Dry	TVDI, NDDI	19 Feb 2015	Visible, NIR, SWIR	–	–
Dry	LST	19 Feb 2015	TIR (band 10)	18-25 Feb 2015	MOD11A2

2.4 Soil Moisture Estimates

2.4.1 Calculation of the TVDI

The LST is the temperature of the Earth's surface as derived from remotely sensed thermal infrared data (Weng, Fu, & Gao, 2014). It depends on the albedo, vegetation cover, and soil moisture. The Landsat 8 LST was computed by fusing images of MODIS LST and Landsat 8 brightness temperature (Tb), provided by Hazaymeh and Hassan (2015). Generating Landsat 8 LST was based on the linear relationship between MODIS LST and Landsat 8 Tb, which were obtained almost simultaneously and under similar atmospheric conditions.

A scatter plot of remotely sensed LST and VI often results in a triangular shape (Price, 1990; Carlson, Gillies, & Perry, 1994) and the "dry" and "wet" edges of the triangle can be used to obtain information on soil moisture content. Figure 2 shows the conceptual TVDI based on the NDVI-LST triangle, where LST is plotted as a function of NDVI. The linear combination of NDVI-LST typically shows a strongly negative relationship and the TVDI can be estimated from the dry and wet edges of the triangle.

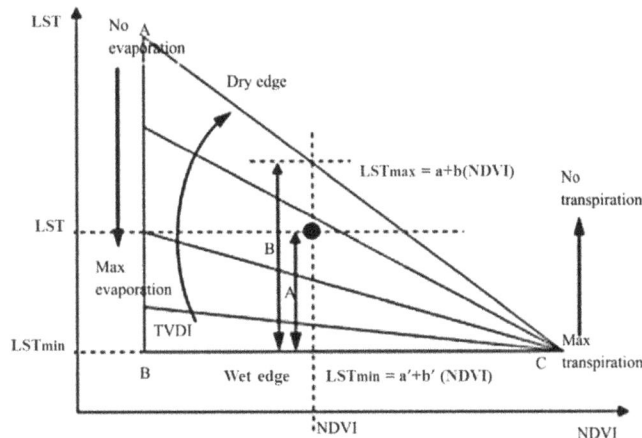

Figure 2. Simplified presentation of TVDI based on the triangular shape of the NDVI-LST relationship (adapted from Sandholt, Rasmussen, & Andersen, 2002)

In the feature space, TVDI is computed based on information about the wet edge representing the minimum LST (LST_{min}, maximum evapotranspiration and thereby, unlimited water access) as a straight line parallel to the NDVI axis. The dry edge, representing the maximum LST (LST_{max}, limited water availability) is linearly correlated with NDVI. Therefore, the TVDI is related to the soil moisture status in that high values indicate dry conditions and low values indicate moist conditions. In this study, the correlations of both NDVI and NDWI to the LST were observed. The TVDI for each pixel can be defined using Equation 3:

$$TVDI = \frac{LST - LSTmin}{LSTmax - LSTmin} \tag{3}$$

Where, LST is the LST (°C) at a given NDVI and NDWI value, LST_{min} is the minimum LST (°C) based on the NDVI and NDWI values along the wet edge, and LST_{max} is the maximum LST (°C) based on the NDVI and NDWI values along the dry edge.

To calculate LST_{max} (dry edge) and LST_{min} (wet edge), we created scatter plots for each NDVI-LST and NDWI-LST pair. Linear regression was applied to scatter plots of the resulting LST_{max} and LST_{min} based on the upper and lower boundary lines of the scatter plots. Positive NDVI and NDWI values were ranked from 0 to 1 and divided into units of 0.01, 0.015, 0.02, 0.025, ... 1. Then, each of the individual values of the scaled NDVI and NDWI were paired with a corresponding LST such as $NDVI_1$, LST_{max1} and $NDVI_1$, LST_{min1} or $NDWI_1$, LST_{max1} and $NDWI_1$, LST_{min1}. Finally, we employed a linear regression approach to fit the point pairs for generating LST_{max} and LST_{min}:

$$LST_{max} = a + b \ (VI) \tag{4}$$

$$LST_{min} = a' + b' \ (VI) \tag{5}$$

Where, a and b are regression coefficients of LST_{max}, a' and b' are regression coefficients of LST_{min}, and VI represents the NDVI and NDWI values. The NDVI is a normalized ratio of the NIR and R reflectance (Tucker, 1979) as described in Equation 6. The NDWI is calculated from NIR and SWIR reflectance (Gao, 1996) as shown in Equation 7:

$$NDVI = \frac{NIR - RED}{NIR + RED} \tag{6}$$

$$NDWI = \frac{NIR - SWIR}{NIR + SWIR} \tag{7}$$

We investigated the NDVI and the NDWI performance and selected the index showing the strongest correlation with LST based on the adjusted R-squared (adj-R^2) of LST_{max} and LST_{min}. The best relationship of the index and LST was later used for TVDI calculation following Equation 3.

2.4.2 Calculation of the NDDI

The NDDI was computed from the NDVI and NDWI values according to the definition proposed by Gu, Brown, Verdin, and Wardlow (2007). The combination of information about both vegetation (NDVI) and water (NDWI) conditions can be used to determine vegetation drought conditions, which reflect the effects of soil moisture. Due to the variation of the NDVI and NDWI within a range from -1 to +1, these values were converted to 8 bits (0-255) for the calculation of the NDDI, which ranges between -1 and +1. Higher NDDI values indicate more severe drought and lower soil moisture. The NDDI is computed as:

$$NDDI = \frac{NDVI - NDWI}{NDVI + NDWI} \tag{8}$$

2.4.3 Soil Moisture Model and Validation

We established a soil moisture estimation model based on a collection of field sampling and remote sensing data. A stepwise multiple regression approach was used to assess the relationship between field soil moisture data and remote sensing data, i.e., TVDI and NDDI were used as independent variables. The model can be computed by a regression formula as follows:

$$Estimated\ soil\ moisture = a + b(TVDI) + b'(NDDI) \tag{9}$$

Where, the estimated soil moisture is given as a percentage (%), and a, b, and b' are the coefficients of the regression lines of the TVDI and NDDI.

The model was validated by ground and remote sensing data. We used the actual soil moisture from the field measurements to evaluate the accuracy of the predictive model by statistical inference: (i) adj-R^2, (ii) root mean squared error (RMSE), (iii) absolute average difference (AAD), and (iv) the precision of the model. The precision (%) of the model is calculated as follows:

$$Precision = \sqrt{\frac{\sum[(Yi - Y'i)/Y'i]^2}{N}} \times 100\% \tag{10}$$

Where, Yi is the actual soil moisture of the field samples (%), Y'i is the estimated soil moisture from remotely sensed data (%), and N is the sample size.

Finally, the validated model was applied to a Landsat 8 image acquired on 19 February 2015 in Sri Lanna National Park (dipterocarp and deciduous forests) in order to estimate and map the spatial soil moisture distribution during the dry season.

2.5 Analysis of the Relationship between Estimated Soil Moisture and Leaf Fuel Moisture

To investigate the relationship between soil moisture estimated from our model and FMC, we performed a correlation analysis using the Pearson correlation and linear regression methods. Estimated soil moisture was extracted from the model at the same locations as were used to measure leaf fuel moisture in the field to determine correlation. We then explored the possibility of applying estimated soil moisture from our model to the prediction of wildfire occurrences.

3. Results and Discussion

Scatter plots of the relationships between NDVI-LST and NDWI-LST are shown in Figure 3. Compared to the NDVI-LST plot, the NDWI-LST relationship shows a clearer triangular shape, following the theoretical triangle of the TVDI. We determined LST_{max} (dry edge) and LST_{min} (wet edge) to highlight linear trends. A comparison of pixels representing LST_{max} and LST_{min} extracted from the NDVI-LST and the NDWI-LST plots indicates a stronger relationship between these pixels in the NDWI-LST space. Based on Figure 3, the LST_{max}, representing the dry edge, shows a strong negative correlation between the NDWI and LST (adj-R^2 = 0.84, p-value < 0.01), and the LST_{min}, representing the wet edge, shows a negative correlation between the NDWI and LST with adj-R^2 = 0.63 at a significant level for p < 0.01. In contrast, NDVI has a lower correlation with LST, with LST_{max} at adj-R^2 = 0.62 (p-value > 0.05) and LST_{min} at adj-R^2 = 0.47 (p-value < 0.01). The results of the collinearity requirement indicate that the NDWI has a stronger negative correlation with the LST than the NDVI, which is why the NDWI was used to calculate TVDI.

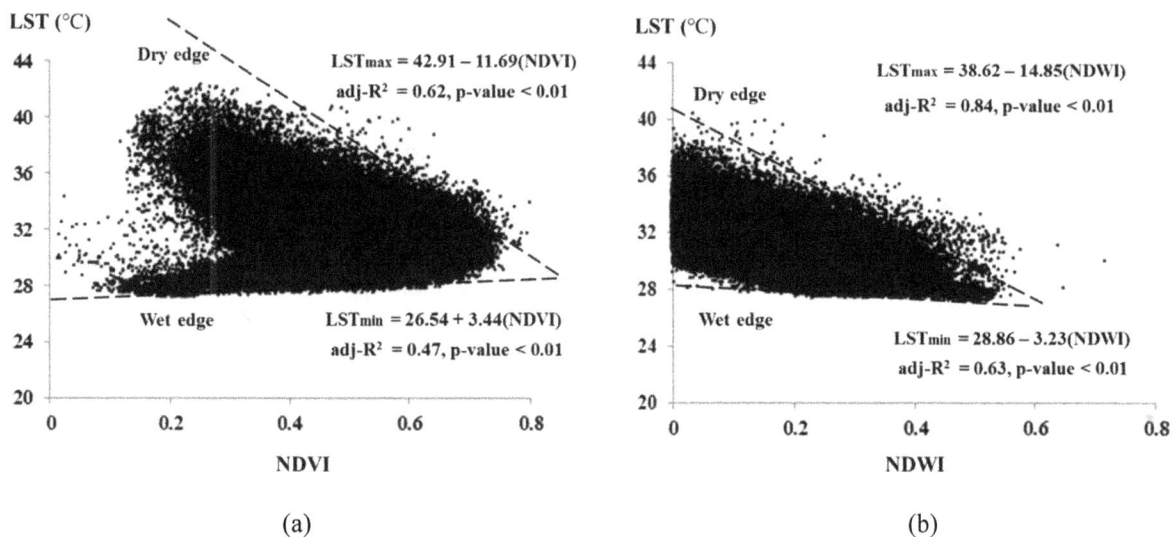

Figure 3. Observed relationships for (a) NDVI-LST and (b) NDWI–LST, based on the conceptual TVDI model

The reason for the better correlation between the NDWI and LST might be that LST is more strongly related to the water content of vegetation (captured by NDWI) than to the chlorophyll content (captured by NDVI). The NDVI measures changes in chlorophyll content (absorption of visible red radiation) and in the leaf spongy mesophyll (reflection of NIR radiation) within the vegetation canopy. Consequently, the NDVI has a limited capability for retrieving vegetation water content information, as it provides information on vegetation greenness (chlorophyll), which is not directly and uniformly related to the quantity of water in the vegetation (Ceccato, Gobron, Flasse, Pinty, & Tarantola, 2002). A change in chlorophyll content detected using the NDVI does not imply a direct change in leaf water content. Conversely, the NDWI is sensitive to changes in leaf water content because the green vegetation spectra in the SWIR region are dominated by water absorption.

The water content in leaves is directly affected by temperature conditions, especially high temperatures. As temperature increases, evaporation from leaves is higher, which affects the water content of the leaves. Evaporation within leaves also causes an increase in heat, and the leaf temperature rises relative to the air temperature or LST. Therefore, NDWI is more sensitive to LST, resulting in a stronger negative correlation with LST. Gu, Brown, Verdin, and Wardlow (2007) found that NDWI values exhibited a quicker response to drought conditions when compared to NDVI values. This is because the NDWI is constructed from the SWIR, which is more sensitive to moisture than other spectra. As a result, the NDWI shows a better correlation with LST and follows more closely the conceptual TVDI model. This result supports our hypothesis that the relationship between the NDWI and LST can be used to improve the calculation of the TVDI.

A TVDI map of the study area extracted from LST_{max} and LST_{min} based on the strong NDWI-LST relationship is shown in Figure 4a, while a NDDI map computed from the NDVI and the NDWI is shown in Figure 4b. Both maps, which show drought conditions during the dry season, can reflect the degree of soil moisture because drought influences the soil moisture status. Extreme drought results in lower soil moisture content. Therefore, both VIs can be used as predictor variables to estimate soil moisture.

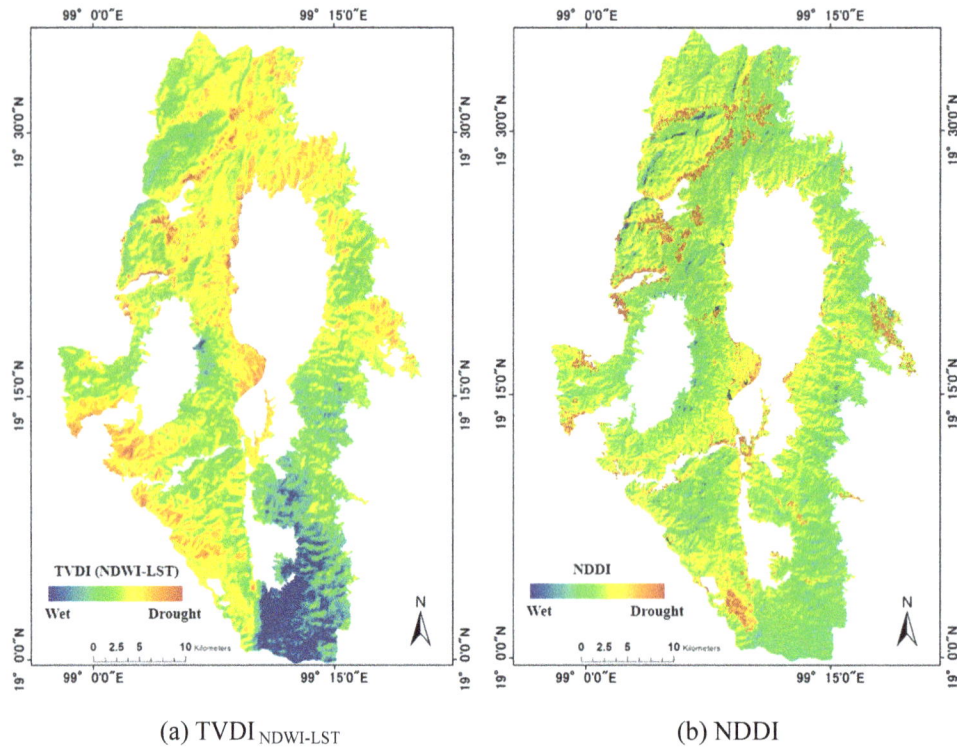

(a) TVDI $_{NDWI-LST}$ (b) NDDI

Figure 4. Extraction of a Landsat 8 image from 19 February 2015 for (a) TVDI$_{NDWI-LST}$ and (b) NDDI

Linear regression models for soil moisture estimation shown in Table 2 were calculated using the modified TVDI$_{NDWI-LST}$ and the NDDI as dependent variables, and field-measured soil moisture content as the independent variable. The model constructed from both indices has the strongest response to the actual soil moisture and likely has a greater ability to accurately estimate soil moisture, based on its high adj-R^2 (0.89, p-value < 0.01) and low RMSE (0.87%) for actual versus estimated soil moisture. In contrast, the model that only uses TVDI$_{NDWI-LST}$ has a lower adj-R^2 (0.72, p-value < 0.01) and a higher RMSE value of 1.39 %. Similarly, the model that only uses the NDDI shows the weakest correlation with an adj-R^2 of 0.52 (p-value < 0.01) and the highest RMSE of 1.82%. Thus, the soil moisture model using both the TVDI$_{NDWI-LST}$ and the NDDI fulfills the collinearity requirements with an increase in the adj-R^2 and a reduced RMSE, which can enhance the efficiency of soil moisture estimation.

Table 2. Comparison of statistical soil moisture models

Predictor variable	Soil moisture model (%)	N	adj-R^2	RMSE (%)
TVDI $_{NDWI-LST}$	10.67 – 12.24(TVDI $_{NDWI-LST}$)	27	0.72*	1.39
NDDI	13.93 – 35.44(NDDI)	27	0.52*	1.82
TVDI $_{NDWI-LST}$, NDDI	14.32 – 9.45(TVDI $_{NDWI-LST}$) – 21.78(NDDI)	27	0.89*	0.87

Note. * is significant at the 0.01 level.

The best model, developed from the combination of the modified TVDI$_{NDWI-LST}$ and the NDDI, was tested for accuracy with regard to field-measured soil moisture, resulting in the statistical parameters shown in Table 3. The model fulfills the statistical requirements. We found a high adj-R^2 of 0.75 with a p-value of < 0.01. We obtained low RMSE and AAD values of 1.22% and 1.06% between the actual and estimated soil moisture, respectively. In addition, the model precision was found it to be 76.65% consistent with the actual and estimated soil moisture. These statistical tests demonstrate that the model generated from the modified TVDI$_{NDWI-LST}$ and the NDDI can provide reliable estimates of soil moisture.

Table 3. Statistical validation between the actual and soil moisture estimated from the model

Soil moisture model (%)	N	adj-R^2	RMSE (%)	AAD (%)	Precision (%)
14.32 − 9.45(TVDI $_{NDWI-LST}$) − 21.78(NDDI)	7	0.75*	1.22	1.06	76.65

Note. * is significant at the 0.01 level.

These results demonstrate that the efficacy of soil moisture estimation can be greatly enhanced using TVDI (modified from NDWI-LST) and NDDI as dependent variables, because both VIs show a strong correlation with soil moisture measured in the field. The reason for this strong correlation is the causal relationship between variations in soil moisture and changes in vegetation; consequently, soil moisture deficits are ultimately tied to drought stress in plants (Gu et al., 2008), which is captured by both the TVDI and NDDI. Based on these results, we applied the model to a Landsat 8 image taken during the dry season to estimate soil moisture (Figure 5). The spatial distribution map shows that the percentage of soil moisture in Sri Lanna National Park is quite low during the dry season at around 0.001% to 31.1%, with a mean value of 15.49%. The degree of estimated soil moisture can indicate drought conditions, which in turn influence the occurrence of wildfires. Areas with lower soil moisture and resulting lower fuel moisture, which influences fire ignition and spread, are more prone to wildfire occurrence.

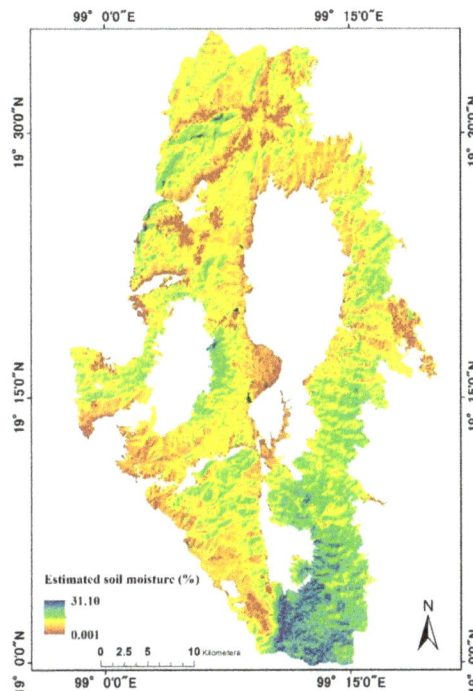

Figure 5. Spatial distribution of soil moisture derived from the model generated by the modified TVDI$_{NDWI-LST}$ and the NDDI in Sri Lanna National Park during the dry season on 19 February 2015

We also investigated the correlation between the estimated soil moisture and leaf fuel moisture determined in the field (Figure 6). Pearson's correlation reveals that leaf fuel moisture shows a statistically significant positive correlation to the estimated soil moisture (Pearson's correlation coefficient = 0.67, p-value < 0.01). Larger values of estimated soil moisture tend to be associated with larger values of leaf fuel moisture. This implies that leaf fuel moisture has a tendency to increase when estimated soil moisture increases and vice versa. The statistical tests also support our hypothesis that the estimated soil moisture is directly related to FMC.

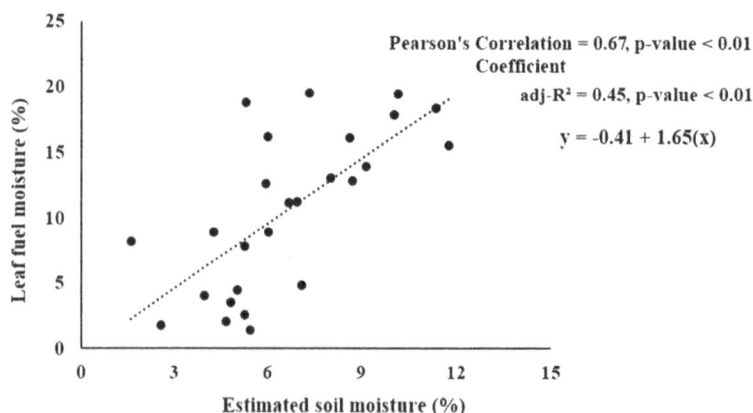

Figure 6. Scatter plot of leaf fuel moisture measured in the field and estimated soil moisture

Moreover, a median adj-R^2 of 0.45 with a p-value of < 0.01 as shown in Figure 6 indicates that estimated soil moisture is a significant variable for predicting leaf fuel moisture. This suggests that soil moisture is a factor that influences FMC since soil moisture condition affects fuel moisture levels which are directly related to wildfire occurrence. At high temperatures during the dry season, soil moisture and FMC are positively correlated, because high temperatures result in low soil moisture, which in turn leads to low FMC. As a result, we can use soil moisture to assess wildfire risks by exploiting the relationship between soil moisture and FMC. When FMC is high, fires do not readily ignite, because heat energy has to be used to evaporate water from plant material before it can burn. During the combustion of the above ground plant material and surface organic layers, the heat energy created is then transferred in the soil (DeBano, Neary, & Ffolliott, 1998). Thus, fuel load with low moisture can transfer more heat into the soil during the combustion of fuel. Soils with higher moisture content tend to absorb more heat energy (DeBano, Neary, & Ffolliott, 1998, 2005); as a result, the intensity of the fire is reduced. In cases where both the FMC and soil moisture are low wildfires will start much easily and spread rapidly resulting in uncontrollable fire condition.

Based on the result, mapping of estimated soil moisture can be used to investigate wildfire risk in large areas. Additionally, soil moisture can give an insight on the dryness of the fuel, which is a crucial parameter for wildfire risk. Therefore, to reduce wildfire risk and intensity, soil moisture should be considered as another indicator for monitoring wildfire prone areas. An analysis of soil moisture could considerable enhance wildfire management, thus in our study we highly recommend estimating soil moisture by remotely sensed data to be used as a complementary dataset for wildlife management in terms of risks and danger assessment.

4. Conclusion

The main goal of this study was to estimate the spatial distribution of soil moisture using TVDI and NDDI derived from Landsat 8 OLI/TIRS data for wildfire risk assessment. Results reveal that an accurate estimate of TVDI can be obtained from the relationship between NDWI, which is more significantly correlated to LST than the NDVI, and LST. This modified TVDI$_{NDWI-LST}$ can be used together with the NDDI to enhance the efficacy of soil moisture estimation. A scatter plot of NDWI-LST shows a linear relationship and is a good match with the theoretical concept of the TVDI, which is characterized by the triangular shape of the NDVI-LST relationship. The good correlation between NDWI and LST fulfills the collinearity requirements for extracting LST$_{max}$ and LST$_{min}$; consequently, the NDWI-LST relationship provides a better estimate of the TVDI than the NDVI-LST relationship.

The soil moisture model generated from a combination of the modified TVDI$_{NDWI-LST}$ and NDDI can improve the accuracy of soil moisture estimates. The accuracy of the model was tested using statistical metrics, and was found to be more than 76% consistent with actual soil moisture and estimated soil moisture derived from our model. We further explored the relationship between estimated soil moisture and wildfire risk by investigating the correlation between estimated soil moisture and leaf fuel moisture measured in the field. Results show that estimated soil moisture is positively correlated to leaf fuel moisture with a Pearson's correlation coefficient of 0.67 (p-value < 0.01). This relationship demonstrates that wildfire-prone areas, which are characterized by low FMC, can be identified through soil moisture estimates, because both soil moisture and FMS show the same or similar behavior under conditions of high temperatures during the dry season.

The model allows to remotely determine the spatial distribution of soil moisture as a complementary dataset for identifying wildfire-prone areas, which is a fundamental step toward involving soil moisture in the assessment of wildfire risk. We therefore recommend soil moisture estimation by remotely sensed model as another indicator for monitoring wildfire risks and intensity. Furthermore, the demonstrated NDWI-LST relationship provides another option for researchers studying soil moisture when the established TVDI based on the NDVI-LST relationship is insufficient. Future studies should address soil moisture as one of the factors used for enhancing estimates of FMC, as soil moisture is shown to be correlated with FMC.

Acknowledgements

The financial supports and the GIS based thematic maps were provided by Department of National Parks, Wildlife and Plant Conservation, Thailand. We truly thank Mr. Yossawat Thiansawat the Sri Lanna National Park superintendent for the field data collection.

References

Carlson, T. N. (2007). An overview of the triangle method for estimating surface evapotranspiration and soil moisture from satellite imagery. *Sensors, 7*, 1612-1629. http://dx.doi.org/10.3390/s7081612

Carlson, T. N., Gillies, R. R., & Perry, E. M. (1994). A method to make use of thermal infrared temperature and NDVI measurements to infer surface soil water content and fractional vegetation cover. *Remote Sensing Reviews, 9*, 161-173.

Ceccato, P., Gobron, N., Flasse, S., Pinty, B., & Tarantola, S. (2002). Designing a spectral index to estimate vegetation water content from remote sensing data: Part 1 Theoretical approach. *Remote Sensing of Environment, 82*, 188-197. http://dx.doi.org/10.1016/S0034-4257(02)00037-8

Chen, S., Wen, Z., Jiang, H., Zhao, Q., Zhang, X., & Chen, Y. (2015). Temperature vegetation dryness index estimation of soil Moisture under different tree species. *Sustainability, 7*, 11401-11417. http://dx.doi.org/10.3390/su70911401

Chmura, D. J., Anderson, P. D., Howe, G. T., Harrington, C. A., Halofsky, J. E., Peterson, D. L., ... Clair, J. B. St. (2011). Forest responses to climate change in the northwestern United States: Ecophysiological foundations for adaptive management. *Forest Ecology and Management, 261*, 1121-1142. http://dx.doi.org/10.1016/j.foreco.2010.12.040

DeBano, L. F., Neary, D. G., & Ffolliott, P. F. (1998). *Fire's effects on ecosystems*. New York, USA: John Wiley & Sons Inc.

DeBano, L. F., Neary, D. G., & Ffolliott, P. F. (2005). Soil physical properties. In D. G. Neary, K. C. Ryan, & L. F. DeBano (Eds.), *Wildland fire in ecosystems effects of fire on soil and water* (pp. 21-55). USA: USDA Forest Service Gen. Tech. Rep.

Department of National Parks, Wildlife and Plant Conservation (DNP). (2003). *Sri Lanna National Park's Area Management Master Plan*. Ministry of Natural Resources and Environment, Thailand.

Desbois, N., Deshayes, M., & Beudoin, A. (1997). Protocol for fuel moisture content measurements. In E. Chuvieco (Ed.), *a review of remote sensing methods for the study of large wildland fires* (pp. 61-72). Spain: Universidad de Alcalá: Alcalá de Henares, Departamento de Geografía.

Eller, H., & Denoth, A. A. (1996). Capacitive soil moisture sensor. *Journal of Hydrology, 185*, 137-146.

Forest Fire Control Division. (2016). *Statistics of fire*. Thailand: Department of national parks, wildlife and Plant conservation. Retrieved from http://www.dnp.go.th/forestfire/2546/firestatistic%20Th.htm

Gao, B. (1996). NDWI–A normalized difference water index for remote sensing of vegetation liquid water from space. *Remote Sensing of Environment, 58*, 257-266.

Gao, Z., Gao, W., & Chang, N. B. (2011). Integrating temperature vegetation dryness index (TVDI) and regional water stress index (RWSI) for drought assessment with the aid of LANDSAT TM/ETM+ images. *International Journal of Applied Earth Observation and Geoinformation, 13*, 495-503. http://dx.doi.org/10.1016/j.jag.2010.10.005

Gardner, W. H. (1986). Water content. In A. Klute (Ed.), *Methods of Soil Analysis. Part 1. Physical and Mineralogical Methods* (2nd ed.). Madison, Wisconsin, USA: Soil Science Society of America, Inc.

Goetz, S. J. (1997). Multisensor analysis of NDVI, surface temperature and biophysical variables at a mixed grassland site. *International Journal of Remote Sensing, 18*(1), 71-94. http://dx.doi.org/10.1080/01431169 7219286

Gouveia, C. M., Bastos, A., Trigo, R. M., & Daamara1, C. C. (2012). Drought impacts on vegetation in the pre- and post-fire events over Iberian Peninsula. *Natural Hazards and Earth System Sciences, 12*, 3123-3137.

Goward, S. N., Xue, Y. K., & Czajkowski, K. P. (2002). Evaluating land surface moisture conditions from the remotely sensed temperature/vegetation index measurements: An exploration with the simplified simple biosphere model. *Remote Sensing of Environment, 79*(2-3), 225-242. http://dx.doi.org/10.1016/S0034-4257(01)00275-9

Gu, Y., Brown, J. F., Verdin, J. P., & Wardlow, B. (2007). A five year Analysis of MODIS NDVI and NDWI for grassland drought assessment over the central great plains of United States. *Geophysical Research Letters, 34*(6). http://dx.doi.org/10.1029/2006GL029127

Gu, Y., Hunt, E., Wardlow, B., Basara, J. B., Brown, J. F., & Verdin J. P. (2008). Evaluation of MODIS NDVI and NDWI for vegetation drought monitoring using Oklahoma Mesonet soil moisture data. *Geophysical Research Letters, 35*, 1-5. http://dx.doi.org/10.1029/2008GL035772

Hazaymeh, K., & Hassan, Q. K. (2015). Fusion of MODIS and Landsat-8 Surface Temperature Images: A New Approach. *PloS One 2015, 10*(3), e0117755. http://dx.doi.org/10.1371/journal.pone.0117755

Hillel, D. (1998). *Environmental soil physics* (1st ed.). San Diego, California, USA: Academic Press.

Ishimura, A., Shimizu, Y., Rahimzadeh, B. P., & Omasa, K. (2011). Remote sensing of Japanese beech forest decline using an improved Temperature Vegetation Dryness Index (iTVDI). *iForest-Biogeosciences and Forestry, 4*, 195-199. http://dx.doi.org/10.3832/ifor0592-004

Kaleita, A. L., Tian, L. F., & Hirschi, M. C. (2005). Relationship between soil moisture content and soil surface reflectance. *American Society of Agricultural Engineers, 48*(5), 1979-1986. http://dx.doi.org/10.13031/2013.19990

Kozlowski, T. T., & Pallardy, S. G. (2002). Acclimation and adaptive responses of woody plants to environmental stresses. *The Botanical Review, 68*(2), 270-334.

Krueger, E. S., Ochsner, T. E., Engle, D. M., Carlson, J. D., Twidwell, D., & Fuhlendorf, S. D. (2015). Soil moisture affects growing-season wildfire size in the Southern Great Plains. *Soil Science Society of America Journal, 79*, 1567-1576.

Mallick, K., Bhattacharya, B. K., & Patel, N. K. (2009). Estimating volumetric surface moisture content for cropped soils using a soil wetness index based on surface temperature and NDVI. *Agricultural and Forest Meteorology, 149*, 1327-1342. http://dx.doi.org/10.1016/j.agrformet.2009.03.004

Nemani, R., Pierce, L., Running, S., & Goward, S. (1993). Developing satellite-derived estimates of surface moisture status. *Journal of Applied Meteorology, 32*, 548-557.

Patel, N. R., Anapashsha, R., Kumar, S., Saha, S. K., & Dadhwal, V. K. (2009). Assessing potential of MODIS derived temperature/vegetation condition index (TVDI) to infer soil moisture status. *International Journal of Remote Sensing, 30*, 23-39. http://dx.doi.org/10.1080/01431160802108497

Price, J. C. (1990). Using spatial context in satellite data to infer regional scale evapotranspiration. *IEEE Transactions on Geoscience and Remote Sensing, 28*, 940-948.

Qi, Y., Dennison, P. E., Spencer, J., & Riaño, D. (2012). Monitoring live fuel moisture using soil moisture and remote sensing proxies. *Fire Ecology, 8*, 71-87. http://dx.doi.org/10.4996/fireecology.0803071

Renza, D., Martinez, E., Arquero, A., & Sanchez, J. (2010). Drought estimation maps by means of multidate Landsat fused images. *Remote Sensing for Science, Education, and Natural and Cultural Heritage*, 775-782.

Rouse, J. W., Haas, H. R., Deering, D. W., Schell, J. A., & Harlan, J. C. (1974). *Monitoring the vernal advancement and retrogradation (green wave effect) of natural vegetation*. NASA/GSFC Type III Final Report, Greenbelt, Maryland, USA. Retrieved from http://ntrs.nasa.gov/archive/nasa/casi.ntrs.nasa.gov/19740004927.pdf

Sandholt, I., Rasmussen, K., & Andersen, J. (2002). A simple interpretation of the surface temperature/vegetation index space for assessment of surface moisture status. *Remote Sensing of Environment, 79*, 213-224. http://dx.doi.org/10.1016/S0034-4257(01)00274-7

Thai Meteorological Department. (2014). *Climatological data for the year 2014 (Chiang Mai)*. Northern Meteorological Center, Chiang Mai, Thailand.

Tucker, C. J. (1979). Red and photographic infrared linear combinations for monitoring vegetation. *Remote Sensing of Environment, 8*, 127-150. Retrieved from http://ntrs.nasa.gov/archive/nasa/casi.ntrs.nasa.gov/19780024582.pdf

Wang, L., Qu, J. J., Zhang, S., Hao, X., & Dasgupta, S. (2007). Soil moisture estimation using MODIS and ground measurements in eastern China. *International Journal of Remote Sensing, 28*(6), 1413-1418. http://dx.doi. org/10.1080/01431160601075525

Weng, Q., Fu, P., & Gao, F. (2014). Generating daily land surface temperature at Landsat resolution by fusing Landsat and MODIS data. *Remote Sensing of Environment, 145*, 55-67. http://dx.doi.org/10.1016/j.rse. 2014.02.003

Yebra, M., Dennison, P. E., Chuvieco, E., Riano, D., Zylstra, P., Hunt Jr., E. R., ... Jurdao, S. (2013). A global review of remote sensing of live fuel moisture content for fire danger assessment: Moving towards operational products. *Remote Sensing of Environment, 136*, 455-469. http://dx.doi.org/10.1016/j.rse.2013. 05.029

Zhang, N., Hong, Y., Qin, Q., & Zhu, L. (2013). Evaluation of the Visible and Shortwave Infrared Drought Index in China. *International Journal of Disaster Risk Science, 4*(2), 68-76. http://dx.doi.org/10.1007/s13753-013-0008-8

17

Manually Produce Clay-Based Housing Materials in Rural Area

Baixin Wu[1], Haifeng Yan[1] & Ao Sun[1]

[1] Hunan Institute of Animal Husbandry and Veterinary Medicine, Changsha, China

Correspondence: Baixin Wu, Hunan Institute of Animal Husbandry and Veterinary Medicine, CN410131, 8#Changlan Route (Quantang), Furong District, Changsha City, China. E-mail: wbx503@163.com

Abstract

Rural housing materials in developing countries (such as African countries) are mostly (crop) straw for roof cover, soil and tree branch for (round) wall. The houses are small with interior dark. In the case of electricity, mechanical and economic conditions are not allowed, farmers hardly know how to improve housing conditions or create economic benefits by applying manual labour, animal power, land, timber and fuel (firewood or coal). In this article, the method of manually producing baked tiles (for roof cover) and bricks (for wall) are described in detail with a set of historic pictures, which aims to inherit Chinese farmers' wisdom and diligence (intangible cultural heritage), arouse farmers in developing countries to improve their housing conditions by self reliance, promote the rise and development of rural industry, at the same time, promote the construction of water conservancy project.

Keywords: mud, mould, tile, brick, kiln, burn

1. Introduction

Rural housing materials mainly contain wall and roof cover materials. The primary function is to withstand wind, rain or snow into the building (Guofeng, 2005; Tiansheng & Hong, 2010). Modern wall materials largely consist of baked bricks and mechanical moulding hollow clay bricks or concrete bricks among which the baked bricks account for 70% of wall materials (Xiaoguang, 1991). Modern roof cover materials mostly refer to baked tiles or glazed tiles. The process flow for the production of these materials mainly consists of earth cutting (with bulldozer), transferring (carry scraper), adding water (running water, hot water or steam), mixing (blender), filtrating, squeezing out (extrusion machine and conveyor belt), cutting unbaked tiles or bricks (cutting machine, plastic compression molding machine), glazing (for glazed tiles), drying, transferring into a tunnel kiln (tractor or conveyor belt) and firing (Situ, 1991; Chengcai, 1994; Xiaoxi, 2000; Zuxian, 2012). In a word, modern production method of the materials needs machine, fuel (diesel oil), electricity, etc. Farmers in developed countries have the economic ability to buy these materials. Most of them have brick & tile structured housing that is spacious and bright.

Many farmers' houses in rural area in developing countries (such as African countries) looked like haystacks. Some dry tree branches or wooden sticks were inserted into ground to form a circular enclosure and become a wall after mud put in around it. Crop straw (or cattle dung, houses of a few nomadic farmers) was covered on a conical (or planiform) framed roof. The houses were small with interior dark, the wall and roof difficult to resist extreme weather such as rainstorm or high temperature (Baixin, 2016). There was no one who did not want to change his own house into spacious and bright. However, most of them could not be affordable to buy, some even had never seen, modern construction materials.

Before 1960s, there were also many houses constructed with naturally dry clay bricks for wall and crop (rice or wheat) straw for roof in rural area in China. The level of rural economy was very low. From then on, farmers forged ahead and manually produced baked tiles and bricks by self-reliance. The housing situation gradually changed. However, the traditional workshop of manually producing clay-based baked tiles and bricks gradually disappeared with the rapid development of economy. In this article, the method of manually producing baked tiles (for roof cover) and bricks (for wall) are described in detail with a set of historic pictures, which aims to inherit Chinese farmers' wisdom and diligence (intangible cultural heritage), arouse farmers in developing countries to improve their housing conditions by self reliance, promote the rise and development of rural industry, at the same time, promote the construction of water conservancy project.

2. Method of Manually Producing Baked Tiles and Bricks

2.1 Preparation of Mud

The main material for making baked tiles or bricks is clay. The silt, shale, coal gangue and flyash (mixed with clay) can also be used. The intensity of burnt clay is stronger than dry clay due to the content of silicon dioxide and aluminium oxide (Table 1, Linwei, 2005). Temporary work shed should be initially built near clay digging place. Removing the topsoil, digging and sprinkling water to make the soil soaked for 2 days. The soaked soil is repeatedly trampled by human feet when its quantity is small or by a cattle (Figure 1) when large. When the soil becomes dough-like mud (Figure 2) it is transferred into the work shed. The mud should be stacked into a number of cuboid shapes (height 28-30 cm and length 82-85 cm) when used for unbaked tile or any shapes when used for unbaked brick. Water should be sprinkled over the mud in the shed to keep certain moisture so as to be easily handled.

Table 1. Main ingredients of clay

Chemical composition	Content (%)	Main effect
Silicon dioxide	55-70	Strength of products
Aluminium oxide	15-20	Strength of products
Iron sesquioxide	4-10	Colorant
Calcium oxide	0-10	Fluxing agent
Magnesium oxide	0-3	Fluxing agent
Sodium oxide/ Burnt potash	a little	Fluxing agent
Organics	3-15	Bad material fall off

Figure 1. A Chinese farmer was guiding the cattle to trample over the soaked soil

Figure 2. The mud transferred to the shed was ready for unbaked tiles or bricks

2.2 Production of Baked Tiles

2.2.1 Preparation of Tile Mould and Unbaked Tiles

Tile mould, 26-30 cm width and 80-82 cm length, is made of planks (1-1.5 cm width and 26-30 cm length) connected by 2 wires. It looks like a hollow cylinder when rolled up (Figure 3). There are 4 raised battens outside the cylinder, which is upright and uniformly spaced. The (future) mud enwrapped here will be relatively thin so that it can be easily divided into 4 pieces when dry (Liyi, 2016). The sheath of tile mould seems to a bag without bottom. A piece of gauze (cloth) is fixed around a round wire (or bamboo) rack. The sheath is used to cover the tile mould and will be covered by mud so that the mould and sheath can be easily removed from the mud.

Figure 3. Tile mould was being covered by its sheath

A piece of mud (length 82-85 cm, width 26-30 cm and thickness 1 cm) is cut by the bow, which is made of an iron wire and a bent tree branch (Figure 4). The mud enwrapped the sheath and mould (Figure 5 left) is sprinkled with water and made smooth by a scraper (Figure 5 right). The top spare mud (25 cm away from the bottom) is cut and removed by a nail in the rectilinear scale, which is 25 cm away from the long end (Figure 6 left). The mould rolled up towards inside from the interface position so as to be easily removed. The sheath is then carefully taken out and a cylinder shape mud ready (Figure 6 right). Another one is made according to the same steps (Figure 7). Each dry cylinder shape mud is carefully broken apart into 4 unbaked tiles and erectly stacked (Figure 8).

Figure 4. A piece of mud was cut by the bow wire

Figure 5. The mud enwrapped the sheath and mould (left) was sprinkled with water and made smooth by a scraper (right)

Figure 6. The top spare mud was cut by a nail in the rectilinear scale (left) and the mould taken out (right)

Figure 7. The man would put down another cylinder shape mud on the ground

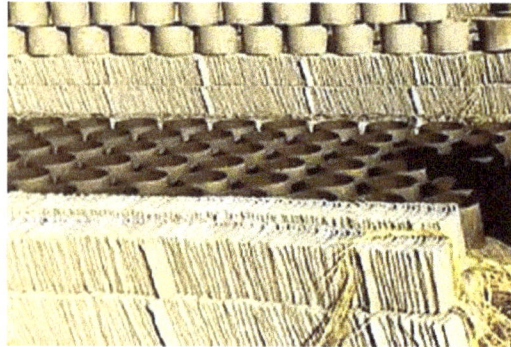

Figure 8. Each dry cylinder shape mud was broken apart into 4 unbaked tiles and erectly stacked

2.2.2 Transferring the Dry Unbaked Tiles into a Kiln and Firing

The dry unbaked tiles are transferred into a kiln (Figure 9), which should be previously prepared, and laid up row by row from the wall (Figure 10). The spacing between every two rows is 8-10 cm for fire passageway. The shape of kiln looks like a cylinder or cellar (Figure 11). There is an outlet flue on the top and one door opening in the bottom for people (unbaked tiles or firewood) getting in and people (baked tiles or ash) out.

Figure 9. The man was transferring dry unbaked tiles into the kiln

Figure 10. The workers were laying up dry unbaked tiles in the kiln

Figure 11. Dry unbaked tiles were being burnt in the kiln

When the unbaked tiles are ready in the kiln, firewood is piled up inside the door opening and burnt. An iron plank closing the door openings is removed a little or more to control fire power. In the initial 9-10 days (first stage), weak fire should be applied to evaporate slowly the moisture in the unbaked tiles. Otherwise the baked tiles will have cracks. 4-8% of artificially added water in the clay gradually evaporates when the temperature is 40-200 $^{\circ}$C (Huaixiang, 1957). In the next 6-7 days (second stage), strong fire is used. The door opening in the bottom should be closed to enlarge the drawing strength of outlet flue. The unbaked tiles will not be well burnt when the fire power is insufficient or deformed when it is excessive. The chemically combined water in the clay ($Al_2O_3 \cdot 2SiO_2 \cdot 2H_2O$) changes into dehydrated clay ($Al_2O_3 \cdot 2SiO_2$) and water ($2H_2O$) when the temperature reaches to 500-600 $^{\circ}$C (Huaixiang, 1957). Lastly (third stage) the fire is stopped. When the temperature reaches to 850-900 $^{\circ}$C, the dehydrated clay ($Al_2O_3 \cdot 2SiO_2$) becomes the mixture of Aluminium oxide (Al_2O_3) and Silicon dioxide ($2SiO_2$) , which is called baked tile (Huaixiang, 1957). It is observed from the outlet flue on top or the door opening in the bottom when the whole height of unbaked tiles descends 4-5 cm or inclines 7-8 cm or the tiles' color has changed into grey (Figure 12), the fire should be stopped. Then about 15 days later, baked tiles can be transferred out and used (sold) for construction (roof cover) material.

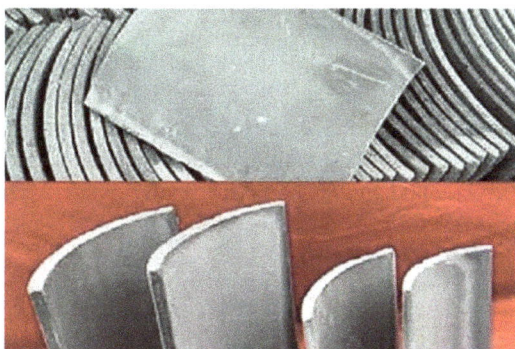

Figure 12. Baked (grey) tiles

2.2.3 Usage of Baked Tiles

The size of a baked tile is generally 25 cm length by 20 cm width) by 1 cm thickness (Jijin, 2015). The support on roof consists of a few wood stretched over two walls and planks fixed on the wood. The width of a plank is about 8 cm. The interval between two planks is approximately 12 cm. Each two intervals among 3 planks are covered with two tiles which are concave upward and both ends contact 4cm of the plank. Every two ends of the tiles on a plank are covered by another tile which is concave downward. The entire roof is covered with tiles according to the same procedures. The function of concave upward tiles is mainly receiving rainy water and making it fall down to ground whereas concave downward tiles are preventing water from dropping into house. Baked bricks should be constructed on the top or both top and bottom of inclined plane so as to fasten the tiles (Figure 13). If leaking inside the house after a much heavy rain, the area of roof should be checked or replaced by new tiles (Figure 14).

Figure 13. A part of roof covered by baked tiles

Figure 14. The aleak site of tiles roof after years was being checked and covered again

2.3 Production of Baked Bricks

2.3.1 Preparation of Brick Mould and Unbaked Brick

Brick mould (Figure 15 right), 24.5 cm length, 12 cm width and 5.8 cm height, is made of two long planks (35 cm length and 5.8 cm width) and two short planks (12 cm length and 5.8 cm width), which the short ones are inserted in the long ones. The two ends of the long are used as handle. The mould, previously soaked in water for a few seconds, sprinkled a little sand on the inside wall and put on a table or ground after taking out, is filled up with mud, compressed, removed erectly and an unbaked brick ready.

Figure 15. One brick mould (right) and four pieces of mould fixed by active button (left)

Four pieces of brick mould can be temporarily fixed together by active button (Figure 15 left), filled up with mud and transferred to outside. The active button is loosened. The gap between two pieces of mould is cut by the bow wire, three pieces of brick mould removed, the first mould taken out, then the second (third and fourth) mould taken out according to the same steps. Hence, 4 pieces of unbaked bricks can be made at the same time (Figure 16).

Figure 16. Many unbaked bricks were staying outside the work shed to become dry

2.3.2 Transferring the Dry Unbaked Bricks into a Kiln and Firing

The kiln (previously prepared) used for burning unbaked bricks in rural area commonly includes small type of kiln which has only one door opening and one outlet flue (Figure 17) and corridor shaped kiln (Figure 18). There are a high chimney for exhaust emission and a few holes for adding fuel (coal) on the top and some door openings in the bottom for people (unbaked bricks in a handcart) getting in and out. The dry unbaked bricks are transferred into the kiln and laid up by one row lying and other row standing. The spacing between every two standing bricks is 5cm for fire passageway. The initial fuel (nubbly coal) is placed between every 4 or 5 rows of brick.

Figure 17. A kind of small kiln used for unbaked bricks

Figure 18. Dry unbaked bricks were being burnt in the corridor shaped kiln

When unbaked bricks and (inside) fuel are ready in the kiln, firewood is piled up in the door openings and burnt. Iron planks closing the door openings are removed a little or more to control fire power. Some (outside) fuel (coal) is added through holes on top during burning period when needed. In the initial 9-10 days (first stage), weak fire should be applied to evaporate slowly the moisture in the unbaked bricks. Otherwise the baked bricks will have cracks. In the next 6-7 days (second stage), strong fire is used. All the door openings in the bottom should be closed to enlarge the drawing strength of chimney. The unbaked bricks will not be well burnt when the fire power is insufficient or deformed when it is excessive. Lastly (third stage) the fire is stopped. It is observed from the top holes or the door openings in the bottom when the whole height of unbaked bricks descends or inclines or the bricks' color become red (Figure 19), the fire should be stopped and all of door openings should be opened to let the hot bricks become cool easily. Then about 15 days later, baked bricks can be transferred out and used (sold) for (wall) construction material.

Figure 19. Baked (red) bricks were being transferred out by the farmers

2.3.3 Usage of Baked Bricks

The standard size of a baked brick is 24 cm by 11.5 cm by 5.3 cm. The error should be within 0.3 cm (Linwei, 2005). The thickness of mortar joint is 1cm. About 128 bricks needed for the construction of per square meter (512 bricks for each cubic meter) of wall (24 cm of thickness). The bricks can be constructed as horizontal type or vertical type. In general, a bearing wall is constructed with horizontal bricks whereas a partition wall with vertical bricks (Figure 20).

Figure 20. A kind of house was being constructed with baked bricks

3. Discussion

It is time-saving, labor saving, size consistent and standardized, high efficient to produce commercial clay-based tiles or bricks with modern equipment. But the investment cost is large and the price of products is higher. It is time consuming, arduous, accurate size uncertain and low efficient to produce the materials manually. However, the equipment needed is simple, the investment cost less and the price of products lower if for commercial purpose (Nana, 2010). Developing countries have resources of land, forest, mineral (coal, iron, etc.), human power and livestock power, etc. Under the premise of unavailable mechanical equipment, fuel (diesel) electricity, etc. and the condition of low economic level, manually producing clay-based tiles and bricks has applicability and feasibility. Firstly, small types of tile kiln and brick kiln are recommended to be promoted. A small amount of tiles and bricks produced for the purpose of improving the producers' own houses. Secondly, large types are tried to be built and a lot of materials produced for commercial purpose so as to improve other farmers' houses. Lastly, the great improvement of rural and urban housing construction will be realized by transition to modern production.

It is affected by climate when producing clay-based materials manually, because earth cutting and drying of unbaked tiles or bricks are handled outdoor. The optimum conditions for the production are that the temperature ≥ 5 °C, 2-4 level of wind and 60-80% of relative humidity, which mass production can be performed. It should be stopped when the lowest temperature $\leq$ -2 °C as the unbaked tiles or bricks will be cracked with cold. The windward side and top of stacked tiles or bricks outdoor should be covered with something when rainfall ≥ 0.5 mm. When it ≥ 5 mm drainage ditch should be dug. When the highest temperature ≥ 30 °C, relative humidity $\leq$ 30% and wind level ≥ 6, the unbaked tiles or bricks should be dried in the morning and evening so as to avoid direct exposure in strong sunshine. When more than 5 consecutive days of low cloud amount are ≥ 8, the drying time should be prolonged (Jinping, 2012).

Yearly total brick production in China was 600 billion, which consumed 1.3 billion m^3 of clay resource, equivalent to 46667 hectares of land destroyed each year that calculated according to 3 m of average depth of excavation (Nana, 2010). Hence, the destruction of arable land should be avoided when producing the materials manually. The production would emit harmful gases and waste water (Haiying, 1994). Therefore, the kiln should be built far away from residential area, animal farm and crop production area.

The development of agricultural production can not be separated from irrigation. There are many rivers and lakes in developing countries that can be used for irrigation. But the land that is far away from a river or lake needs irrigation canal built to get irrigation. Under the condition of limited national financial resources, construction of irrigation channel is very difficult. If guiding and administrating the construction material producers to cut earth from rivers (lakes) nearby to the area where irrigation is needed, construction of irrigation channel would be probably achieved. Assuming the earth cutting is 3 m in width and 3 m in depth, 0.144 billion meters [= 1.3/(3×3)] or 144 thousands kilometers length of canal would be finished after 600 billion bricks produced.

References

Baixin, W. (2016). Better living facilities for African farmers. *African Journal of Science and Research, 5*(4), 48-49. Retrieved from http://ajsr.rstpublishers.com

Chengcai, Y. (1994). Producing tiles with clay in which has fly ash. *Bricks and Tiles*. Retrieved from http://www.brick-tile.com

Guofeng, S. (2005). Study on rainproof performance of roofing tiles. *Bricks and Tiles*. Retrieved from http://www.brick-tile.com

Haiying, W. (1994). Red brick production in Taiwan tends to shrink. *Information of Building Materials Industry, 3*(14). Retrieved from http://www.cnki.net

Huaixiang, L. (1957). Discussion on the basic theory of clay tiles during the course of baking. *Building Materials Industry*. Retrieved from http://www.cnki.net

Jijin, D. (2015). *Disappearing traditional process of producing unbaked tiles*. Retrieved from http://www.360doc.com

Jinping, L. (2012). The meteorological conditions and suggestions for red brick production. *Animal Husbandry and Feed Science*. Retrieved from http://www.cnki.net

Linwei, Z. (2005). The traditional firing craft of red brick kiln in Southern Fujian. *The Chinese Journal for the history of Science and Technology, 26*(3), 249-256. Retrieved from http://www.doc88.com

Liyi, H. (2016). *Tiler-Let the clay reborn in the water and fire*. Retrieved from http://news.beiww.com

Nana, X. (2010). Study on the destruction of solid clay brick production to land resources and the countermeasures. *Journal of Henan University of Urban Construction, 19*(3). Retrieved from http://www.cnki.net

Situ. (1991). Strictly control the quality of red bricks production. *Construction Workers, 2*(24). Retrieved from http://www.cnki.net

Tiansheng, C., & Hong, C. (2010). Analyzing the reasons for the slow development of China ancient building masonry structure. *Jiangsu Construction*, 47-49. Retrieved from http://www.cnki.net

Xiaoguang. (1991). Expectation of red bricks production in Beijing. *China Building Materials, 1*, 35-36. Retrieved from http://www.cnki.net

Xiaoxi, Z. (2000). Liquid static pressure forming and device for clay tiles. *Hebei Ceramics, 28*(3). Retrieved from http://www.cnki.net

Zuxian, C. (2012). Production process of color clay tiles. *Bricks and Tiles*. Retrieved from http://www.brick-tile.com

PERMISSIONS

The contributors of this book come from diverse backgrounds, making this book a truly international effort. This book will bring forth new frontiers with its revolutionizing research information and detailed analysis of the nascent developments around the world.

We would like to thank all the contributing authors for lending their expertise to make the book truly unique. They have played a crucial role in the development of this book. Without their invaluable contributions this book wouldn't have been possible. They have made vital efforts to compile up to date information on the varied aspects of this subject to make this book a valuable addition to the collection of many professionals and students.

This book was conceptualized with the vision of imparting up-to-date information and advanced data in this field. To ensure the same, a matchless editorial board was set up. Every individual on the board went through rigorous rounds of assessment to prove their worth. After which they invested a large part of their time researching and compiling the most relevant data for our readers.

The editorial board has been involved in producing this book since its inception. They have spent rigorous hours researching and exploring the diverse topics which have resulted in the successful publishing of this book. They have passed on their knowledge of decades through this book. To expedite this challenging task, the publisher supported the team at every step. A small team of assistant editors was also appointed to further simplify the editing procedure and attain best results for the readers.

Apart from the editorial board, the designing team has also invested a significant amount of their time in understanding the subject and creating the most relevant covers. They scrutinized every image to scout for the most suitable representation of the subject and create an appropriate cover for the book.

The publishing team has been an ardent support to the editorial, designing and production team. Their endless efforts to recruit the best for this project, has resulted in the accomplishment of this book. They are a veteran in the field of academics and their pool of knowledge is as vast as their experience in printing. Their expertise and guidance has proved useful at every step. Their uncompromising quality standards have made this book an exceptional effort. Their encouragement from time to time has been an inspiration for everyone.

The publisher and the editorial board hope that this book will prove to be a valuable piece of knowledge for researchers, students, practitioners and scholars across the globe.

LIST OF CONTRIBUTORS

Mark Anglin Harris
College of Natural & Applied Sciences, Northern Caribbean University, Mandeville, WI, Jamaica

Roland Clement Abah
College of Agriculture and Environmental Sciences, Department of Environmental Sciences, University of South Africa, Pretoria, South Africa
National Agency for the Control of AIDS, Abuja, Nigeria

Brilliant Mareme Petja
College of Agriculture and Environmental Sciences, Department of Environmental Sciences, University of South Africa, Pretoria, South Africa

M. R. Ramasubramaniyan
National Agro Foundation, Chennai, Tamilnadu, India

J. Vasanthakumar
Faculty of Agriculture, Annamalai University, Tamilnadu, India

B. S. Hansra
School of Agriculture, Indira Gandhi National Open University, New Delhi, India

Kuniaki Sato and Tsugiyuki Masunaga
Faculty of Life and Environmental Science, Shimane University, Matsue, Japan

Linca Anggria
Faculty of Life and Environmental Science, Shimane University, Matsue, Japan
Indonesian Soil Research Institute, Bogor, Indonesia

Husnain
Indonesian Soil Research Institute, Bogor, Indonesia

Baraka B. Mdenye
Masasi District Council, Masasi, Mtwara, Tanzania
College of Agriculture and Veterinary Sciences, Faculty of Agriculture, University of Nairobi, Nairobi, Kenya

Josiah M. Kinama, Florence M. Olubayo and James W. Muthomi
College of Agriculture and Veterinary Sciences, Faculty of Agriculture, University of Nairobi, Nairobi, Kenya

Benjamin M. Kivuva
Kenya Agricultural and Livestock Research Organization (KALRO), Kenya

Hanna Jaworska and Katarzyna Matuszczak
Department of Soil Science and Soil Protection, UTP University of Science and Technology, Bydgoszcz, Poland

Anetta Siwik-Ziomek
Department of Biochemistry, UTP University of Science and Technology, Bydgoszcz, Poland

Dongliang Qi and Tiantian Hu
College of Water Resources and Architectural Engineering, Northwest A&F University, Yangling, China Key Laboratory of Agricultural Soil and Water Engineering in Arid and Semiarid Areas of Ministry of Education, Northwest A&F University, Yangling, China

Alice M. Mweetwa, Gwen Chilombo and Brian M. Gondwe
Department of Soil Science, School of Agricultural Sciences, University of Zambia, Lusaka, Zambia

Frank E. Johnson II and & Peter P. Motavalli
Department of Soil, Environmental and Atmospheric Sciences, University of Missouri, Columbia, Missouri, USA

Kelly A. Nelson
Division of Plant Sciences, University of Missouri, Novelty, Missouri, USA

Plinio L. Kroth and Clesio Gianello
Soil Science Department, Federal University of Rio Grande do Sul, Porto Alegre, Brazil

Leandro Bortolon and Elisandra S. O. Bortolon
Embrapa Pesca e Aquicultura, Palmas, Brazil

Jairo A. Schlindwein
Soil Science Department, Federal University of Rondônia, Porto Velho, Brazil

Mark Anglin Harris
College of Natural & Applied Sciences, Northern Caribbean University, Mandeville, WI, Jamaica
Correspondence: Mark Anglin Harris, College of Natural & Applied Sciences, Northern Caribbean University

Amanda O. Andrade, Maria A. P. da Silva, Alison H. de Oliveira, Marcos Aurelio F. dos Santos, Lilian C. S. Vandesmet, Maria E. M. Generino, Helen K. R. C. Coelho, Hemerson S. Landim, Ana C. A. M. Mendonça and Natália C. da Costa
Programa de Pós-graduação em Bioprospecção Molecular, Departamento de Ciências Biológicas, Universidade Regional do Cariri, Rua Cel. Antônio Luis, Crato, Ceará, Brazil

E. T. Sebetha
Crop Science Department, School of Agriculture, Science and Technology, North-West University, Mafikeng Campus, Mmabatho, South Africa

A. T. Modi
Crop Science, School of Agriculture, Earth and Environmental Sciences, University of KwaZulu-Natal, Scottsville, South Africa

Kopano Conferance Phefadu
Department of Plant Production, Soil Science and Agricultural Engineering, University of Limpopo, Sovenga, South Africa

Funso Raphael Kutu
Food Security & Safety Niche Area Research Group, Department of Crop Science, School of Agricultural Sciences, North-West University, Mafikeng Campus, Mmabatho, South Africa

Prosper I. Massawel, Linus K. Munishi and Patrick A. Ndakidemi
Department of Sustainable Agriculture and Biodiversity Management, The Nelson Mandela African Institution of Science and Technology, Arusha, Tanzania

Kelvin M. Mtei
Department of Water and Environmental Sciences, The Nelson Mandela African Institution of Science and Technology, Arusha, Tanzania

Kansuma Burapapol
The United Graduate School of Agricultural Sciences, Tottori University, Tottori, Japan

Ryota Nagasawa
Faculty of Agriculture, Tottori University, Tottori, Japan

Baixin Wu, Haifeng Yan and Ao Sun
Hunan Institute of Animal Husbandry and Veterinary Medicine, Changsha, China

Index

www.ingramcontent.com/pod-product-compliance
Lightning Source LLC
Chambersburg PA
CBHW080254230326
41458CB00097B/4446